AF598119

Methods in Molecular Biology™

Series Editor
John M. Walker
School of Life Sciences
University of Hertfordshire
Hatfield, Hertfordshire, AL10 9AB, UK

For further volumes:
http://www.springer.com/series/7651

Mobile Genetic Elements

Protocols and Genomic Applications

Second Edition

Edited by

Yves Bigot

Physiologie de la Reproduction et des Comportements, UMR INRA-CNRS 6175, Nouzilly, France

Editor
Yves Bigot
Physiologie de la Reproduction et des Comportements
UMR INRA-CNRS 6175
Nouzilly, France

ISSN 1064-3745 e-ISSN 1940-6029
ISBN 978-1-61779-602-9 e-ISBN 978-1-61779-603-6
DOI 10.1007/978-1-61779-603-6
Springer New York Dordrecht Heidelberg London

Library of Congress Control Number: 2012931007

Printed on acid-free paper

Humana Press is part of Springer Science+Business Media (www.springer.com)

Preface

This second edition of *Mobile Genetic Elements: Protocols and Genomic Applications* gathers into a highly practical, single source volume, a wide array of strategies and protocols for identifying transposable elements (TEs) and their evolutionary derivatives, and for studying genome structure, function, and evolution. The overlaps between both editions have been limited in order to set a second volume depicting in silico tools, strategies, and protocols that are complementary of those described in the first.

In front of the exponential increase of sequence genome data and the "easiness" to obtain them, Chapters 2–7 provide a series of complementary approaches in silico to identify, to name, and to classify TEs, but also to follow the consequences of their mobility between datasets obtained from genomes resequenced. Chapters 8 and 9 described TE-derived techniques that have been used successfully in vitro to detect genome polymorphisms. Chapters 10–15 are focused on TE-based technologies to make mutagenesis or gene delivery in vivo in a wide range of organisms ranging from bacteria to mammals, including nematods, insects, and plants. Chapter 16 is devoted to an in vitro method to define the insertion profile of neo-inserted TEs in the genomes of a cell population. Chapters 17 and 18 concerns novel TE-based technologies used for cancer and cell biology purposes.

I hope that this second edition of *Mobile Genetic Elements: Protocols and Genomic Applications* will appeal to those scientists and students intending to use TEs as genetic tools for dissecting the function of a specific gene and elaborating on mechanisms leading to genetic change and diversity, as well as to those studying the evolutionary impact of mobile DNA on the biology and evolution of organism.

Nouzilly, France *Yves Bigot*

Contents

Contributors

ISMAHEN AMMAR • *Max Delbruck Center for Molecular Medicine, Berlin, Germany*
AHMED ARNAOTY • *GICC, UMR CNRS 6239, Université François Rabelais, UFR des Sciences et Technques, Tours, France*
CYNTHIA C. BARTHOLOMAE • *Department of Translational Oncology, National Center of Tumor Diseases and German Cancer Research Center, Heidelberg, Germany*
BENOIT BATEAU • *InCellArt, Nantes, France*
SUSANTA K. BEHURA • *Department of Biological Sciences, Eck Institute for Global Health, Univerity of Notre Dame, Galvin Life Sciences Center, Notre Dame, IN, USA*
EYAYU BELAY • *Division of Gene Therapy & Regenerative Medicine, Free University of Brussels (VUB), Faculty of Medicine & Pharmacy, University Medical Center – Jette, Laarbeeklaan 103, B-1090 Brussels, Belgium*
HUGO J. BELLEN • *Department of Molecular and Human Genetics, Howard Hughes Medical Institute Program in Developmental Biology, Baylor College of Medicine, Houston, TX, USA*
YVES BIGOT • *Physiologie de la Reproduction et des Comportements, UMR INRA-CNRS 6175, Nouzilly, France*
SOLENNE BIRE • *GICC, UMR CNRS 6239, Université François Rabelais, UFR des Sciences et Technques, Tours, France*
M. CHANDLER • *Centre National de la Recherche Scientifique, LMGM, Toulouse, France*
CRISTIAN CHAPARRO • *UMR LGDP, CNRS/UPVD. Université de Perpignan Via Domitia, Perpignan Cedex, France*
MARINEE K.L. CHUAH • *Division of Gene Therapy & Regenerative Medicine, Free University of Brussels (VUB), University Medical Center – Jette, Brussels, Belgium; Department of Molecular Cardiovascular Medicine, University of Leuven, University Hospital Campus, Gasthuisberg, Belgium*
T.H. NOEL ELLIS • *Institute of Biological, Environmental & Rural Sciences, Aberystwyth University, Aberystwyth, Ceredigion, UK*
MARIO DI MATTEO • *Division of Gene Therapy & Regenerative Medicine, Free University of Brussels (VUB), University Medical Center – Jette, Brussels, Belgium; Department of Molecular Cardiovascular Medicine, University of Leuven, University Hospital Campus, Gasthuisberg, Belgium*
TEWODROS FIRDISSA • *Division of Gene Therapy & Regenerative Medicine, Free University of Brussels (VUB), University Medical Center – Jette, Brussels, Belgium; Department of Molecular Cardiovascular Medicine, University of Leuven, University Hospital Campus, Gasthuisberg, Belgium*
ANDREW J. FLAVELL • *Division of Plant Sciences, University of Dundee at JHI, Invergowrie, UK*

Timothée Flutre • *Unité de Recherches en Génomique Info URGI (UR1164) – INRA – Centre de Versailles, Versailles cedex, France*
Pascal Gantet • *Université Montpellier 2, UMR DIADE, Montpellier Cedex 5, France; Department of Biotechnology and Pharmacology, University of Sciences and Technology of Hanoi, Institute of Agricultural Genetics, Laboratoire Mixte International Rice Functional Genomics and Plant Biotechnology, Ha Noi, Vietnam*
Eric Gilson • *Laboratory of Biology and Pathology of Genomes, CNRS UMR 6267, Institut National de la Santé et de la Recherche Médicale U998, University of Nice Sophia-Antipolis, Nice, France; Department of Medical Genetics, Centre Hospitalier Universitaire of Nice, Nice, France*
Hanno Glimm • *Department of Translational Oncology, National Center of Tumor Diseases and German Cancer Research Center, Heidelberg, Germany*
Emmanuel Guiderdoni • *CIRAD, UMR AGAP, Montpellier Cedex 5, France*
Béatrice Horard • *Centre de Génétique et de Physiologie Moléculaire et Cellulaire-UMR CNRS 5534 / Université LYON 1, Villeurbanne, France*
Zoltan Ivics • *Max Delbruck Center for Molecular Medicine, Berlin, Germany; University of Debrecen, Debrecen, Hungary*
Zsuzsanna Izsvák • *Max Delbruck Center for Molecular Medicine, Berlin, Germany; University of Debrecen, Debrecen, Hungary*
Thierry Lecomte • *GICC, UMR CNRS 6239, Université François Rabelais, UFR des Sciences et Technques, Tours, France; Department of Hepatogastroenterology and Digestive Oncology, University Hospital of Tours, Tours, France*
Janka Mátrai • *Division of Gene Therapy & Regenerative Medicine, Free University of Brussels (VUB), University Medical Center – Jette, Brussels, Belgium; Department of Molecular Cardiovascular Medicine, University of Leuven, University Hospital Campus, Gasthuisberg, Belgium*
Jacques Nicolas • *IRISA, INRIA centre de recherche Rennes-Bretagne Atlantique, Campus Universitaire de Beaulieu, Rennes Cedex, France*
Etienne Paux • *Genetics, Diversity & Ecophysiology of Cereals, INRA Clermont-Ferrand – Theix, Clermont-Ferrand, France*
Emmanuelle Permal • *Unité de Recherches en Génomique Info – URGI (UR1164) – INRA – Centre de Versailles, Versailles cedex, France*
J. Perochon • *Centre National de la Recherche Scientifique, LMGM, Toulouse, France*
Mathieu Picardeau • *Unité de Biologie des Spirochètes, Institut Pasteur, Paris Cedex 15, France*
Bruno Pitard • *IRTUN, Institut du thorax – UMR INSERM 915, NANTES Cedex 1, France; InCellArt, Nantes, France*
Hadi Quesneville • *Unité de Recherches en Génomique Info – URGI (UR1164) – INRA – Centre de Versailles, Versailles cedex, France*
Valérie J.P. Robert • *Laboratory of Molecular and Cellular Biology, Ecole Normale Supérieure de Lyon, Lyon, France*
Florence Rouleux-Bonnin • *GICC, UMR CNRS 6239, Université François Rabelais, UFR des Sciences et Technques, Tours, France*

FRANCOIS SABOT • *UMR DIADE, IRD/UM2, Montpellier Cedex 2, France*
MANFRED SCHMIDT • *Department of Translational Oncology, National Center of Tumor Diseases and German Cancer Research Center, Heidelberg, Germany*
ALAN H. SCHULMAN • *Plant Genomics, MTT Agrifood Research Finland, Jokioinen, Finland; MTT/BI Plant Genomics Laboratory, University of Helsinki, Institute of Biotechnology, Helsinki, Finland*
P. SIGUIER • *Centre National de la Recherche Scientifique, LMGM, Toulouse, France*
LEYLA SLAMTI • *Unité de Biologie des Spirochètes, Institut Pasteur, Paris Cedex 15, France*
SÉBASTIEN TEMPEL • *IBISC EA 4526, University of Evry/Genopole, Evry, France*
THIERRY VANDENDRIESSCHE • *Division of Gene Therapy & Regenerative Medicine, Free University of Brussels (VUB), University Medical Center – Jette, Brussels, Belgium; Department of Molecular Cardiovascular Medicine, University of Leuven, University Hospital Campus, Gasthuisberg, Belgium*
A. VARANI • *Centre National de la Recherche Scientifique, LMGM, Toulouse, France*
KOEN J.T. VENKEN • *Department of Molecular and Human Genetics, Howard Hughes Medical Institute, Baylor College of Medicine, Houston, TX, USA*
CHRISTOF VON KALLE • *Department of Translational Oncology, National Center of Tumor Diseases and German Cancer Research Center, Heidelberg, Germany*

Chapter 1

Transposable Elements as Tools for Reshaping the Genome: It Is a Huge World After All!

Solenne Bire and Florence Rouleux-Bonnin

Abstract

Transposable elements (TEs) are discrete pieces of DNA that can move from one site to another within genomes and sometime between genomes. They are found in all major branches of life. Because of their wide distribution and considerable diversity, they are a considerable source of genomic variation and as such, they constitute powerful drivers of genome evolution. Moreover, it is becoming clear that the epigenetic regulation of certain genes is derived from defense mechanisms against the activity of ancestral transposable elements. TEs now tend to be viewed as natural molecular tools that can reshape the genome, which challenges the idea that TEs are natural tools used to answer biological questions. In the first part of this chapter, we review the classification and distribution of TEs, and look at how they have contributed to the structural and transcriptional reshaping of genomes. In the second part, we describe methodological innovations that have modified their contribution as molecular tools.

Key words: Transposable elements, Genome evolution, Genome network, Transgenesis, Tool box, Epigenetic

1. Introduction

It is more than 30 years since autonomous transposable elements (TEs) were first described as selfish sequences, acting as parasites in the genome that increased the amount of "junk" DNA in living organisms. In the interim, considerable evidence has accumulated suggesting that transposable elements are a reservoir of genetic innovation and that they can shape the genome, and has shown how the genome defends itself against these invaders.

1.1. Distribution and Classification Among Branches of Life

Bacteria and archaea genomes contain the simplest forms of transposable elements that code only the information required for their mobility flanked by inverted repeats. They are known as insertion sequences (ISs), and 20 families have been described to date (1).

Yves Bigot (ed.), *Mobile Genetic Elements: Protocols and Genomic Applications*, Methods in Molecular Biology, vol. 859, DOI 10.1007/978-1-61779-603-6_1, © Springer Science+Business Media, LLC 2012

Most of these sequences are complete, but internally deleted forms have also been described, the smallest corresponding to typical "miniature inverted-repeat transposable elements" (MITEs) (2) and solo inverted repeats. In a given genome, the distribution of ISs is patchy: some regions are IS rich, perhaps as a result of horizontal transfers, whereas others display IS exclusion or differential IS extinction (3, 4). Moreover, the number of ISs varies greatly among prokaryotic species. In bacterial chromosomes, the density of ISs is generally below 3%, whereas 20% of organisms, such as the bacterial model *Bacillus subtilis* (5), are completely devoid of IS; so ISs must be distributed very unevenly. In bacteria, transposons can jump from chromosomal DNA to plasmid DNA and back, allowing the transfer and permanent addition of genes, such as those encoding antibiotic resistance. In natural plasmids smaller than 20 kb, no IS has been detected, but plasmids above this size threshold contain between 5 and 15% of ISs, and in the extreme case of pW100 ISs make up more than 40% of the total length (6).

Simple transposons are similar to IS elements. They contain DNA segments flanked by short inverted repeat sequences. The DNA segments carry additional genes for functions other than transposition, often for antibiotic resistance (7). Some IS elements flanked a DNA segment at each end, and they are known as "composite transposons." In other words, instead of each IS element's moving independently, they act in concert and move together, along with the intervening DNA (8). The *Mu* transposon is an unusual bacteriophage of *Escherichia coli* that inserts itself into the host genome by a specific transposition mechanism (9).

In eukaryotes, TEs are widespread and persistent entities in metazoans, fungi, plants, and unicellular organisms. The majority of them are divided into two major classes that are distinguished by their mechanism of transposition. On the one hand, there are the class I elements, consisting of the retrotransposons. They are characterized by having a specific mechanism of transposition using an RNA intermediate obtained after transcription of the element. This particular mechanism, known as the "copy-and-paste" mode of transposition, is confined to eukaryotes. Class I elements include the long terminal repeat (LTR) retrotransposons that undergo reverse transcription in virus-like particles and the non-LTR retrotransposons, such as LINEs (long interspersed elements) and SINEs (short interspersed elements) for which RNA copies are carried back into the nucleus, where their reverse transcription and integration occur (10). On the other hand, the "cut-and-paste" mode of transposition is restricted to class II elements, which transpose via a DNA intermediate, which is cut out from the host DNA prior to insertion. "Conventional" DNA transposons mobilize dsDNA (double-stranded DNA) by the interaction of their own transposase with their terminal inverted repeats located at both ends of the transposon. Examples of class II elements are the *Ac/Ds* element of maize, the first transposable element identified by Barbara McClintock in the 1940s, the *P* element of *Drosophila*, the *Tc1* element of *Caenorhabditis elegans*, and the *mariner* family, which is found in a large group of organisms ranging from protozoans to vertebrates. Other class II elements include Helitrons, which mobilize single-stranded DNA through a replicative, rolling-circle mechanism (11), and Mavericks, also known as Polintons, which might be related to certain dsDNA viruses (12).

Both class I and class II transposons comprise autonomous and nonautonomous elements. Only autonomous elements are equipped with all the sequences required to encode the proteins essential for their propagation in the host genome. In contrast, nonautonomous elements have to be mobilized by enzymatic activities that are supplied in trans by autonomous elements. MITEs are internally truncated forms of autonomous elements, the terminal recognition sequences of which have been retained.

1.2. Amount and Distribution of TEs in the Genome

One way to increase genome plasticity is to increase the genome size. However, genome size as measured by the "C value" (genome size (bp) $= (0.921 \times 10^9) \times$ DNA content (pg)) is not correlated to organism complexity or to the number of protein-coding genes (13). Most of the differences in genome size between species result from noncoding parts. This variable part of the genome consists mostly of repetitive sequences, such as satellite DNAs and TEs. We first focus on TEs in different species, and then go on to look at the variable distribution patterns of different classes of TEs between organisms.

In prokaryotes, there is some correlation between genome size and IS number, i.e., the larger the genome, the more ISs it contains; however, most genome size variation is not caused by transposable elements, but results from horizontal gene transfer (14). Touchon and Rocha (15) have shown that in the particular case of pathogen prokaryotes genome size is the most important variable, accounting for 40% of the variance in IS abundance, but that the type of ecological association also accounts for half of this variance (i.e., whether they are free living, commensal, pathogens, or mutualists). Nevertheless, no association has been found between pathogenicity and the presence or absence of ISs.

There are great differences in genome structure between prokaryotes and eukaryotes, and the latter show a positive correlation between genome size and transposable element abundance. However, there are discrepancies between unicellular and multicellular organisms. Several unicellular eukaryotic genomes have no TEs (red alga *Cyanidioschyzon merolae* (16) and protozoan parasites, such as *Cryptosporidium hominis* (17)). However, in mammalian species, TEs make up a large fraction of the genome. TEs account for 45% of the human genome (18), 37.5% of the mouse genome (19), and 55% of the opossum genome (20). Nevertheless, the fraction of the genome accounted by TEs differs dramatically in nonmammalian vertebrates and invertebrates. TEs account for approximately 8% in chicken, 26% in zebra fish, 10% in *Drosophila*, and 9% in *Caenorhabditis elegans* genomes. For plant species with relatively small genomes, such as *Brachypodium distachyon*, DNA derived from TEs usually constitutes 20–30% of the genome. Species with larger genomes have commensurately higher proportions of TE-derived DNA, which constitutes more than 85% of the maize and barley genomes, for instance (21).

Since retrotransposons and transposons have experienced differing degrees of success with regard to genome invasion, the proportion of the classes of TEs, in terms of their total number, composition and location diverge from one genome to another (22). For instance, class I retrotransposons are the only TEs present in the genomes of *Saccharomyces cerevisiae* or *Entamoeba dispar*. Their abundance is also relatively high in human, mouse, and *Drosophila erecta*, where they represent roughly 85% of the total TE population (23). In several plant genomes, they have been shown to constitute about 40–60% of total TEs (24, 25).

Class II elements also contribute to differences in genome size, but to a lesser extent, as they are ten times less abundant than class I TEs, except in *Trichomonas vaginalis* or *Entamoeba moshkovskii*, in which 100% of TEs are DNA transposons (23). In general, DNA TEs constitute only a small fraction (2–3%) of the human, rice, and maize genomes (24, 25). In comparison, the abundance of class II TE is relatively high in wheat genomes (almost 12%) (26). The prevalence of MITEs (74% of the total number of DNA TEs) over transposase-encoding elements is particularly striking in the human genome. The human genome contains MITEs derived from different evolutionary groups: the hAT, *pogo*, *mariner*, *piggyBac*, *Merlin*, and *Mutator* superfamilies (27, 28). They are also a characteristic of the DNA transposon population of plants and nematodes. Fifty-six thousand MITEs have been identified in sorghum (29) and 73,500 in rice (30).

The propensity for rapid changes in genome size between species and individuals as a result of TE amplifications shows that an understanding of TE dynamics is critical for understanding the evolution of genome size, structure, and function. Comparative studies of the distribution of TEs among genomes are essential if we are to understand the balance between the evolutionary forces that have resulted in this distribution.

1.3. Life Cycle

The life cycle of a TE family is akin to a birth-and-death process (31). The scenario is likely to proceed in three distinct phases. First, the initial phase of the TE invasion of a population (see the following Subheading 1.4) does not generally proceed at a low and uniform rate, but tends to occur in sudden episodic bursts (32). If excessive disruption of a genome were to occur, this could lead to the extinction of the individual or lineage. This is the phase during which the genome is highly unstable and generates new combinations of genes and new regulatory networks (see the following Subheading 2.3). Second, after this invasion, there is a period of time during which the TE family proliferates within the population accompanied by mutations that render some elements inactive. Most TE insertions are tolerated (see the following Subheading 2.2) as a result of the repression of other members of the TE family; a kind of host defense mechanism. This period is a stage during which the genome is considered to be dynamic, with copy amplification and loss of copies roughly in balance. Finally, during a more quiescent state, transposition continues at a reduced rate, and selection against insertions is still required to stabilize the copy number. During this final phase, the TE could be extinguished as a result of the accumulation of mutations within TE copies, and some copies could be "recruited" by the genome for new functions through the domestication process (33). There are two main ways for TEs to escape extinction: by horizontal transfer to a new host genome prior to inactivation (34) and by inflicting of minimal harmful effects by having a low replication rate so as to evade detection by the host.

1.4. Transposon Activity and Speciation

Characterization of the repetitive landscape in mammalian model organisms initially suggested a disparity between class I and class II transposable elements in terms of their prevalence and activity levels. Little is known about the absolute rates of transposition because few of the elements described have been observed to be active in their host. It has, therefore, been difficult to characterize transposition rates experimentally. One of the first signs that demonstrate that a transposable element is potentially active is finding complete elements in the genome. However, completeness is not a demonstration of activity in itself. A rough analysis reveals that most TE-derived sequences seem to be nonfunctional. Moreover, the proportion of active copies varies considerably among species. Less than 5 and 20% of the elements of *Schizosaccharomyces pombe* (35) and *Drosophila melanogaster* (36), respectively, appear to be of full length. In humans, Alu and LINE-1 (L1) elements were first identified during the late 1960s, but active Alu and L1 copies were not recognized until the 1980s, when they were found to transpose into genes causing human diseases (37). Only 1% of these *L1* elements are complete, and only 1% of these full-length *L1s* are thought to be active; this is 100 times fewer than in the mouse. Active human transposons have been estimated to generate about 1 new insertion per 10–100 live births. At the opposite extreme, there is no evidence of DNA TE activity in the past 50 My in the human genome (18), whereas class II TE activity can be detected in other eukaryotic lineages. Numerous studies have also noted a connection between TE transcription activity and abiotic or biotic stress (38–41). The analysis of the genetic distance between two organisms, the crosses between species and the phylogeny, has revealed interesting evolutionary pathways of TEs. Increases in TE activity in response to physiological stress may provide the basis for the punctuated equilibrium model of evolutionary change (42). Several researchers have found examples of concordant timing between bursts of transposition or massive TE extinction and speciation events (43). As many TEs are lineage specific; this suggests that TEs have contributed to the process of speciation either as a cause or an effect (44–46). For instance, the timing of the bursts of *Tc1-like* and *piggyBac* transposon replication activity coincides with that of the radiation of the *Salmoninae* to *Salmo*, *Oncorhynchus*, and *Salvelinus* (47). In the plant genus *Helianthus*, interspecies hybrids may show macroevolutionary driving by a burst of *Ty3/gypsy*-like LTR retrotransposons (48). In Mills et al., the time line for speciation of humans and chimpanzees is compared with the time line for the generation of transposon insertions (49). This shows that a burst of the specific Alu family occurred in the human genome about 6 Myr ago. During this period, the brain tripled in mass in our lineage. In his review, Britten et al. developed arguments that TE insertions, and particularly Alu, were involved in the establishment of the human brain size (50).

2. TEs Are Natural Tools Coming from the Dark Ages

2.1. TEs as Natural Tools that Reshape the Genome

The emergence of increasingly complex life forms was accompanied by bursts of TEs not only leading to deleterious genomic effects, but also enhancing genome complexity by creating new genes by the process of exaptation and elaborating sophisticated gene regulatory pathways.

The direct effect of an insertion depends on the precise locus, where the element is inserted (intergenic sequences, exons, introns, promoters). An accumulation of active TEs can be observed in neutral sites, whereas deleterious events lead to preferential elimination. Of course, these extreme cases are modulated by mutation events in TEs. It is noteworthy that the most influential effects of TEs can arise and persist long after transposition has occurred. In fact, even dead TEs can lead to chromosomal rearrangements; when high TE content is coupled with low TE diversity, it can promote ectopic recombination between homologous sequences located within TEs (51, 58). Genome modifications due to transposition events are responsible for several human diseases (52). Alu or L1 insertions have both been linked to 15 diseases, such as hemophilia A (53, 54), glycogen storage disease, and duplication of the beta globin gene (55, 56). However, in fact, less than 0.2%

of known disease-causing mutations are due to gene inactivation as a result of TE insertion; and 0.3% of pathogenic mutations in the human genome appear to result from ectopic recombination between TEs. The positive effects of TEs are to increase genome complexity and speed up the adaptation process (57, 58). These positive effects occur at three molecular levels. First, TEs contribute to various macrostructural aspects of the organization of the chromatin within the nucleus, such as the centromere and telomere. They are also source of chromatin loop structures, since in the human genome 55% of scaffold/matrix attachment regions (S/MARs) are derived from TEs. Abundant TEs in S/MARs have also been observed in yeast, *Drosophila*, and plants (59). Other loop regions contain TE-derived sequences and are defined as locus control regions. These two kinds of sequence play a substantial role in gene regulation by partitioning the genome into distinct transcriptional foci.

Second, at the gene scale, duplication and amplification of genes can occur indirectly as a result of retrotransposition. One consequence of segmental duplications is the appearance of new coding sequences as a result of exon shuffling. DNA transposons can also mediate exon shuffling and gene duplication, as demonstrated by *mutator-like* and *helitron-like* elements in plants (60, 61). TEs have also been co-opted by the host genome via an evolutionary process known as "molecular domestication." They give rise to what have been called "neogenes." About 50 neogenes are known in the human genome; for example, the centromere-binding protein (CENP-B), which is an ancient descendant of a *Pogo-like* transposase (62). The most striking example of TE domestication in vertebrates is the immune system's recombination mechanism. The V(D)J antibody recombination system is dependent on recombinase-activating genes 1 and 2 (RAG1 and RAG2) (63). These genes originated from a single class II element, *Transib*. One of the transposase-derived proteins to have emerged most recently in primates is the protein SETMAR, which results from the fusion of a preexisting SET histone methyltransferase gene to the transposase gene of the *Hsmar1* transposon (64). This protein has been implicated in the nonhomologous end joining of double-stranded DNA breaks. Domestication of TEs also occurs in nonmammalian eukaryotes. For instance, in *Drosophila*, the telomerase function does not exist, but this function is carried out by the non-LTR retrotransposons TART and HeT-A that have accumulated in multiple copies at the chromosome ends. Similarly, in the silkworm, *Bombyx mori*, and the green seaweed, *Chlorella vulgaris*, elements of the LINE type are located at the telomere level, even though these species do have the telomerase function (65).

And thirdly, their effects take place at a more molecular level. Indeed, TEs contribute directly to gene regulation since they supply *cis*-regulatory elements, such as basal promoters, enhancers, insulators, alternative splicing sites, polyadenylation signals, and a plethora of transcription binding sites. Nearly 4% of human genes have TE-derived sequences embedded in their coding sequences, and about 25% of human promoter regions contain a TE-derived sequence (46). Among other examples, five transcription factors, ESR1, TP53, POU5F1, SOX2, and CTCF, are embedded in different families of transposable elements (66).

2.2. DNA Methylation Landscape and Epigenetic Control of TEs

We accept that TE activity within a host genome could be catastrophic. However, in practice, all organisms have evolved cellular TE control measures to minimize the potential harm caused to their somatic cells. A number of epigenetic gene-silencing mechanisms, such as cytosine methylation, genomic imprinting, and heterochromatinization, are thought to have evolved as defense mechanisms against transposition. DNA methylation contributes to the formation of heterochromatin, particularly in TE sequences, in order to silence them and so to block their activity. Land plants and vertebrates retain extensive DNA methylation to repress TEs in their somatic cells. Moreover, they reproduce primarily by sexual outcrossing, a process during which TE aggressiveness is reactivated, because outcrossing and gametogenesis partially break the link between host and TE fitness (67–69). TEs become transcriptionally active in response to dramatic hypomethylation of DNA. However, there are exceptions to this link between transposons and DNA methylation, which call into question the conservation of transposon methylation as Susuki and Bird point out in their review "DNA methylation landscapes: provocative insights from epigenomics" (70).

In a similar way, RNA interference by siRNAs or miRNA is considered to have originated to silence TEs in plants. In animals, another class of small RNA, the piRNAs, plays a crucial role in TE control in the germ line. RNA interference defines a posttranscriptional gene-silencing mechanism that mediates destruction, translational repression, and transcriptional silencing. Many small RNAs originating from transposable elements and MITEs have been identified in the human genome, and could potentially contribute to the autoregulation and regulation of thousands of host genes (71).

Naturally, the question that arises is: "How can TEs invade a genome if they are trapped in an inviolable prison?" In fact, we know that TEs often do transpose, albeit at a very low rate, suggesting that the prison is not, after all, entirely inviolable. Consequently, we need to identify the causes of genome-wide epigenetic modifications and subsequent TE activation. Interspecies crosses induce genomic stress, i.e., changes in genomic stability (chromatin alteration, density of repeats…) and organization (DNA recombination, TE replication) that could have an impact on epialleles and provoke TE activation (42). Cell stressors, including heat shock, genetic damage, oxidative stress, translational inhibition, and viral infection, are known to activate TEs, particularly SINEs (38, 40, 41, 72–74) or retrotransposons in plants (75, 76). So, in response to biotic or environmental challenges, an increased level of TE transposition might be advantageous, accelerating the rate of genome restructuring and promoting potentially useful genetic variability (58). One appealing possibility is that TEs may actually be part of the normal physiological response of the cell to stress, with putative functional roles in DNA double-strand break (DSB) repair (77) and the regulation of protein translation (78).

2.3. TE or Not TE That Is the Network?

In this review, we have collated some of the facts available about transposable elements, and seen that genomes contain a huge number and variety of TEs. Most of them have been dead for a long time, but they are still reminiscent in the various genomes of the entire tree of life. They contribute to the macroevolution of the genome and to the regulation of gene expression, since they are source of new regulatory sequences, they initiate and are subject to epigenetic mechanisms, and most of them are characteristic of specific lineages. However, it has also been reported that some DNA regions are completely devoid of transposons. These transposon-free regions have been maintained through vertebrate evolution. So, how can we envisage the role of TEs in the process of evolution? Are they just " junk DNA" or "parasite elements," as Doolittle thinks (79)?

TE-free regions or TFRs are not restricted to mammals, but are a common feature in vertebrate genomes. These regions are resistant to transposons of both major classes, and most of them have been maintained at the same locus in different species. This hypothesis is supported by the observation that in cancer-associated retroviral screens the integration of retroviruses, which occurs by a mechanism somewhat similar to retrotransposon integration (80), appears to be uninhibited within TFRs (81). However, the molecular and genetic mechanisms that prevent these extended regions from tolerating transposon sequences are still unclear. So, it looks as though TE-rich regions and TFRs are implicated in different regulatory networks, leading to the linking of sets of genes. This linking or wiring would allow genes belonging to the same network to respond simultaneously to environmental changes (10, 42, 58). TFRs are significantly associated with genes encoding developmental and transcriptional regulators, whereas TE-rich regions are involved in transport and metabolism (82). This idea was first put forward by Barbara McClintock, and further developed by Britten and Davidson (83, 84). For instance, multiple p53 binding sites in the human genome are located in SINE/Alu repeats, and many of these Alu sites constitute regulatory transcriptional regions of host genes (85). So, the binding of p53 induces Alu silencing and brings multiple genes simultaneously under the control of p53 through its association with different copies of the same transposon elements.

This constitutes a specific transcriptional network of p53-regulated genes. However, this network is not established once and for all. New insertions or mutations in the existing elements can appear, and cause changes in the network by the addition or loss of *cis*-regulatory sequences.

Furthermore, the distribution of TEs in the genome is conserved through evolution. This could be explained by different selection pressures acting on TE-free and TE-rich regions. Zeh (42) proposed that the equilibrium between host genomes and transposable elements may result from an evolutionary tug-of-war mediated through the epigenome.

3. Traveling Through the Expanding Universe of Transposon Technologies

Many successful technical applications have been derived from the use of TEs. This technology was made possible by the wide distribution of TEs in the living world and also by better understanding how they operate. What is new is that it is now realized that this mode of operation is not limited to the species in which the TEs were originally described, but displays a broad operating potential regardless of the species or the differentiation stage of the cells involved. The natural abilities of TEs as mutagenic agents or as gene carriers have made it possible to develop tools to provide answers to many biological questions, including the characterization of new genes and their involvement in regulatory networks, and also the development of new integrative vectors for treating diseases or producing specific biomolecules. Obviously, to achieve mastery of these new biological tools, we have to develop new tools to store, manage, and analyze new transposons in order to provide a better understanding of their contribution to the evolution of genomes.

3.1. The Dense Forest of Bioinformatic Tools

Genome sequences have accumulated for almost all the phyla of the tree of life. All this progress has been achieved as a result of the improvement in genome-sequencing techniques. The correct assemblage and analysis of the genomes encounter recurrent difficulties owing to the huge number of repeated sequences, such as transposon elements. They have been present in the genome for a very long time, and even if they are similar in sequence they are not identical as a result of point mutations or the presence of nested copies. Due to their repeated nature, their wide distribution throughout the genome, and their mode of propagation, they can lead to confusion in gene annotation, in determining boundaries, and in the process of genome assembly, three steps that are essential for the full characterization of coding and repeated sequences. In this respect, several computational tools have been developed for the ab initio identification and automatic annotation of TE sequences recently reviewed by Lerat (86). Some of them have been presented in Chapters 2–6. The main drawback highlighted in various publications is that we only look for what we already know, and we only find what we are looking for. Abrusan et al. have developed a tool for the annotated classification of unknown eukaryotic transposable elements (87). In the case of taxonomic groups that have so far received less attention, newly identified TEs frequently show no clear similarity to known repeats, and other approaches therefore have to be adopted to classify them. For example, no IS has been identified in the *Nanoarchaeota* or in the *Thermoproteales* because a smaller portion of their genomes has been sequenced. This limited background information may lead to bias (6). Programs, such as Repbase and Repeat Masker, have been developed to standardize the classification of repeats and to expand tools for the systematic annotation of repeated sequences (88, 89). They have proved to be extremely valuable by "masking" transposable elements in query sequences during homology searches so that the presence of a common transposon does not lead to spurious matches. Some browsers are dedicated to specific organisms, such as the ISbrowser for bacterial species (90), specific classes of transposons, such as MITE-Hunter (91), or a specific functionality, such as the exonization of TEs, like SERpredict (92). Nevertheless, in order to provide high-quality gene annotation, models must be produced by combining multiple independent sources of computational evidence.

3.2. TEs as Genomic and Modern Phylogenetic Markers

The huge amount of genomic data and the myriad applications of modern molecular techniques, such as molecular cloning, sequencing, and polymerase chain reactions of DNA, have resulted in enhanced resolution of phylogenetic analysis. The study of evolution frequently requires resolution of the relationships and the gene flow between species. The identification of these patterns relies extensively on the use of genetic markers, and the importance of the choice of marker in phylogenetic studies has long been stressed. However, certain properties of the molecular markers have to be taken into consideration when the results of the molecular phylogenetic analyses are discussed, notably issues regarding potential sources of homoplasy (a character shared by a set of species, but not present in their common ancestor), lack of neutrality of genetic markers, or uneven chromosomal distribution. Analysis of nuclear genomes using only truly orthologous genes is required for phylogeny reconstruction, but it may face problems in distinguishing genes from pseudogenes. The most widely used markers are satellite DNA used, for example, as taxonomic markers in nematodes (93) and mitochondrial DNA (mtDNA) sequences. Recently, Hurst and Jiggins concluded that mtDNA is inappropriate for use as a sole marker in studies of the recent history of arthropods (94).

Mobile elements have been recognized as powerful tools for phylogenetic and population-level analyses since they have ubiquitous distribution, high copy number, and wide chromosomal dispersion. Moreover, precise excision (95), insertion hot spots (96), and incomplete lineage sorting (97) of retroposed elements are thought to be extremely rare events in mammalian evolution. Due to the virtually homoplasy-free nature of retroposons, the analysis of retroposon presence/absence patterns avoids the pitfalls of other molecular methodologies and provides a rapid unequivocal way of revealing the evolutionary history of organisms. For example, retropositions have provided conclusive evidence for therian mammals (98) and early placental mammalian divergences (99), as well as revealed relationships within other mammalian orders (100, 101). Gu et al. demonstrated that IS256 can be used as molecular marker to discriminate invasive strains from commensal strains of *Staphylococcus epidermidis* (102). The only significant limitation of this method is that nodes that are difficult to resolve by sequence data are also rarely supported by the presence/absence patterns of retroposed elements (103). To overcome this drawback, Kriegs et al. have developed several strategies relying on the presence of a given retroposon in related taxa (99). This makes interpreting retroposon markers relatively simple and straightforward: the presence of one of these elements in the orthologous genomic loci of two species signals a common ancestry while its absence in another species signals a prior divergence (104). The use of presence/absence analyses to reconstruct the systematic biology of mammals depends on the availability of retroposed elements that were actively integrated before a particular species diverged. In Chapters 7 and 8, current trends in the application of DNA marker techniques in diverse domains of insects or plants have contributed significantly to progress toward understanding the genetic basis of organism diversity, and for mapping the expression profile of a domesticated transposase in cancer (see also "Genome regulation: TEs are part of the epigenetic journey"). There are various methods (RAPD-PCR, IS-PCR, RELP, ISSR-PCR) available for revealing insertion polymorphism of retrotransposons; Behura, Jones et al., and Kalendar et al. discussed their advantages and shortages (105–107). A careful analysis of the conformities and contradictions between different data sets and looking for congruent conclusions deduced from different characters are the most fruitful ways to advance phylogenetic development.

3.3. Genome Exploration: The Adventure of TE-Based Mutagenesis

Because of their intrinsic ability to disrupt genes and cross-species boundaries, TEs act as natural mutagenic agents in various genomes. The possibility of engineering these elements makes transposon-based mutagenesis a powerful approach for gene identification and analysis in species as disparate as insects, protozoans, bacteria, plants, worms, and mammals, especially for the identification of new genes of unknown function or to identify genes involved in a metabolic process or pathogenic mechanism (108).

Although prokaryotic genomes contain numerous active bacterial transposons, the main drawbacks are that these elements often require host factors and the DNA targets of the transposition events are not totally random, since insertions can occur in different hot spots in the genome, as demonstrated for *Tn917* in different bacteria (109). Similarly, some regions of the genome are unaffected by transposition events. To circumvent these issues, TEs showing a broad tropism should be used so that the whole genome can be targeted. Members of the *mariner* family are eukaryotic elements known to present no host range restrictions and to transpose both in eukaryotes and prokaryotes with no specific sequence requirements. As reviewed in ref. 109, *Mos1* from *D. melanogaster* and *Himar1* from *H. irritans* are *mariner* transposons widely used to generate mutant libraries in bacteria and archaea to screen for genes implicated in various aspects of metabolism and physiology, as well as in pathogenic characters. Insertional mutagenesis could be performed both in vivo and in vitro. The in vivo transposition system relies on a plasmid containing an antibiotic resistance gene flanked by transposon terminal inverted repeats and a source of transposase. This resistance gene makes it possible to positively select host cells bearing random insertions of the transposon. In vitro transposition requires an additional step to move the mutated DNA back into the bacterial cell in order to study the effect of the transposon insertion in vivo. The major advantage of these techniques is that the transposon insertion acts as a molecular hallmark, allowing the rapid identification of the mutated gene for forward genetic screens. Judson and Mekelanos (110) and Kim (111) pinpointed essential genes in *Vibrio cholerae* and *Salmonella enteritidis* using the TnAraOut *mariner*-based transposon system. This system, which can be used for high-throughput mutant screening, contains an outward-facing arabinose promoter. Mutant-harboring TnAraOut insertions in the promoter region of an essential gene display an arabinose-dependent growth phenotype. Signature-tagged mutagenesis, initially described in *Salmonella typhimurium* with *Tn5* (112) and further explored with *Himar1*-based vectors (113), is a method generating random mutants with various tags in order to distinguish simultaneously individual attenuated mutants in a mixed population of mutants.

Genomic analysis and mapping by in vitro transposition or the GAMBIT technique is performed in vitro in PCR products, which are then transformed and studied by genetic footprinting (114). GAMBIT provides a powerful system for identifying essential genes in a pool of mutants. Recently, a refinement of the microarray technology has allowed the emergence of Transposon Site Hybridization (TraSH) method for large-scale genomic studies. This technique uses DNA microarrays to locate transposon insertions in a mutant library and identify genes necessary for in vitro growth and survival after exposure to particular environments. Chapter 9 describes one of the methods and protocols for mutagenesis in bacteria based on TEs. This method, principally using *Himar1 mariner* elements, has led to the development of genetic molecular tools that provide a better understanding of bacterial species to identify protein structure–function (115), and of several putative virulence factors (116, 117) to provide a promising approach to validate new antibacterial drug targets.

While prokaryotic TEs often have short sequences and are not numerous in bacterial genomes, TEs are major components of most eukaryotic genomes and are particularly abundant in plants. Chapter 10 sets out to describe various protocols used to carry out TE mutagenesis in plants, and particularly in crop species, since they constitute an economically important line of research. Knowing the genes involved in plant growth or reproduction is the first step before any attempt can be made to manipulate the genome in the hope of enhancing crops and especially yields. This has been done in *Arabidopsis thaliana*, the reference model, and in other species essentially using T-DNA. DNA transposons, such as *En/Spm-I/dSpm*, *Ac/Ds*, or *nDart1*, and retroelements, such as *Tos17*, were employed in forward or reverse genetic approaches to relate a gene to its function, particularly in rice (118). Rice is one of the world's most important food crops, and is the first cereal to have its genome sequenced. Even so, the function of a large number of these putative genes remains to be elucidated. Insertional mutagenesis can be attempted with foreign elements, but another approach is to reactivate the endogenous *Tos17* retroelement using various biotic or abiotic stressors (119). However, some of these systems may be associated with somaclonal variation (120). One way round this is to use endogenous active DNA transposons, which are free from somaclonal variation. These transposons have been extensively used for gene tagging in maize, snapdragon, petunia, morning glories, and recently in rice.

Whereas endogenous transposons appear to be a preferential option for genomic analyses in plants, such a strategy is hindered in the nematode *Caenorhabditis elegans* because transposition of endogenous elements like *Tc1* and *Tc3* is tightly repressed in the germ line of standard laboratory strains by an RNAi mechanism. Only specific strains, such as *Mutator*, offer backgrounds in which transposition can be achieved as a result of the derepression of all DNA transposons. However, this implies that the overall transposition cannot be controlled and is unspecific, making it difficult to perform studies in these strains. Consequently, exogenous DNA transposons, including *Himar*, *Sleeping beauty*, *Minos*, or *Tol1*, were assayed in *Caenorhabditis elegans*, but no or few transposition events were observed in the germ line. To date, only the *Mos1 mariner* element has been efficiently mobilized in both somatic and germ cells. In the nematode, transposition is also used to trigger DNA DSBs at the transposon excision site (121). Imperfect repair by the host machinery could lead to chromosomal lesions, and therefore to the inactivation of genes. Induced DSBs can also serve to "instruct" a desired transgenic template to manipulate *Caenorhabditis elegans* loci by homologous recombination. In this context, a new generation of genetic tool based on *Mos1* and known as "Mos1 excision-induced Transgene-Instructed gene Conversion" or MosTIC was developed to perform mutagenesis, genome engineering, and for integrating single-copy transgenes into the genome of *Caenorhabditis elegans.* This technique is described in Chapter 11, and was recently used as a basis for the development of the MosSCI, the *Mos1*-mediated Single-Copy Insertion technique used to insert single-copy transgenes at defined positions in the *Caenorhabditis elegans* genome (122).

The *Tc1-mariner* family is not only widely used for nematode mutagenesis, but it is also often utilized for protozoan genomic studies. This is done in an attempt to identify new drug targets or vaccination strategies, and could provide important benefits since protozoan parasites affect millions of people worldwide (400 million people are affected by *Plasmodium*, and 12 million by *Leishmania*), and no vaccine or efficient treatment is yet available. Now that the *Plasmodium*, *Leishmania*, and *Trypanosoma* genomes are available, genome-wide strategies to study gene function are necessary. At present, the *Mos1 mariner* transposon is the transposon toolkit most often used in gain- or loss-of-function studies and gene-trapping assays, especially in *Leishmania* (123). The *piggyBac* and *Tn5* transposons have been shown to transpose in *Plasmodium* (124, 125), and *Tn5* was used in a shuttle mutagenesis protocol (126). *piggyBac* is ideal for trapping a gene promoter due to its biased tendency to insert within the 5′UTR of plasmodia genes. Some of the transformation and mutagenesis systems using TEs in protozoans have all proved useful genetic tools for studying parasite drug resistance and virulence (127).

Most of parasitic protozoans, such as *Plasmodium*, *Leishmania*, and *Trypanosomia*, and other pathogenic agents, such as the yellow fever virus, are carried by an insect vector. Although the very first transposon-based, genome-wide approach to the study of gene function in insects was developed in the 1980s in *Drosophila* with the *P* element, TE-based insertional mutagenesis systems are increasingly being engineered for genomic analyses of insects that constitute a public health issue, as well as for medical or agronomic purposes (128). Studies can be performed in insect cell lines or conducted directly in vivo in the entire insect. One of the major challenges is to create a heritable mutant by inserting an element in or near a specific gene to provide critical information about insect biology and development.

3.4. Genome Manipulation: Jumping to Public Health Issues and Gene Therapy with TE-Based Transgenesis

Once the insect genome and biology have been deciphered, TEs can be considered to be genetic resources for insect transgenesis in species of economic and ecological impact, such as mosquitoes (129), *Lepidoptera* (130), and *Diptera* (131), with *Drosophila melanogaster* as the gold standard (132). Even though only a few cases have been reported, the generation of stable transformants of the *Bombyx mori* silkworm to improve its qualities is of economic importance. In this context, *piggyBac* has been used to confer enhanced resistance to nucleopolyhedrovirus on silkworms (133), which could be beneficial for the silk industry. In a different way, insect transgenesis provides a valuable genetic tool for modifying and thereby controlling insect pest species ranging from crop or fruit fly pests to vectors of human diseases. The sterile insect technique is one of the major components of agricultural pest management programs, based on the massive release of radiation-sterilized insects to compete with wild individuals for reproduction. Transgenesis has been suggested as a way to sterilize natural populations (134), albeit posing the problem of mass rearing. Secondly, insect transgenesis can be used to mark the released flies with genetic reporters to make it easier to identify the modified insects, and lastly a transgenic approach can be applied to SIT programs by leading to the production of only male progeny (135). Another major focus of TE applications in insect transformation is to prevent the spread of insect-borne pathogens by engineering mosquitoes so as to confer artificial refractoriness to pathogen transmission. This promising concept was developed after naturally occurring refractoriness had been shown to be under the control of the immune system genes, and could be successfully selected for laboratory strains of *Anopheles gambiae*.

To date, insect germline transformation systems allowing stable and vertical transmission are all based on DNA transposons isolated from insect species. There are at least four gene-vector systems that can be employed to generate transgenic nondrosophilid insects: *Hermes* (136), *Mos1* (137), *Minos* (138), and *piggyBac* (139). These systems and their applications in insects are well described in ref. 140. They have been used to transform 15 species of insect, including *Diptera*, *Hymenoptera*, *Lepidoptera*, and *Coleoptera* (135). Nevertheless, the efficiency and ease of handling of each of these elements are variable depending on the insect targeted (141). Consequently, in spite of their broad host ranges, TEs must be tested in the studied lab model. For example, more than 50% of *Drosophila melanogaster* mutant offspring can be obtained with *Hermes*, whereas this transposon yields less than 10% of transgenic *Aedes aegypti* (142, 143). *Mos1* moves efficiently in *Drosophila mauritiana*, but is almost immobile in the closely related *Drosophila melanogaster* (144, 145). *piggyBac* is inefficient in *Anopheles gambiae*, whereas mutants of *Anopheles albimanus* have been observed (146, 147). The *mariner Himar1* is completely inactive in insects

(148), whereas the bacterial transposon *Tn5* appears to be an efficient insect gene vector (149). Chapter 12 presents some approaches to manipulating flies by generating mutant libraries with the *P* element and the ϕC31 integrase (150).

Gene transfer is no longer limited to insects, plants, and worms, but is widely applied to create transgenic vertebrates in the hope of establishing gene therapy protocols. In the first approach, viral vectors were identified that could be used to integrate a desired transgene into animal cell lines. Nevertheless, immunity issues, low cargo capacity, and the oncogenic risks associated with their insertion pattern prompted a search for nonviral integrative methods, including site-specific recombination systems, phage integrases, and transposons (150–153). Various transposon families have already been exploited to perform vertebrate transgenesis in both somatic and germ culture cell lines or directly in vivo, including attempts with retro- or DNA elements in mammals (154–156), zebra fish (157), and chicken (158). This book focuses in particular on the "tool boxes" of two of the most promising elements, *Sleeping Beauty* (*SB*) and *piggyBac* (*PB*) in Chapters 13 and 14, respectively. These DNA transposons are mainly engineered for effective gene delivery applications in mammals, including in vitro and in vivo murine and human studies. The *Sleeping Beauty* transposon is a synthetic transposable element, molecularly reconstructed by reverse engineering from defective copies of salmonid genomes (159), and hyperactive transposase mutants are now available to address the problem of weak transformation levels. The activity of this transposon in a broad range of vertebrate models has now been demonstrated. This element can transpose and operate gene delivery in zebra fish (160), *Xenopus* (161), and pigs (162), but attention has focused particularly on mouse and human trials (159, 163). In mice and rats, *SB* was assayed for transgenesis in somatic tissue, germ line, embryonic stem cells, and a one-cell embryo (164–167). Most importantly, a protocol for in vivo gene delivery in order to replace a defective gene has been successfully used to treat tyrosinemia (168). On the basis of these first preclinical models, *SB* has been investigated for stable transgene expression in human cells, including primary T cells (169) and hematopoietic stem and progenitor cells (170). This pioneering work paved the way for a still ongoing human trial in the USA (171). Another DNA transposon recently used in vertebrates is the *piggyBac* transposon system derived from the cabbage looper moth *Trichoplusia ni* (172). Like its counterpart *SB*, *piggyBac* is mobile in many different species, and has been tested for efficient transgenesis and long-term expression in human cells and mice (173, 174), as well as in the chicken (175). However, *PB* shows a significantly higher transposition rate in mammalian cell lines and higher cargo capacity than *SB* or *Tol2* (176–178). Moreover, this element mostly excises itself precisely upon transposition and leaves no footprint behind (172), thus allowing seamless reversible remobilization (179). *PB* has also been engineered to generate inducible (180) or hyperactive transposases (181). Another important trait of *PB* transposase is the possibility of using it to fuse site-specific DNA-binding domains without any loss of activity, unlike *SB*, in order to confer site integration specificity (163, 182). Its unique features and the emergence of clinical applications of *PB* (183) have made this transposon a promising and attractive alternative to *SB* for the purposes of genetic study and gene therapy (184). In a different area of clinical research, the *Sleeping Beauty* and *piggyBac* transposition systems have been explored with a view to reprogramming fibroblasts into induced pluripotent stem cells (185, 186) and to coaxed differentiation of these iPS cells into myogenic progenitor cells (187). This novel approach consolidates the position of transposable elements in regenerative medicine, and cellular and gene therapy.

The use of integrating vectors for clinical gene therapy studies necessitates determination of vector integration loci to assess vector biosafety and to monitor the fate of gene-corrected cells. Protocols for high-throughput integration-site profiling have been established in recent years. Insertion site patterns can be performed by various PCR-based methods, such as inverse PCR (188), ligation-mediated PCR (LM-PCR) (189), or linear amplification-mediated PCR (LAM-PCR) (190). LAM-PCR is a reliable and robust method to detect and sequence unknown DNA sequences flanking the transposon at the single-cell level (191). Such protocols have been used to study clonal behavior and the physiology of gene-modified hematopoietic (192) and T cells (193). LAM-PCR is currently the most sensitive PCR-based method available, but it depends on restriction enzymes, which can hinder identification of the entire pool of integration sites. To bypass the use of restriction enzymes, Parunzynski et al. developed nonrestrictive LAM-PCR (nrLAM-PCR) coupled with mass sequencing for universal genome-wide and comprehensive integrome analysis (190). Chapter 15 proposes a detailed protocol for integration-site profiling using LAM-PCR. Other efforts have been made in this field, and have resulted in the Transposon Insertion Site Profiling chip (TIP-chip) assay. This strategy is based on using a tiling microarray to search for transposons either throughout an entire genome or solely in a region of interest. The proof of concept was originally described in *Saccharomyces cerevisiae* (194) and it was subsequently used to map L1 retrotransposons in the human genome (195).

3.5. Genome Regulation: TEs Are Part of the Epigenetic Journey

A parallel application of integrome analysis is the investigation of the potential effects of a transposon on the neighboring genes flanking the integration site. Active TEs, especially endogenous retroelements, are highly mutagenic and, depending on their location in the genome, can influence surrounding genes by altering splicing or polyadenylation signals or by acting as enhancers or promoters (196). For example, Zhu demonstrated that the presence of the *SB* transgene affected the expression of neighboring host genes at distances of >45 kb (197). To prevent potential harmful effects, three major epigenetic defense mechanisms against TEs have emerged during evolution, consisting of posttranscriptional silencing by RNAi, of histone modification, and more importantly of DNA methylation. These mechanisms are effective in most of the differentiated tissues of animals and plants, including the germ line (198). But what is the impact of the epigenetic silencing of TEs upon regulation of nearby genes and, by extension, upon the genome? Alteration of gene expression due to methylation of nearby TEs was expected to have deleterious effects on gene and genome functions, resulting in a loss of methylated TEs from gene-rich chromosomal regions. This hypothesis was recently confirmed in *Arabidopsis thaliana* (199). Numerous studies were carried out intended to pinpoint silencing effectors of TEs, and they succeeded in demonstrating that many cell cycle components, such as E2F–Rb (200), p53 (201), and c-Myc (202), and that chromatin-remodeling proteins, such as Lsh (lymphoid-specific helicase) (203), interact with transposable elements, and especially with SINEs, LINEs, and LTR retroelements, to control them and ensure that genomic integrity is preserved.

Since TEs are able to interact with factors of epigenetic regulation and are indeed a preferential target of these factors, the development of methods and protocols for epigenetic and histone profiling of the genome in a TE context appears to be a promising approach to elucidating differences in the methylation patterns in germ and somatic lines (204), or between healthy and cancerous cells (205), thus making it possible to identify candidate cancer genes. Indeed, some genes are frequently found to be methylated in cancer, principally due to retroelement silencing (206). Genome-wide analysis predicted that 22% of human genes implied in developmental processes would be methylation prone, i.e., down-regulated, in cancer. More and more studies highlighted this link between the methylation of TEs and their immediate environment during cancer development (207). Conversely, a lack or loss of methylation suppressing TE activity in somatic cells also drives tumorigenesis, as a result of the reactivation of endogenous active retroelements, leading to insertional mutagenesis (208) or chromosomal rearrangements (209). For instance, de novo L1 insertions occur at high frequency in human lung tumor genomes (210). To date, the real impact of TEs in oncogenesis is not completely clear. TEs tend to be viewed as triggers of tumorigenesis following external stimuli (heat shock, UV radiation, infection), but on the other hand a high level of reactivation of endogenous TEs can be the consequence of malignancy (211). According to the second hypothesis, reactivated retroelements act as enhancers of tumor progression by increasing DNA damage (212). In both cases, the potential effects of currently used cancer therapies on TE activity must be further investigated, since many cancer therapies include chromatin-modifying and methylation-reducing agents (209). The methylation state of LINE is now considered to be a potential biomarker for predisposition to many human cancers, such as colorectal (213), lung (214), familial testicular (215), and bladder cancers (216). In Chapter 16, several methods and protocols for characterizing tumors by investigating the methylation profile of TEs are described. Recently, a genomic tool called RepArray, which is used to generate repeat-specific methylation maps, has demonstrated that repeats can be used as markers of genome-wide methylation changes (217). In addition to being used as hallmarks for cancer predisposition, transposons can be employed to identify oncogenes or study molecular mechanisms driving cancer development by creating animal models of cancers. For instance, the *Sleeping Beauty* and *piggyBac* systems have been used to model a wide variety of human cancers in mice (218).

The potential uses and impact of TEs in oncology do not stop here. Domesticated transposons could also have a role in carcinogenesis. Some of the human transposon-derived neogenes encode proteins that are directly or indirectly implicated in genome stability (cell cycle, chromatin modification, regulation of transcription…). The neogenic recombinase SETMAR, or Metnase, has been shown to be involved in replication and genome repair machineries (219). Consequently, neogenes are an important source of genetic variability and could be involved in genetic instability mechanisms that lead to tumorigenesis or tumor progression. A deeper analysis of the impact of neogenes in cancer appears to be called for. However, most of the neogenes encode proteins that have not been well characterized and that have biological functions that are poorly understood. Chapter 17 describes an innovative protocol to produce antibodies directed against neogenic recombinases. These antibodies are generated through DNA vaccination of model mice, and have the interesting characteristic of targeting the entire protein expressed in its native conformation with high affinity. The antibody tool will make it possible to carry out expression and analysis studies of neogenes in both cancerous and noncancerous tissues, and to determine their real impact on tumor formation.

4. Conclusion

Sequencing various prokaryotic and eukaryotic genomes has revealed that transposable elements are ubiquitous among plants, bacteria, fungi and animals, and that they exhibit very diverse biological features. Their massive dominance in genomes means that they have been heavily involved in genome evolution, as they shape genome architecture and contribute to genetic innovations. Transposable elements that are currently available can be used to elucidate many biological, medical, and genetic questions. TEs form a large and still-expanding series of molecular tools that can be used to perform insertional mutagenesis and genome manipulation for functional genomic studies, and they are also used in transgenesis and gene therapy protocols. In addition to their technical applications, TEs provide an interesting way to elucidate some of the genome regulatory mechanisms. Genome-sequencing projects are still in their infancy, and there is no doubt that new elements with innovative applications will be soon discovered from a variety of life forms.

References

1. Chandler M, Mahillon J, (2002) Insertion sequences revisited. In: Craig N, Craigie R, Gellert M, Lambowitz A (ed.), Mobile DNA, vol. 2. ASM Press, Washington, DC
2. Blount Z, Grogan D, (2005) New insertion sequences of Sulfolobus: functional properties and implications for genome evolution in hyperthermophilic archaea. Mol Microbiol 55:312–325
3. Wagner A, (2006) Periodic extinctions of transposable elements in bacterial lineages: evidence from intragenomic variation in multiple genomes. Mol Biol Evol 23:723–733
4. Siguier P, et al (2006) ISfinder: the reference centre for bacterial insertion sequences. Nucl Acids Res 34:D32–D36
5. Kunst F, et al (1997) The complete genome sequence of the Gram-positive bacterium Bacillus subtilis. Nature 390:249–256
6. Filée J, Siguier P, Chandler M. (2007) Insertion sequence diversity in archaea. Microbiol Mol Biol Rev 71:121–157
7. Parks A, Peters J, (2009) Tn7 elements: engendering diversity from chromosomes to episomes. Plasmid 61:1–14
8. Reznikoff W, (2008) Transposon Tn5. Annu Rev Genet 42:269–286
9. Chaconas G, Harshey R, (2002) Transposition of phage Mu DNA. Craig N, Craigie R, Gellert M, Lambowitz A (ed.), Mobile DNA, vol. 2. ASM Press, Washington, DC
10. Feschotte C, Pritham E, (2007) DNA transposons and the evolution of eukaryotic genomes. Annu Rev Genet 41:331–368
11. Kapitonov V, Jurka J, (2001) Rolling-circle transposons in eukaryotes. Proc Natl Acad Sci USA 98:8714–19
12. Pritham E, Putliwala T, Feschotte C, (2007) Mavericks, a novel class of giant transposable elements widespread in eukaryotes and related to DNA viruses. Gene 390:3–17
13. Biémont C, Vieira C, (2006) Genetics: junk DNA as an evolutionary force. Nature 443:521–524
14. Frost L, et al (2005) Mobile genetic elements: the agents of open source evolution. Nat Rev Microbiol 3:722–732
15. Touchon M, Rocha E, (2007) Causes of insertion sequences abundance in prokaryotic genomes. Mol Biol Evol 24:969–981
16. Misumi O, et al (2005) Cyanidioschyzon merolae genome. A tool for facilitating comparable studies on organelle biogenesis in photosynthetic eukaryotes. Plant Physiol 137:567–585
17. Xu P, et al (2004) The genome of Cryptosporidium hominis. Nature 431: 1107–1112
18. Lander E, et al (2001) Initial sequencing and analysis of the human genome. Nature 409:860–921
19. Waterston R, et al (2002) Initial sequencing and comparative analysis of the mouse genome. Nature 420:520–562
20. Gentles A, et al (2007) Evolutionary dynamics of transposable elements in the short-tailed opossum *Monodelphis domestica*. Genome Res 17:992–1004
21. Tenaillon M, Hollister J, Gaut B (2010) A triptych of the evolution of plant transposable elements. Trends Plant Sci 15: 471–478
22. Sela N, Kim E, Ast G. (2010) The role of transposable elements in the evolution of

non-mammalian vertebrates and invertebrates. Genome Biol 11:R59
23. Pritham E, (2009) Transposable elements and factors influencing their success in eukaryotes. J Hered 100:648–655
24. Hawkins J, et al (2006) Differential lineage-specific amplification of transposable elements is responsible for genome size variation in Gossypium. Genome Res 16:1252–1261
25. Piegu B, et al (2006) Doubling genome size without polyploidization: dynamics of retrotransposition-driven genomic expansions in Oryza australiensis, a wild relative of rice. Genome Res 16:1262–1269
26. Wicker T, et al (2003) CACTA transposons in Triticeae. A diverse family of high-copy repetitive elements. Plant Physiol 132:52–63
27. Pace J, Feschotte C, (2007) The evolutionary history of human DNA transposons: evidence for intense activity in the primate lineage. Genome Res 17:422–432
28. Feschotte C, Jiang N, Wessler S, (2002) Plant transposable elements: where genetics meets genomics. Nat Rev Genet 3:329–341
29. Paterson A, et al (2009) The Sorghum bicolor genome and the diversification of grasses. Nature 457:51–556
30. Oki N, et al (2008) A genome-wide view of miniature inverted-repeat transposable elements (MITEs) in rice, Oryza sativa ssp. japonica. Genes Genet Syst 83:321–329
31. Blumenstiel J, (2011) Evolutionary dynamics of transposable elements in a small RNA world. Trends Genet 27:23–31
32. Ray D, et al (2008) Multiple waves of recent DNA transposon activity in the bat, Myotis lucifugus. Genome Res 18:717–728
33. Sinzelle L, Izsvák Z, Ivics Z. (2009) Molecular domestication of transposable elements: from detrimental parasites to useful host genes. Cell Mol Life Sci 66:1073–1093
34. Schaack S, Gilbert C, Feschotte C (2010) Promiscuous DNA: horizontal transfer of transposable elements and why it matters for eukaryotic evolution. Trends Ecol Evol 25:537–546
35. Bowen N, et al (2003) Retrotransposons and their recognition of pol II promoters: a comprehensive survey of the transposable elements from the complete genome sequence of Schizosaccharomyces pombe. Genome Res 13:1984–1997
36. Quesneville H, Nouaud D, Anxolabéhère D, (2003) Detection of new transposable element families in Drosophila melanogaster and Anopheles gambiae genomes. J Mol Evol 57 Suppl 1:S50-S59
37. Mills R, et al (2007) Which transposable elements are active in the human genome? Trends Genet 23:183–91
38. Liu W, et al (1995) Cell stress and translational inhibitors transiently increase the abundance of mammalian SINE transcripts. Nucleic Acids Res 23:1758–1765
39. Grandbastien M, et al (1997) The expression of the tobacco Tnt1 retrotransposon is linked to plant defense responses. Genetica 100:241–252
40. Li T, Schmid C, (2001) Differential stress induction of individual Alu loci: implications for transcription and retrotransposition. Gene 276:135–141
41. Kimura R, et al (2001) Stress induction of Bm1 RNA in silkworm larvae: SINEs, an unusual class of stress genes. Cell Stress Chaperones 6:263–272
42. Zeh D, Zeh J, Ishida Y, (2009) Transposable elements and an epigenetic basis for punctuated equilibria. Bioessays 31:715–726
43. Rebollo R, et al (2010) Jumping genes and epigenetics: towards new species. Gene 454:1–7
44. Marino-Ramirez L, et al (2005) Transposable elements donate lineage-specific regulatory sequences to host genomes. Cytogenet Genome Res 110:333–341
45. Le Rouzic A, Boutin T, Capy P, (2007) Long-term evolution of transposable elements. Proc Natl Acad Sci USA 104:19375–19380
46. Böhne A, et al (2008) Transposable elements as drivers of genomic and biological diversity in vertebrates. Chromosome Res 16: 203–215
47. de Boer J, et al (2007) Bursts and horizontal evolution of DNA transposons in the speciation of pseudotetraploid salmonids. BMC Genomics 8:422
48. Ungerer M, Strakosh S, Zhen Y (2006) Genome expansion in three hybrid sunflower species is associated with retrotransposon proliferation. Curr Biol 16:R872–R873
49. Mills R, et al (2006) Recently mobilized transposons in the human and chimpanzee genomes. Am J Hum Genet 78:671–679
50. Britten R (2010) Transposable element insertions have strongly affected human evolution. Proc Natl Acad Sci USA 107:19945–19948
51. Ling A, Cordaux R, (2010) Insertion sequence inversions mediated by ectopic recombination between terminal inverted repeats. PLoS One 5:e15654
52. Kazazian H, et al (1988) Haemophilia A resulting from *de novo* insertion of L1 sequences represents a novel mechanism for mutation in man. Nature 332:164–166

53. Deininger P, Batzer M, (1999) Alu repeats and human disease. Mol Genet Metab 67:183–193
54. Chen J, et al (2005) A systematic analysis of LINE-1 endonuclease-dependent retrotranspositional events causing human genetic disease. Hum Genet 117:411–427
55. Burwinkel B, et al (1998) Mutations in the liver glycogen phosphorylase gene (PYGL) underlying glycogenosis type VI. Am J Hum Genet 62:785–791
56. Fitch D, et al (1991) Duplication of the gamma-globin gene mediated by L1 long interspersed repetitive elements in an early ancestor of simian primates. Proc Natl Acad Sci USA 88:7396–7400
57. Feschotte C (2008) Transposable elements and the evolution of regulatory networks. Nat Rev Genet 9:397–405
58. Oliver K, Greene W (2009) Transposable elements: powerful facilitators of evolution. Bioessays 31:703–714
59. Jordan I, et al (2003) Origin of a substantial fraction of human regulatory sequences from transposable elements. Trends Genet 19:68–72
60. Jiang N, et al (2004) Pack-MULE transposable elements mediate gene evolution in plants. Nature 431:569–573
61. Morgante M, et al (2005) Gene duplication and exon shuffling by helitron-like transposons generate intraspecies diversity in maize. Nat Genet 37:997–1002
62. Kipling D, Warburton P, (1997) Centromeres, CENP-B and Tigger too. Trends Genet 13:141–145
63. Zhou L, et al (2004) Transposition of hAT elements links transposable elements and V(D)J recombination. Nature 432:995–1001
64. Liu D, et al (2007) The human SETMAR protein preserves most of the activities of the ancestral Hsmar1 transposase. Mol Cell Biol 27:1125–1132
65. George J, et al (2010) Evolution of diverse mechanisms for protecting chromosome ends by Drosophila TART telomere retrotransposons. Proc Natl Acad Sci USA 107:21052–21057
66. Bourque G, et al (2008) Evolution of the mammalian transcription factor binding repertoire via transposable elements. Genome Res 18:1752–1762
67. Bestor T, (1999) Sex brings transposons and genomes into conflict. Genetica 107:289–295
68. Malone C, et al. (2009) Specialized piRNA pathways act in germline and somatic tissues of the Drosophila ovary. Cell 137:522–535
69. Malone C, Hannon G, (2009) Small RNAs as Guardians of the Genome. Cell 136:656–668
70. Suzuki M, Bird A, (2008) DNA methylation landscapes: provocative insights from epigenomics. Nat Rev Genet 9:465–476
71. Piriyapongsa J, Jordan I, (2007) A family of human microRNA genes from miniature inverted-repeat transposable elements. PLoS One 2:e203
72. Rudin C, Thompson C, (2001) Transcriptional activation of short interspersed elements by DNA-damaging agents. Genes Chromosomes Cancer 30:64–71
73. Cam H, et al (2008) Host genome surveillance for retrotransposons by transposon-derived proteins. Nature 451:431–436
74. Evans LH, et al (2009) Mobilization of endogenous retroviruses in mice after infection with an exogenous retrovirus. J Virol. 83(6):2429–2435
75. Maumus F, et al (2009) Potential impact of stress activated retrotransposons on genome evolution in a marine diatom. BMC Genomics. 10:624
76. Grandbastien MA, et al. (2005) Stress activation and genomic impact of Tnt1 retrotransposons in Solanaceae. Cytogenet Genome Res. 110(1–4):229–241
77. Eickbush T, (2002) Repair by retrotransposition. Nat Genet 31:126–127
78. Hâsler J, Strub K, (2006) Alu RNP and Alu RNA regulate translation initiation in vitro. Nucleic Acids Res 34:2374–2385
79. Doolittle W, Sapienza C, (1980) Selfish genes, the phenotype paradigm and genome evolution. Nature 284:601–603
80. Leib-Mosch C, Seifarth W, (1995) Evolution and biological significance of human retroelements. Virus Genes 11:133–145
81. Simons C, et al (2006) Transposon-free regions in mammalian genomes. Genome Res 16:164–172
82. Mortada H, Vieira C, Lerat E, (2010) Genes devoid of full-length transposable element insertions are involved in development and in the regulation of transcription in human and closely related species. J Mol Evol 71: 180–91
83. McClintock B, (1950) The origin and behavior of mutable loci in maize. Proc Natl Acad Sci USA 36:344–355
84. Britten R, Davidson E, (1976) DNA sequence arrangement and preliminary evidence on its evolution. Fed Proc 35:2151–2157
85. Cui F, Sirotin M, Zhurkin V, (2011) Impact of Alu repeats on the evolution of human p53 binding sites. Biol Direct 6:2
86. Lerat E, (2010) Identifying repeats and transposable elements in sequenced genomes: how

to find your way through the dense forest of programs. Heredity 104:520–533

87. Abrusan G, et al (2009) Teclass – a tool for automated classification of unknown eukaryotic transposable elements. Bioinformatics 25:1329–1330
88. Chen N, (2004) Using RepeatMasker to identify repetitive elements in genomic sequences. Curr Protoc Bioinformatics Chapter 4:Unit 4.10
89. Jurka J, et al (2005) Repbase Update, a database of eukaryotic repetitive elements. Cytogenet Genome Res 110:462–7
90. Kichenaradja P, et al (2010) ISbrowser: an extension of ISfinder for visualizing insertion sequences in prokaryotic genomes. Nucleic Acids Res 38(Database issue):D62–68
91. Han Y, Wessler S, (2010) MITE-Hunter: a program for discovering miniature inverted-repeat transposable elements from genomic sequences. Nucleic Acids Res 38:e199
92. Mersch B, et al (2007) SERpredict: detection of tissue- or tumor-specific isoforms generated through exonization of transposable elements. BMC Genet 8:78
93. Grenier E, Castagnone-Sereno P, Abad P, (1997) Satellite DNA sequences as taxonomic markers in nematodes of agronomic interest. Parasitol Today 13:398–401
94. Hurst G, Jiggins F, (2005) Problems with mitochondrial DNA as a marker in population, phylogeographic and phylogenetic studies: the effects of inherited symbionts. Proc Biol Sci 272:1525–1534
95. van de Lagemaat L, et al (2005) Genomic deletions and precise removal of transposable elements mediated by short identical DNA segments in primates. Genome Res 15:1243–1249
96. Ludwig A, et al (2005) An unusual primate locus that attracted two independent Alu insertions and facilitates their transcription. J Mol Biol 350:200–214
97. Salem A, et al (2003) Alu elements and hominid phylogenetics. Proc Natl Acad Sci USA 100:12787–12791
98. Warren W, et al (2008) Genome analysis of the platypus reveals unique signatures of evolution. Nature 453:175–183
99. Kriegs J, et al (2006) Retroposed elements as archives for the evolutionary history of placental mammals. PLoS Biol 4:e91.
100. Farwick A, et al (2006) Automated scanning for phylogenetically informative transposed elements in rodents. Syst Biol 55:936–948
101. Nilsson M, et al (2010) Tracking marsupial evolution using archaic genomic retroposon insertions. PLoS Biol 8:e1000436
102. Gu J, et al (2005) Bacterial insertion sequence IS256 as a potential molecular marker to discriminate invasive strains from commensal strains of Staphylococcus epidermidis. J Hosp Infect 1:342–348
103. Nishihara H, et al (2005) A retrospon analysis of Afrotherian phylogeny. Mol Biol Evol 22:1823–1833
104. Shedlock A, Okada N, (2004) SINEs of speciation: tracking lineages with retroposons. Trends Ecol Evol 19:545–553
105. Behura S, (2006) Molecular marker systems in insects: current trends and future avenues. Mol Ecol 15:3087–3113
106. Jones N, et al (2009) Markers and mapping revisited: finding your gene. New Phytol 183:935–966
107. Kalendar R, et al (2011) Analysis of plant diversity with retrotransposon-based molecular markers. Heredity 106:520–530
108. Reznikoff W, Winterberg K, (2008) Transposon-Based Strategies for the Identification of Essential Bacterial Genes. In: Osterman A, and Gerdes Y (ed) Methods in Molecular Biology, vol. 416: Microbial Gene Essentiality, Totowa, NJ
109. Picardeau M, (2010) Transposition of fly mariner elements into bacteria as a genetic tool for mutagenesis. Genetica 138:551–558
110. Judson N, Mekalanos J, (2000) TnAraOut, a transposon-based approach to identify and characterize essential bacterial genes. Nat Biotechnol 18:740–745
111. Kim J, Youm G, Kwon Y, (2008) Essential genes in Salmonella enteritidis as identified by TnAraOut mutagenesis. Curr Microbiol 57:391–394
112. Hensel M, et al (1995) Simultaneous identification of bacterial virulence genes by negative selection. Science 269:400–403
113. Grant A, et al (2005) Signature-tagged transposon mutagenesis studies demonstrate the dynamic nature of cecal colonization of 2-week-old chickens by Campylobacter jejuni. Appl Environ Microbiol 71:8031–8041
114. Akerley B, et al (1998) Systematic identification of essential genes by in vitro mariner mutagenesis. Proc Natl Acad Sci USA 95:8927–8932
115. Reznikoff W, (2006) Tn5 transposition: a molecular tool for studying protein structure-function. Biochem Soc Trans 34:320–323
116. Aviat F, et al (2010) Expanding the genetic toolbox for Leptospira species by generation of fluorescent bacteria. Appl Environ Microbiol 76:8135–8142
117. Bourhy P, et al (2005) Random insertional mutagenesis of Leptospira interrogans, the

agent of leptospirosis, using a mariner transposon. J Bacteriol. 187:3255–3258
118. Upadhyaya N, Zhu Q, Bhat R, (2011) Transposon insertional mutagenesis in rice. Methods Mol Biol 678:147–177
119. Wang H, et al (2010) Transpositional reactivation of two LTR retrotransposons in rice-Zizania recombinant inbred lines (RILs). Hereditas 147:264–277
120. Takagi K, et al (2010) Transposition and target preferences of an active nonautonomous DNA transposon nDart1 and its relatives belonging to the hAT superfamily in rice. Mol Genet Genomics 284:343–355
121. Robert V, Bessereau JL, (2007) Targeted engineering of the Caenorhabditis elegans genome following Mos1-triggered chromosomal breaks. EMBO J 26:170–183
122. Frokjaer-Jensen C, et al (2008) Single-copy insertion of transgenes in Caenorhabditis elegans. Nat Genet 40:1375–1383
123. Laurentino E, et al (2007) The use of Tn5 transposable elements in a gene trapping strategy for the protozoan Leishmania. Int J Parasitol 37:735–742
124. Balu B, et al (2005) High-efficiency transformation of Plasmodium falciparum by the lepidopteran transposable element piggyBac. Proc Natl Acad Sci USA 102:16391–16396
125. Balu B, et al (2009) piggyBac is an effective tool for functional analysis of the Plasmodium falciparum genome. BMC Microbiol 9:83
126. Damasceno J, Beverley S, Tosi L, (2010) A transposon toolkit for gene transfer and mutagenesis in protozoan parasites. Genetica 138:301–311
127. Garraway L, et al (1997) Insertional mutagenesis by a modified in vitro Ty1 transposition system. Gene 198:27–35
128. Renyu X, et al. (2009) Elementary research into the transformation BmN cells mediated by the piggyBac transposon vector. J Biotechnol 144:272–278
129. Sethuraman N, et al (2007) Post-integration stability of piggyBac in Aedes aegypti. Insect Biochem Mol Biol 37:941–951
130. Mohammed A, Coates C, (2004) Promoter and piggyBac activities within embryos of the potato tuber moth, P hthorimaea operculella, Zeller (Lepidoptera: Gelechiidae). Gene 342:293–301
131. Raphaël K, et al (2010) Germ-line transformation of the Queensland fruit fly, Bactrocera tryoni, using a piggyBac vector in the presence of endogenous piggyBac elements. Genetica 139:91–97
132. Mathieu J, et al (2007) A Sensitized PiggyBac-Based Screen for Regulators of Border Cell Migration in Drosophila. Genetics 176:1579–1590
133. Kanginakudru S, et al (2007) Targeting ie-1 gene by RNAi induces baculoviral resistance in lepidopteran cell lines and in transgenic silkworms. Insect Mol Biol 16:635–644
134. Alphey L, (2002) Re-engineering the sterile insect technique. Insect Biochem Mol Biol 32:1243–1247
135. Robinson A, Franz G, Atkinson P, (2004) Insect transgenesis and its potential role in agriculture and human health. Insect Biochem Mol Biol 34:113–120
136. Smith R, Atkinson P, (2010) Mobility properties of the Hermes transposable element in transgenic lines of Aedes aegypti. Genetica 139:7–22
137. Coates C, et al (2000) Purified mariner (Mos1) transposase catalyzes the integration of marked elements into the germ-line of the yellow fever mosquito, Aedes aegypti. Insect Biochemistry and Molecular Biology 30:1003–1008
138. Catteruccia F, et al (2000). Toward Anopheles transformation: Minos element activity in anopheline cells and embryos. Proc Natl Acad Sci USA 97:2157–2162
139. Wang N, et al (2010) Using chimeric piggyBac transposase to achieve directed interplasmid transposition in silkworm Bombyx mori and fruit fly Drosophila cells. J Zhejiang Univ-Sci B 11:728–734
140. O'Brochta D, (2003) Gene vector and transposable element behavior in mosquitoes. J Exp Biol 206:3823–3834
141. Atkinson P, Pinkerton A, O'Brochta D, (2001) Genetic transformation systems in insects. Ann Rev Entomol 46:317–346
142. Jasinskiene N, Coates C, James A, (2000) Structure of Hermes integrations in the germline of the yellow fever mosquito, Aedes aegypti. Insect Mol Biol 9:11–18
143. O'Brochta D, et al (1996) Hermes, a functional non-drosophilid insect gene vector from Musca domestica. Genetics 142:907–914
144. Bryan G, Jacobson J, Hartl D, (1987) Heritable somatic excision of a Drosophila transposon. Science 235:1636–1638
145. Lidholm D, Lohe A, Hartl D, (1993) The transposable element mariner mediates germline transformation in Drosophila melanogaster. Genetics 134:859–868
146. Grossman G, et al (2001) Germline transformation of the malaria vector, Anopheles gambiae, with the piggyBac transposable element. Insect Mol Biol 10:597–604

147. Perera O, Harrell R, Handler A, (2002) Germline transformation of the South American malaria vector, Anopheles albimanus, with a piggyBac/EGFP transposon vector is routine and highly efficient. Insect Mol Biol 11:291–297
148. Lampe D, et al (2000) Genetic engineering of insects with mariner transposons. In: Handler A, James A (ed) Transgenic Insects: Methods and Applications Boca Raton: CRC
149. Rowan K, et al (2004) Tn5 as an insect gene vector. Insect Biochem Mol Biol 34:695–705
150. Venken K, et al. (2006) P(acman): A BAC Transgenic Platform for Targeted Insertion of Large DNA Fragments in D. melanogaster. Science 314:1747–1751.
151. Gorman C, Bullock C, (2000) Site-specific gene targeting for gene expression in eukaryotes. Curr Opin Biotechnol 11:455–460
152. Groth A, Calos M, (2004) Phage integrases: biology and applications. J Mol Biol 335:667–678
153. Ivics Z, Izsvak Z, (2006) Transposons for gene therapy! Curr Gene Ther 6:593–607
154. Zhang L, et al (1998) The Himar1 mariner transposase cloned in a recombinant adenovirus vector is functional in mammalian cells. Nucleic Acids Res 26:3687–3693
155. Keravala A, et al (2006) Hyperactive Himar1 transposase mediates transposition in cell culture and enhances gene expression in vivo. Hum Gene Ther 17:1006–1018
156. Huang X, et al (2010) Gene transfer efficiency and genome-wide integration profiling of Sleeping Beauty, Tol2, and piggyBac transposons in human primary T-cells. Mol Ther 18:1803–1813
157. Kawakami K, Shima A, Kawakami N, (2000) Identification of a functional transposase of the Tol2 element, an Ac-like element from the Japanese medaka fish, and its transposition in the zebrafish germ lineage. Proc Natl Acad Sci USA 97:11403–11408
158. Sherman A, et al (1998) Transposition of the Drosophila element mariner into the chicken germ line. Nat Biotechnol 16:1050–1053
159. Ivics Z, et al (1997) Molecular reconstruction of Sleeping Beauty, a Tc1-like transposon from fish, and its transposition in human cells. Cell 91:501–510
160. Davidson E, et al (2003) Efficient gene delivery and gene expression in zebrafish using the Sleeping Beauty transposon. Dev Biol 263:191–202
161. Yergeau D, et al (2009) Transgenesis in Xenopus using the Sleeping Beauty transposon system. Dev Dyn 238:1727–1743
162. Jakobsen J, et al (2011) Pig transgenesis by Sleeping Beauty DNA transposition. Transgenic Res 20:533–545
163. Horie K, et al (2001) Efficient chromosomal transposition of a Tc1/mariner-like transposon Sleeping Beauty in mice. Proc Natl Acad Sci USA 98:9191–6
164. Dupuy A, Fritz S, Largaespada D, (2001) Transposition and gene disruption in the male germline of the mouse. Genesis 30:82–88
165. Carlson C, et al (2003) Transposon mutagenesis of the mouse germline. Genetics 165:243–256
166. Kitada K, et al (2009) Generating mutant rats using the Sleeping Beauty transposon system. Methods 49:236–42
167. Dupuy A, et al (2002) Mammalian germ-line transgenesis by transposition. Proc Natl Acad Sci USA 99:4495–4499
168. Wilber A, et al (2006) RNA as a source of transposase for Sleeping Beauty-mediated gene insertion and expression in somatic cells and tissues. Mol Ther 13:625–630
169. Mátés L, et al (2009) Molecular evolution of a novel hyperactive Sleeping Beauty transposase enables robust stable gene transfer in vertebrates. Nat Genet 41:753–761
170. Xue X, et al (2009) Stable gene transfer and expression in cord blood-derived CD34+ hematopoietic stem and progenitor cells by an hyperactive Sleeping Beauty transposon system. Blood 114:1319–1330
171. Williams A, (2008) Sleeping Beauty vector system moves toward human trials in the United States. Mol Ther 16:1515–1516
172. Fraser M, et al (1995) Assay for movement of Lepidopteran transposon IFP2 in insect cells using a baculovirus genome as a target DNA. Virology 211:397–407
173. Nakanishi H, et al (2010) piggyBac transposon-mediated long-term gene expression in mice. Mol Ther 18:707–714
174. Saridey S, et al (2009) piggyBac transposon-based inducible gene expression in vivo after somatic cell gene transfer. Mol Ther 17:2115–2120
175. Lu Y, Lin C, Wang X, (2009) piggyBac transgenic strategies in the developing chicken spinal cord. Nucleic Acids Res 37:e141
176. Wu S, (2006) piggyBac is a flexible and highly active transposon as compared to sleeping beauty, Tol2, and Mos1 in mammalian cells. Proc Natl Acad Sci USA 103:15008–15013
177. Ding S, et al (2005) Efficient transposition of the piggyBac (PB) transposon in mammalian cells and mice. Cell 122:473–483
178. Wang W, et al (2008) Chromosomal transposition of piggyBac in mouse embryonic stem

cells. Proc Natl Acad Sci USA 105:9290–9295

179. Lacoste A, Berenshteyn F, Brivanlou A, (2009) An efficient and reversible transposable system for gene delivery and lineage-specific differentiation in human embryonic stem cells. Cell Stem Cell 5:332–342
180. Cadiñanos J, Bradley A, (2007) Generation of an inducible and optimised piggyBac transposon system. Nucleic Acids Res 35:e87
181. Yusa K, et al (2010) A hyperactive piggyBac transposase for mammalian applications. Proc Natl Acad Sci USA 108:1531–1536
182. Maragathavally K, Kaminski J, Coates C, (2006) Chimeric Mos1 and piggyBac transposases result in site-directed integration. FASEB J 20:1880–1882
183. Raja Manuri P, et al (2010) piggyBac transposon/transposase system to generate CD19-specific T cells for the treatment of B-lineage malignancies. Hum Gene Ther 21:427–437
184. Feschotte C, (2006) The piggyBac transposon holds promise for human gene therapy. Proc Natl Acad Sci USA 103:14981–14982
185. Woltjen K, et al (2009) piggyBac transposition reprograms fibroblasts to induced pluripotent stem cells. Nature 458:766–770
186. Kaji K, et al (2009) Virus-free induction of pluripotency and subsequent excision of reprogramming factors. Nature 458:771–5
187. Belay E, et al (2010) Novel hyperactive transposons for genetic modification of induced pluripotent and adult stem cells: a nonviral paradigm for coaxed differentiation. Stem Cells 28:1760–1771
188. Silver J, Keerikatte V, (1989) Novel use of polymerase chain reaction to amplify cellular DNA adjacent to an integrated provirus. J Virol 63:1924–1928
189. Mueller P, Wold B, (1989) In vivo footprinting of a muscle specific enhancer by ligation mediated PCR. Science 246:780–786
190. Paruzynski A, et al (2010) Genome-wide high-throughput integrome analyses by nrLAM-PCR and next-generation sequencing. Nat Protoc 5:1379–1395
191. Schmidt M, et al (2007) High-resolution insertion-site analysis by linear amplification-mediated PCR (LAM-PCR). Nat Methods 4:1051–1057
192. Schmidt M, et al (2003) Clonality analysis after retroviral-mediated gene transfer to CD34+ cells from the cord blood of ADA-deficient SCID neonates. Nat Med 9:463–468
193. Cattoglio C, et al (2010) High-Definition Mapping of Retroviral Integration Sites Defines the Fate of Allogeneic T Cells After Donor Lymphocyte Infusion. PLoS One 5:e15688
194. Wheelan S, et al (2006) Transposon insertion site profiling chip (TIP-chip). Proc Natl Acad Sci USA 103:17632–7
195. Huang C, et al (2010) Mobile interspersed repeats are major structural variants in the human genome. Cell 141:1171–1182
196. Huda A, et al (2011) Epigenetic regulation of transposable element derived human gene promoters. Gene 475:39–48
197. Zhu J, et al (2010) High-Level Genomic Integration, Epigenetic Changes, and Expression of Sleeping Beauty Transgene. Biochemistry 49:1507–1521
198. Saze H, Kakutani T, (2011) Differentiation of epigenetic modifications between transposons and genes. Curr Opin Plant Biol 14:81–87
199. Hollister J, Gaut B, (2009) Epigenetic silencing of transposable elements: a trade-off between reduced transposition and deleterious effects on neighboring gene expression. Genome Res 19:1419–1428
200. Montoya-Durango D, et al (2009) Epigenetic control of mammalian LINE-1 retrotransposon by retinoblastoma proteins. Mutat Res 665:20–28
201. Harris C, et al (2009) p53 responsive elements in human retrotransposons. Oncogene 28: 3857–3865
202. Wang J, et al (2009) A c-Myc regulatory subnetwork from human transposable element sequences. Mol Biosyst 5:1831–1839
203. Huang J, et al (2004) Lsh, an epigenetic guardian of repetitive elements. Nucleic Acids Res 32:5019–5028
204. Muramoto H, et al (2010) Enrichment of short interspersed transposable elements to embryonic stem cell-specific hypomethylated gene regions. Genes Cells 15:855–865
205. Phokaew C, et al (2008) LINE-1 methylation patterns of different loci in normal and cancerous cells. Nucleic Acids Res 36: 5704–5712
206. Yates P, et al (1999) Tandem B1 elements located in a mouse methylation center provide a target for de novo DNA methylation. J Biol Chem 274:36357–36361
207. Estécio M, et al (2010) Genome architecture marked by rétrotransposons modulates predisposition to DNA methylation in cancer. Genome Res 20:1369–1382
208. Howard G, et al (2008) Activation and transposition of endogenous retroviral elements in hypomethylation induced tumors in mice. Oncogene 27:404–408

209. Belancio V, et al (2010) All y'all need to know 'bout retroelements in cancer. Semin Cancer Biol 20:200–210
210. Iskow R, et al (2010) Natural Mutagenesis of Human Genomes by Endogenous Retrotransposons. Cell 141:1253–1261
211. Romanish M, Cohen C, Mager D, (2010) Potential mechanisms of endogenous retroviral-mediated genomic instability in human cancer. Semin Cancer Biol 20:246–253
212. Xu T, Deng K, (2002) Transposable elements and tumor progression. Med Hypotheses 58:293–6
213. Kawakami K, (2010) Long interspersed nuclear element-1 hypomethylation is a potential biomarker for the prediction of response to oral fluoropyrimidines in microsatellite stable and CpG island methylator phenotype-negative colorectal cancer. Cancer Sci 102:166–174
214. Saito K, et al (2010) Long Interspersed Nuclear Element 1 Hypomethylation Is a Marker of Poor Prognosis in Stage IA Non–Small Cell Lung Cancer. Clin Cancer Res 16: 2418–2426
215. Mirabello L, et al (2010) Line-1 methylation is inherited in familial testicular cancer kindreds. BMC Med Genet 11:77
216. Wilhelm C, et al (2010) Implications of LINE1 Methylation for Bladder Cancer Risk in Women. Clin Cancer Res 16:1682–1689
217. Horard B, et al (2009) Global analysis of DNA methylation and transcription of human repetitive sequences. Epigenetics 4:339–350
218. Dupuy A, et al (2009) A modified Sleeping Beauty transposon system that can be used to model a wide variety of human cancers in mice. Cancer Res 69:8150–8156
219. Beck B, et al (2008) Human Pso4 is a metnase (SETMAR)-binding partner that regulates metnase function in DNA repair. J Biol Chem 283:9023–9030

Chapter 2

Using and Understanding RepeatMasker

Sébastien Tempel

Abstract

RepeatMasker is a program that screens DNA sequences for interspersed repeats and low-complexity DNA sequences. In this chapter, we present the procedure to routinely use this program on a personal computer.

Key words: Repeats, Mobile genetic elements, Transposon, Retrotransposon, In silico

1. Introduction

RepeatMasker is a free application created by A.R.A Smith in 1998. The latest version 3-2-9 is available since January 7, 2010. RepeatMasker screens one or more genomic sequences in FASTA format and detects transposable elements, satellites and low-complexity DNA sequences. It is the most use d software for the detection of transposable elements. Although it was not the subject of scientific publication, it is particularly used for genome annotation and for studying families of multiple transposable elements.

RepeatMasker is written in PERL. It uses a search engine like Basic Local Alignment Search Tool (BLAST; (1)) with a library of transposable elements as the query and the input sequence as the database. RepeatMasker reads "raw results" from the alignment tool, manipulates them based on some criteria and writes them in the RepeatMasker standard outputs. There are currently two versions of RepeatMasker: RepeatMasker that runs on the user's computer (UNIX or LINUX computer) and WEBRepeatMasker that runs on the Institute for System Biology server. WEBRepeatMasker is a Web interface of RepeatMasker (Fig. 1). They both accept the same input sequences, have some options and outputs in common, but the RepeatMasker has more outputs and options.

Yves Bigot (ed.), *Mobile Genetic Elements: Protocols and Genomic Applications*, Methods in Molecular Biology, vol. 859, DOI 10.1007/978-1-61779-603-6_2, © Springer Science+Business Media, LLC 2012

Fig. 1. Web interface of RepeatMasker: WEBRepeatMasker.

2. Installation of RepeatMasker

RepeatMasker is not limited to one input sequence and there is no limitation of the sequence size. WEBRepeatMasker has a limit of 10 Mbp for deferred processing with an email reply and 100 kbp for immediate processing. The user needs to install RepeatMasker on his personal computer if he wants to use it on long sequences (longer than 10 Mbp) or other output formats. The user can download the last version of RepeatMasker at http://www.repeatmasker.org/RMDownload.html and uncompress it. RepeatMasker runs only on Unix and Linux systems.

Before configuring the software, the user also needs to install a library and a minimum of three pre-requirements: PERL, Tandem Repeat Finder (TRF; (2)), and a search engine sotfware. All these files must be installed before the first run and the configuration of RepeatMasker.

The user must download the RepeatMasker library file (repeatmaskerlibraries-XXX.tar.gz, XXX corresponds to the date when the last version was created) from the GIRI Web site (http://www.girinst.org/repbase/index.html) after the user has requested an account opening. Uncompress the file in the "Libraries" subdirectory inside RepeatMasker directory. RepeatMasker needs PERL 5.8.0 or one of its update for the installation. This is usually installed by default on Unix systems. Tandem Repeat Finder detects the tandem repeat in a FASTA sequence. After the software installed, copy the binary file in the same directory than RepeatMasker binaries and rename it to "trf." The search engine library is one of the four alignment tools: AB-BLAST (new name for WU-BLAST, http://www.advbiocomp.com/blast.html), Cross_match (http://www.phrap.org/phredphrapconsed.html), Decypher (http://www.timelogic.com/decypherblast.html), or RMBLAST (http://www.repeatmasker.org/RMBlast.html).

The four alignment tools use the BLAST algorithm at the core of their methods.

3. Configuration of RepeatMasker

After the installation of Tandem Repeat Finder (2) and the alignment software, the user must configure RepeatMasker before the first run. There are two ways to configure it: manually editing the configuration file of RepeatMasker (RepeatMaskerConfig.pm) or using the PERL script 'configure' in RepeatMasker directory that automatically writes the configuration file. Here, I assume that the user will use the PERL script. First, the script asks the user to enter some necessary paths as follows:

```
**PERL INSTALLATION PATH**

Enter path [ /usr/bin/perl ]:
```

Confirm the path to the PERL binaries. The path inside the brackets corresponds to the path written in RepeatMaskerConfig.pm. Usually, PERL is installed at '/usr/bin/perl', it is only necessary to confirm it (*press Enter*).

```
**REPEATMASKER INSTALLATION PATH**

Enter path [ /usr/local/bin/RepeatMasker ]:
```

Enter the complete path to where you install RepeatMasker. If the user does not know the complete path of RepeatMasker, he can use the Unix command 'pwd' in the software's directory.

```
**TRF INSTALLATION PATH**

Enter path [ ]:
```

Enter the complete path to where you put the TRF binary. Normally it is the same path as RepeatMasker.

Add a Search Engine:

```
1. CrossMatch: [ Un-configured ]

2. RMBlast-NCBI Blast with RepeatMasker extensions: [
Un-configured ]

3. WUBlast/ABBlast (required by DupMasker): [ Un
configured ]

4. DeCypher (TimeLogic): [ Un-configured ]

5. Done
```

Enter Selection: 3

```
**ABBlast/WUBlast BLASTP INSTALLATION PATH**

Enter path [ ]: /usr/local/bin/WUBLAST

Do you want ABBlast/WUBlast to be your default
search engine for Repeatmasker? (Y/N) [ Y ]:
```

In this part of the script, choose which alignment tool you want use. In this example, I choose WUBLAST (number 3). The script prompts for the path of AB-BLAST. Confirm AB-BLAST as the default alignment tool. The script will return to the Search Engine menu. Enter another search engine or finish the configuration.

4. Running RepeatMasker with Basic Options

On WEBRepeatMasker, load the sequence file (maximum 100 kbp) by entering the sequence into the indicated text field or browsing the file with the 'Load' button (Fig. 1). Choose the results format: 'html' or 'tar file'. The 'html' choice shows the results as Web page and the 'tar file' choice sends a compress file that contains all output files. Select an appropriate genome for your sequence with the "DNA source" button (Fig. 1).

If the user chooses to use the RepeatMasker from the command line, it works as follows:

```
RepeatMasker [-options] <seqfiles(s) in FASTA format>
```

Enter the RepeatMasker name following by the options and the name of the input file. The output files are automatically created. The different options of RepeatMasker (and WEB RepeatMasker) are presented in the Subheading 5.

4.1. Input Format Sequence

The input sequence must be in FASTA format. FASTA format looks like this:

```
>Sequence1

ACGTGCGCGATCGCCTGCTAGGCGTACGTCGCAGGCGATCGATGTGCTAGATCAG

ATGACA

>Sequence2 human

GGGCTAGATTAGCACCACATACATCGCTCA
```

The FASTA format of a sequence always start with the symbol '>'. This symbol is the beginning of the sequence name. This name can be one letter or number, one word or one sentence. The sequence name is limited to one line (it finishes with return symbol). The size of the sequence is not limited and could be written in one or more lines. The FASTA sequence finishes with a blank line or a new start symbol '>'.

4.2. Output Results

WEBRepeatMasker and RepeatMasker returns three output files for each query: a masked file, a map file and an overview table file.

4.2.1. Map File

The map file is a list of all repeats identified in the query sequence by the alignment software. Figure 2 shows the map file format. The format of map file is the cross_match summary lines (Figs. 2 and 3). RepeatMasker converts the other search engine output into this format. The map file lists all best matches (above a minimum score) between the query sequence and any of the sequences in the repeat database or with low-complexity DNA. A match of a repeat A could be inserted in the match of a repeat B. If this inserted match has a lower score than its container, RepeatMasker hides it. The matches in the list masked in the map file are masked in masked sequence file (see Subheading 4.2.2). In the map file, matches are ranked by position from the start of the alignment.

4.2.2. Masked Sequence

RepeatMasker returns a .masked file containing the query sequence(s) with all identified repeats and low-complexity sequences masked. All nucleotides contained in a repeat in the map file are replaced by "N" in the sequences (Fig. 4). Some options allow modifying the format and the category of repeats that are masked (described in Subheading 5).

±	score	% div.	% del.	% ins.	query sequence	----position in query---- begin	end	(left)	C matching + repeat		repeat class/family	-position in repeat-- (left) begin	end end	begin (left)	linkage id/graphic
±	282	14.5	0.0	7.2	chr1_Arabidopsis	1	105	(3499825)	C	ATREP18	DNA	(1142)	649	561	1
±	201	9.4	0.0	0.0	chr1_Arabidopsis	1066	1097	(3498833)	+	(C)n	Simple_repeat	1	32	(0)	2
±	26	34.6	0.0	0.0	chr1_Arabidopsis	3305	3330	(3496600)	+	AT_rich	Low_complexity	1	26	(0)	3 *
±	29	88.4	0.0	0.0	chr1_Arabidopsis	3308	3350	(3496580)	+	AT_rich	Low_complexity	1	43	(0)	4
±	267	7.9	0.0	0.0	chr1_Arabidopsis	4291	4328	(3495602)	+	(TA)n	Simple_repeat	2	39	(0)	5
±	279	0.0	0.0	0.0	chr1_Arabidopsis	8669	8699	(3491231)	+	(TC)n	Simple_repeat	1	31	(0)	6
±	221	16.9	1.6	3.3	chr1_Arabidopsis	9970	10030	(3489900)	+	(TA)n	Simple_repeat	2	61	(0)	7
±	297	10.9	0.0	0.0	chr1_Arabidopsis	11915	11960	(3487970)	+	(CAT)n	Simple_repeat	2	47	(0)	8
±	186	4.3	0.0	0.0	chr1_Arabidopsis	13346	13368	(3486562)	+	(GA)n	Simple_repeat	2	24	(0)	9
±	26	81.5	0.0	0.0	chr1_Arabidopsis	15266	15319	(3484611)	+	AT_rich	Low_complexity	1	54	(0)	10
±	26	57.7	0.0	0.0	chr1_Arabidopsis	15296	15321	(3484609)	+	AT_rich	Low_complexity	1	26	(0)	11
±	180	0.0	0.0	0.0	chr1_Arabidopsis	16804	16823	(3483107)	+	(TTG)n	Simple_repeat	2	21	(0)	12
±	1523	6.5	0.4	0.8	chr1_Arabidopsis	17010	17256	(3482674)	+	ATREP3	RC/Helitron	1	246	(1851)	13 *
±	8985	6.0	0.8	1.7	chr1_Arabidopsis	17256	18642	(3481288)	+	ATREP20	RC/Helitron	27	1471	(673)	14
±	552	4.4	0.0	0.0	chr1_Arabidopsis	18661	18728	(3481202)	+	ATREP20	RC/Helitron	2077	2144	(0)	14
±	28	87.6	0.0	0.0	chr1_Arabidopsis	18796	18900	(3481030)	+	AT_rich	Low_complexity	1	105	(0)	15
±	198	0.0	0.0	0.0	chr1_Arabidopsis	20510	20531	(3479399)	+	(TA)n	Simple_repeat	2	23	(0)	16
±	198	0.0	0.0	0.0	chr1_Arabidopsis	22621	22642	(3477288)	+	(TTTTA)n	Simple_repeat	6	27	(0)	17
±	333	0.0	0.0	0.0	chr1_Arabidopsis	37736	37772	(3462158)	+	(GAA)n	Simple_repeat	1	37	(0)	18
±	225	0.0	0.0	0.0	chr1_Arabidopsis	41361	41385	(3458545)	+	(TA)n	Simple_repeat	2	26	(0)	19
±	34	77.1	0.0	0.0	chr1_Arabidopsis	42444	42491	(3457439)	+	AT_rich	Low_complexity	1	48	(0)	20
±	35	73.2	0.0	0.0	chr1_Arabidopsis	44763	44818	(3455112)	+	AT_rich	Low_complexity	1	56	(0)	21
±	209	19.1	0.0	0.0	chr1_Arabidopsis	46523	46564	(3453366)	+	CT-rich	Low_complexity	139	180	(0)	22
±	271	24.7	0.0	0.0	chr1_Arabidopsis	50433	50517	(3449413)	+	(CAG)n	Simple_repeat	1	85	(0)	23
±	5236	6.5	0.7	1.1	chr1_Arabidopsis	55676	56576	(3443354)	+	SIMPLEHAT1	DNA/hAT	1	1059	(0)	24
±	234	0.0	0.0	0.0	chr1_Arabidopsis	62347	62372	(3437558)	+	(GAA)n	Simple_repeat	2	27	(0)	25

Fig. 2. An example map file from the start of *Arabidopsis thaliana* chromosome 1.

282	= Smith-Waterman score of the match, (complexity adjusted, dependent on repeat age and GC level)
14.5	= % of divergence = mismatches/(matches+mismatches)
0.0	= % of deletions in the query sequence (deleted bp)
7.2	= % of insertions in the query sequence (inserted bp)
chr1_Arabidopsis	= name of query sequence
1	= starting position of match in query sequence
105	= ending position of match in query sequence
(3499825)	= number of bases in query sequence past the ending position of match
C	= match is with the Complement of the repeat consensus sequence
ATREP18	= name of the matching interspersed repeat
DNA	= the class of the repeat, in this case a DNA transposon
(1142)	= number of bases in (complement of) the repeat consensus sequence prior to beginning of the match (0 means that the match extended all the way to the end of the repeat consensus sequence)
649	= starting position of match in repeat consensus sequence
561	= ending position of match in repeat consensus sequence
1	= unique identifier for individual insertions

An asterisk (*) following the final column (see below example) indicates that there is a higher-scoring match whose domain partly (<80%) includes the domain of the current match.

Fig. 3. The meaning of each column in the map file.

```
>chr1_Arabidopsis
NNNNNNNNNNNNNNNNNNNNNNNNNNNNNNNNNNNNNNNNNNNNNNNNNNNNNNNNNNNNNNNNNNNNNNNNNNNNNNNNNNNNNNNNNNNNNNNNNNNN
NNNNNACCCGAAACCGGTTTCTCTGGTTGAAAATCATTGTGTATATAATGATAATTTTATCGTTTTTATGTAATTGCTTATTGTTGTGTGTAGATTTTTT
AAAAATATCATTTGAGGTCAATACAAATCCTATTTCTTGTGGTTTTCTTTCCTTCACTTAGCTATGGATGGTTTATCTTCATTTGTTATATTGGATACAA
GCTTTGCTACGATCTACATTTGGGAATGTGAGTCTCTTATTGTAACCTTAGGGTTGGTTTATCTCAAGAATCTTATTAATTGTTTGGACTGTTTATGTTT
GGACATTTATTGTCATTCTTACTCCTTTGTGGAAATGTTTGTTCTATCAATTTATCTTTTGTGGGAAAATTATTTAGTTGTAGGGATGAAGTCTTTCTTC
GTTGTTGTTACGCTTGTCATCTCATCTCTCAATGATATGGGATGGTCCTTTAGCATTTATTCTGAAGTTCTTCTGCTTGATGATTTTATCCTTAGCCAAA
AGGATTGGTGGTTTGAAGACACATCATATCAAAAAAGCTATCGCCTCGACGATGCTCTATTTCTATCCTTGTAGCACACATTTTGGCACTCAAAAAAGTA
TTTTTAGATGTTTGTTTTGCTTCTTTGAAGTAGTTTCTCTTTGCAAAATTCCTCTTTTTTTAGAGTGATTTGGATGATTCAAGACTTCTCGGTACTGCAA
AGTTCTTCCGCCTGATTAATTATCCATTTTACCTTTGTCGTAGATATTAGGTAATCTGTAAGTCAACTCATATACAACTCATAATTTAAAATAAAATTAT
GATCGACACACGTTTACACATAAAATCTGTAAATCAACTCATATACCCGTTATTCCCACAATCATATGCTTTCTAAAAGCAAAAGTATATGTCAACAATT
GGTTATAAATTATTAGAAGTTTTCCACTTATGACTTAAGAACTTGTGAAGCAGAAAGTGGCAACANNNNNNNNNNNNNNNNNNNNNNNNNNNNNNNNAAA
TTGAGAAGTCAATTTTATATAATTTAATCAAATAAATAAGTTTATGGTTAAGAGTTTTTTACTCTCTTTATTTTTCTTTTTCTTTTTGAGACATACTGAA
AAAAGTTGTAATTATTAATGATAGTTCTGTGATTCCTCCATGAATCACAT
```

Fig. 4. Example of a masked sequence. This sequence is the beginning of Arabidopsis chromosome 1.

Summary:

```
==================================================
file name: RM2_temp.fna_1298535356
sequences:            1
total length:    3499930 bp  (3499930 bp excl N/X-runs)
GC level:         37.31 %
bases masked:      91778 bp ( 2.62 %)
==================================================
               number of      length   percentage
               elements*    occupied  of sequence
--------------------------------------------------
Retroelements             24         7427 bp    0.21 %
   SINEs:                  1          869 bp    0.02 %
   Penelope                0            0 bp    0.00 %
   LINEs:                 12         4125 bp    0.12 %
    CRE/SLACS              0            0 bp    0.00 %
     L2/CR1/Rex            0            0 bp    0.00 %
     R1/LOA/Jockey         0            0 bp    0.00 %
     R2/R4/NeSL            0            0 bp    0.00 %
     RTE/Bov-B             0            0 bp    0.00 %
     L1/CIN4              12         4125 bp    0.12 %
   LTR elements:          11         2433 bp    0.07 %
     BEL/Pao               0            0 bp    0.00 %
     Ty1/Copia             8         1810 bp    0.05 %
     Gypsy/DIRS1           3          623 bp    0.02 %
       Retroviral          0            0 bp    0.00 %

DNA transposons           44        25896 bp    0.74 %
   hobo-Activator          9         4483 bp    0.13 %
   Tc1-IS630-Pogo          6         2074 bp    0.06 %
   En-Spm                  1           67 bp    0.00 %
   MuDR-IS905             16        17023 bp    0.49 %
   PiggyBac                0            0 bp    0.00 %
   Tourist/Harbinger       7         1687 bp    0.05 %
   Other (Mirage,          0            0 bp    0.00 %
    P-element, Transib)

Rolling-circles            0            0 bp    0.00 %

Unclassified:             35        22581 bp    0.65 %

Total interspersed repeats:         55904 bp    1.60 %


Small RNA:                 0            0 bp    0.00 %

Satellites:               13         4286 bp    0.12 %
Simple repeats:          273         9639 bp    0.28 %
Low complexity:          472        21949 bp    0.63 %
```

Fig. 5. Table of summary result. RepeatMasker sums the number of occurrences and the number of nucleotides of each "superfamily" detected. A superfamily is defined here by A.R.A Smith and corresponds to the definition of Wicker et al. (3).

4.2.3. Summary Results

RepeatMasker returns a .tbl file containing a plain text in the form of a table. The file summarizes the number of identified repeats clustered in superfamilies and sums the nucleotides that belong to each superfamily (Fig. 5). A superfamily is mainly defined as described by Wicker et al. (3). Because it is extremely difficult to distinguish which repeat fragments are derived from the same insertion event of a transposable element, there is a slight overestimate of the copy numbers. There are also differences in the number of "bases masked" and the sum of the bases annotated in this summary file. For example, the fragments shorter than 10 bp are not annotated but are masked. Note that the classification is not based on the target size duplication sizes (for example, a TA for a Tc1/Mariner element), but on the proteins coded by the repeat or, whether these are not known in the classification, on the extremity matches (i.e., inverted terminal repeat (ITR) or long terminal repeat (LTR). I also noted the classification of some superfamilies are more precise than others, especially mammalian repeats are very well annotated. These differences are more precisely described in Subheading 5.4 about algorithm identification of repeats. We observed this classification was not updated at least since 2006, especially, it does not mention the Polinton/Marverick superfamily (4).

5. RepeatMasker Main Functions

Before describing how to use the options of RepeatMasker, the main functions are presented during a standard run (i.e., without option). DateRepeats and DupMasker are two binaries installed with RepeatMasker and not used in standard run. The user can use them separately.

Figure 6 shows the main PERL scripts involved in Repeat Masker execution and the chronological order they are used (the numbers adjacent to the arrows). The execution of RepeatMasker can be mainly split into seven different steps:

- Verify hit point to the valid alignment tool.
- Read and check the input sequence(s).
- Check the RepeatMasker library or the user transposable elements library.
- Split the sequence(s) into fragments and prepare a list of execution on the different fragments.
- Launch the alignment tool on the sequences.
- Change the search engine output into RepeatMasker standard output.
- Merge the fragment sequences and merge the fragmented hits of transposable elements.

5.1. RepeatMasker

The first task of RepeatMasker is to verify its own configuration. RepeatMasker loads and checks all binary paths found in the module RepeatMaskerConfig.pm (number 1 in Fig. 7). RepeatMasker

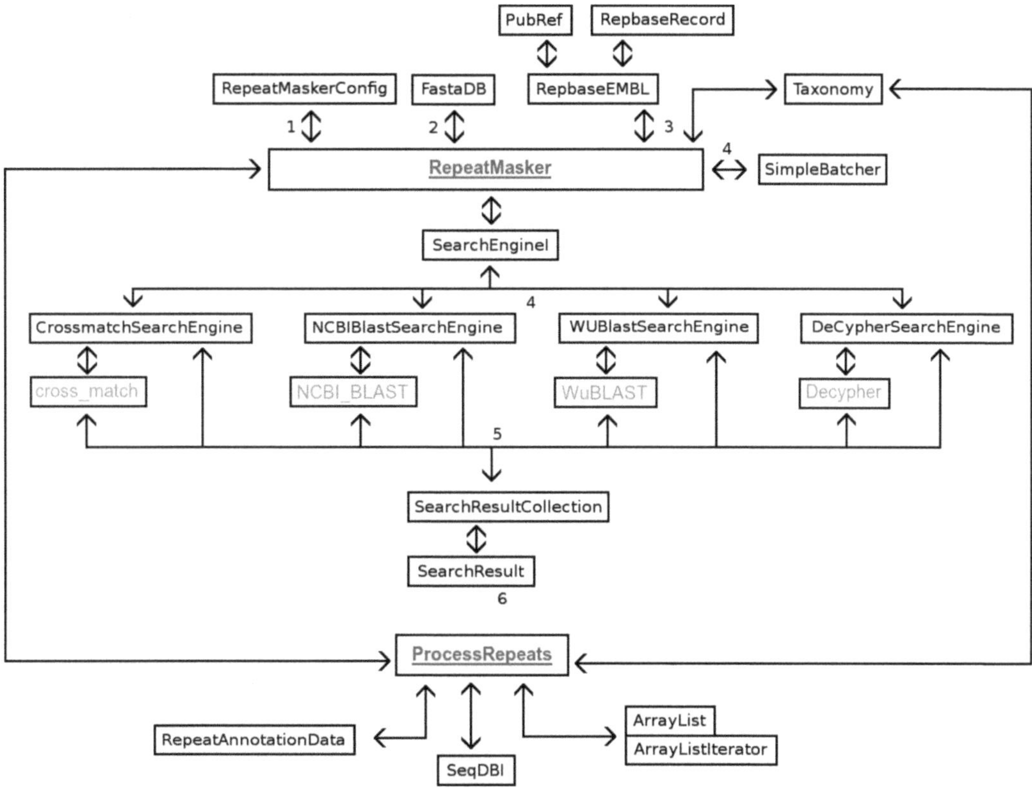

Fig. 6. Main organization of RepeatMasker binaries and libraries. These files are used in all RepeatMasker executions. The *arrows* correspond to the relationship between two files. The *numbers* indicate the main chronological order where the files are used. The binary files of RepeatMasker are written in *gray underline color*, the search engine binaries are written in *light gray color*.

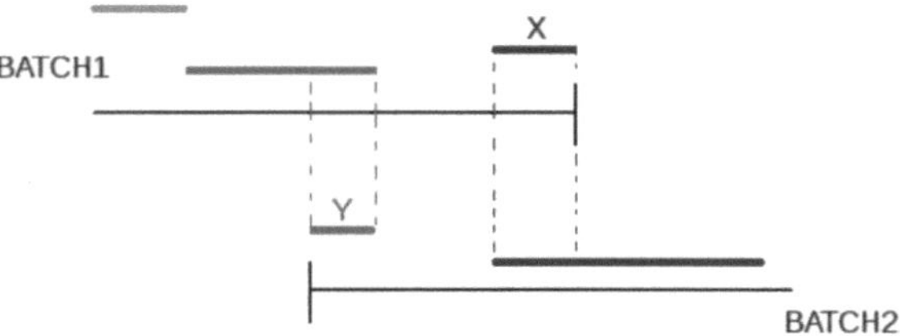

Fig. 7. Example of duplicate hits on two fragmented batch sequences.

initializes some parameters as the size of fragment that will run with the alignment tool and loads the date for creating the output directory for each run.

The second task of RepeatMasker is to check the input sequence (number 2 in Fig. 7). The FASTADB.pm module loads and checks many parameters: the number of FASTA sequences, the length, the number of G and C nucleotides (calculates the GC ratio in the FASTA sequence), and the number of X and/or N nucleotides.

The module checks whether the FASTA sequence is valid and splits the sequences into short overlapping fragments (their length depends on the search engine parameters). The length of the overlap is 2,000 bp.

The third task of RepeatMasker is to check the database file that contains the repeat elements (transposable elements and satellites; number 3 and 4 in Fig. 7). The user can use the RepeatMasker library downloaded from Genetic Information Research Institute (http://www.girinst.org) or can use its own library. If the user provides a custom database, RepeatMasker checks the format of the database. With the use of custom library, RepeatMasker cannot fill the summary table, except the value of GC percentage and the length of the sequence. If the user uses the RepeatMasker library, the RepbaseEMBL.pm, RepbaseRecord.pm and PubRef.pm modules extract the information about the repeats. Section 4.2 details the information extracted from the database. In this case, the user must specify to which species its sequences belong (option '-specie') or the taxonomy order, if RepeatMasker does not recognize the specified species. For example, the option '-specie bear' does not work because the beard genome is not recognized but the user can enter '-specie mammal' instead.

The fourth task of RepeatMasker creates and writes the list of command batch of each fragmented sequence (SimpleBatcher.pm) according to the search engine chosen. Five PERL modules convert the option of RepeatMasker into Search Engine options and write them in the command batch: SearchEngineI.pm, CrossmatchSearchEngine.pm, DeCypherSearchEngine.pm, NCBI BlastSearchEngine.pm, and WUBlastSearchEngine.pm. Their name is correlated to the search engine they control. Note that the search engine parameters cannot be changed directly by the user, but are calculated based on RepeatMasker parameters. For example, the minimal size of the exact words found by Decypher, NCBI-BLAST and WU-BLAST is 14., and WU-BLAST uses BLASTP with "`-warnings -T=1000000 -p=1 -hspmax=2000000000 -gi V=0 B=100000000 Q=12 R=2 W=14 S=30 gapS2=30 S2=15 X=30 gapX=60`" parameters.

After the runs of search engine, the fifth task of RepeatMasker is to convert the different outputs in a standard output: cross_match output (Figs. 2 and 3). The only task left to RepeatMasker is to run ProcessRepeats.

5.2. ProcessRepeats

ProcessRepeats follows the RepeatMasker application. It organizes and processes the results by assembling and sorting the fragmented repeats. Many repeat subfamilies are similar to each other and could be close in the sequence, thus the main difficulty of ProcessRepeats algorithm is to merge fragmented repeats and to assign appropriate subfamily names to the search engine output.

The main algorithm of ProcessRepeats is:

- Read the options
- Load the information data (specie, GC content…)
- Remove duplicated hit present in many fragment sequences
- Assemble the sequence fragmented by RepeatMasker
- Sort all repeats hits
- Join the fragmented repeats subject to some parameters
- Removing insignificant fragments
- Write the (optional) annotation output files

The first task of ProcessRepeats is to read and check the parameters users. These parameters will define the ProcessRepeats tasks. For example, the option '-gff' will write a gff output file and the option '-noint' will skip all tasks involving transposable elements.

The second task loads and checks the information about the query sequence and the library. For example, if the user works on Drosophila genomes and enter it with the '-specie' option, ProcessRepeats will not merge of the mammalian long interspersed nuclear elements (LINEs). The GC content (%) calculated by RepeatMasker is a very important information which determines the minimal threshold of many repeats and adjusts the Smith-Watermann score of each hit.

The third task of ProcessRepeats is to remove the duplicated hit present in two batch sequences (Fig. 7): when the same repeat is present in two hits on two batch files, ProcessRepeats removes the one on the edges of the batch file.

The fourth task is the most difficult. It consists in merging the fragmented repeats. This task is subdivided based on the A.R.A. Smith defined superfamilies: DNA transposons, short interspersed nuclear elements (SINEs), LTR retrotransposons, LINEs, and recombinant LINEs.

The fifth task removes the insignificant hits. The parameters and the information data allow ProcessRepeats to calculate the minimal threshold. The insignificant hits are those that have a Smith-Watermann score lower than this threshold. This threshold is different for each superfamily.

Finally, the last task of ProcessRepeats is to write the assembled repeats in the different output files according to the options.

5.3. DateRepeats

The goal of the script DateRepeats is to know if a repeat, present in one specific species, is expected to be present in another one. Actually, DateRepeats works only with the mammal species. DateRepeats takes a RepeatMasker .out file. The user must choose the species of the repeats in RepeatMasker .out file and choose the compared species. For example, if a user wants to know whether the repeats in the RepeatMasker map file from mouse genome are also present in the rat and the human genome, it will be able to process as follows:

```
DateRepeats MouseRepeat.out -q mouse -c rat -c human
```

DateRepeats analyses the MouseRepeat.out file that contains mouse transposable elements and checks if they exist in rat and human genome. DateRepeats prints out the following output file:

```
12436 19.0 1.7 7.7 chr3_400     9   536 (4464) + L1_Mur2
LINE/L1 1850 2408 (3469) 1  X 0
2728    5.2 0.0 3.9 chr3_400   537   896 (4104) + ORR1A0
LTR/MaLR   1  346    (0) 2  0 0
12436 19.0 1.7 7.7 chr3_400   897 2441 (2559) + L1Mur2
LINE/L1 2408 3665 (2212) 1  X 0
5229    7.2 0.0 1.1 chr3_400 2442 3134 (1866) + L1Md_F2
LINE/L1 5877 6580    (2) 3  0 0
```

DateRepeats adds the two last columns for the rat and the human respectively. The X indicates that the repeat is expected to be present at orthologous sites, while the O predicts an absence.

5.4. DupMasker

DupMasker annotates segmental duplications in query sequence. It analyses this file against a library known segmental duplications. If the map file does not exist, DupMasker will launch RepeatMasker to create this file. DupMasker writes a '.duplicons' output file.

5.5. RepeatMasker Library

The user can use the RepeatMasker library downloaded from Genetic Information Research Institute (http://www.girinst.org) or can use its own library. The RepeatMasker library has an EMBL-like format. In addition to the sequence, the EMBL format gives some important information. Each lane starts with a 'two-letter code' that defines the type of line. An example from the library is written below:

```
ID AluJb repbase; DNA; PRI; 283 BP.
XX
AC .
XX
DT 20-AUG-1998 (Rel. 1, Created)
DT 20-AUG-1998 (Rel. 1, Last updated, Version 1)
XX
DE Alu-Jb subfamily - a consensus.
```

```
XX
KW SINE1/7SL; SINE; Non-LTR Retrotransposon;
Transposable Element;
KW Alu-J; Alu-Jb; AluJ; AluJb; Repetitive sequence.
XX
OS Primates
OC Eukaryota; Metazoa; Chordata; Craniata;
Vertebrata; Euteleostomi;
OC Mammalia; Eutheria; Euarchontoglires.
XX
RN [1]
RA Jurka J.;
RT "Origin and evolution of Alu repetitive
elements.";

RL Molecular Biology Intelligence Unit:The impact of
short
RL interspersed elements (SINEs) on the host genome
(ed. Richard J.
RL Maraia), R.G. Landes Company, Austin, pp.25-41
(1995).
XX
DR [1] (Consensus)
XX
SQ Sequence 283 BP; 59 A; 82 C; 98 G; 44 T; 0 other;
ggccgggcgc ggtggctcac gcctgtaatc ccagcacttt
gggaggccga ggcgggagga 60
tcacttgagc ccaggagttc gagaccagcc tgggcaacat
ggtgaaaccc cgtctctaca 120
aaaaatacaa aaattagccg ggcgtggtgg cgcgcgcctg
tagtcccagc tactcgggag 180
gctgaggcag gaggatcgct tgagcccggg aggtcgaggc
tgcagtgagc cgtgatcgcg 240
ccactgcact ccagcctggg cgacagagcg agaccctgtc tca 283
```

The first line, starting with the 'ID' identifier, gives the name of the repeats (here AluJb). This line also gives the size of the consensus sequence. The fourth line, starting with 'DT' (Date), gives the date when the repeat was available in Repbase database (5). The 'DE' line gives a short definition of the repeat and the KW (KeyWord) informs about the class and superfamily of the repeat. The 'OS' and the 'OC' lines give the organism where this repeat was first found and the taxonomy of this specie. 'RA', 'RT,' and 'RL' lines inform about the author and the article where it was first mentioned. The 'SQ' line indicates the size and the ratio of each nucleotide in the sequence. The following lines are the sequence itself.

5.6. Search Engine Software

5.6.1. AB-BLAST (Old WU-BLAST)

Basic Local Alignment Search Tool (BLAST) was developed by Stephen Altschul, Warren Gish, and David Lipman at the U.S. National Center for Biotechnology Information (NCBI), Webb Miller at the Pennsylvania State University, and Gene Myers at the University of Arizona (1). The software WU-BLAST, that is an alternative implementation of BLAST, was acquired from Washington University by Warren R. Gish. The right to license the software to the community was acquired by Advanced Biocomputing, LLC in 2009 (http://blast.advbiocomp.com/) (6). The user can use the old WU-BLAST software, but it is no more available on AB-BLAST Web site. AB-BLAST is free for academic uses. AB-BLAST uses the same core algorithm of WU-BLAST: the binaries kept the same name and the same options. For information, the main difference between AB-BLAST and WU-BLAST is the match/mismatch scores M=+1 N=−3 with gap penalties Q=7 R=2; whereas WU-BLAST uses M=+5 N=−4 with gap penalties Q=10 R=10 by default. This difference changes the hits score and could identify new repeats that the old version misses and vice versa.

5.6.2. RMBLAST

RMBlast is a RepeatMasker compatible version of the standard NCBI BLAST (7). It is a free software that can be downloaded at the RepeatMasker Web site (http://www.repeatmasker.org/RMBlast.html). RMBlast supports some new features such as custom matrices (without KA-Statistics) and cross_match-like complexity adjusted scoring.

5.6.3. Cross_Match

Cross_Match was implemented by Phil Green (http://www.phrap.org/phredphrapconsed.html). This software is part of the Phred/Phrap/Consed package and it is free for academic use but the user must email the information requested in the academic user agreement to David Gordon and receive an email reply before downloading the search engine. Cross_Match is the search engine reference in output format, and some options are specific to it.

5.6.4. Decypher

DeCypher, so-called DeCypherSW, is a software that belongs to "TimeLogic biocomputing solutions." If it could be found in RepeatMasker configuration, it is not listed on RepeatMasker Web site as a search engine tool for RepeatMasker.

6. Options

RepeatMasker's author classifies the options in five categories: Species, Contamination, Masking, Output, and Speed options. I will precise which options are also present on WEBRepeatMasker.

```
-h(elp) Detailed help
```

This option prints all options present in the RepeatMasker command line.

6.1. Species Options

The two following options select the custom library or the part of RepeatMasker library using by RepeatMasker. The goal of these two options is limited to the number of unnecessary alignment against the input sequence. For example, it is useless to search for human Alu in plant genomes.

```
-species <query species>
Specify the species or clade of the input sequence. The
species name must be a valid NCBI Taxonomy Database
species name and be contained in the RepeatMasker
repeat database. Some examples are:
-species human
-species rattus
-species "ciona savignyi"
Other commonly used species:
mammal, carnivore, rodentia, rat, cow, pig, cat, dog,
chicken, fugu,danio, "ciona intestinalis" drosophila,
anopheles, elegans, diatoaea, artiodactyl, arabidopsis,
rice, wheat, and maize
```

This option is present on WEBRepeatMasker. The user must choose the species (or the clade) to which the input data belongs. For example, if the input data is a human chromosome, the user can type "human," "homo sapiens," or "mamma." RepeatMasker does not distinguish between uppercase and lowercase words. The user can use multiple words to define a species like "homo sapiens." Quotation marks must be used before and after the multiple words. The PERL script queryRepeatDatabase.pl in the util directory checks whether the specie's name that the user enters is available in RepeatMasker library.

```
-lib [filename]
Allows use of a custom library (e.g. from another
species)
```

The user can enter a custom library with this option, especially if the studied specie is not covered by RepeatMasker. The library sequence must be in FASTA format. Information of the class repeat can be added after the repeat name: `>repeatname#class`. The symbol '#' indicates the name of the repeat class is after the repeat name. The user can also combine the custom sequence and RepeatMasker library to create a custom library.

6.2. Contamination Options

By default, RepeatMasker looks for bacterial insertion sequences (IS elements). Here, the options control the research of contamination. I noticed the search of IS elements generally takes less than one second per fragment sequence.

```
-no_is
Skips bacterial insertion element check
```

This option skips the research of IS elements in the query sequence. The search of interspersed repeats does not change.

```
-no_is
Only clips E coli insertion elements out of FASTA and
.qual files
```

This option limits the run of RepeatMasker to check only the IS elements, it does not search the other repeats.

```
-no_is
Clips IS elements before analysis (default: IS only
reported)
```

This option removes the IS elements found in the input sequence before the standard runs of RepeatMasker. It writes out a '.withoutIS' file that correspond to the unmasked query sequence without the IS elements. Note the coordinates of repeat elements in the map file correspond to the withoutIS file and not to the original query.

```
-rodspec
Only checks for rodent specific repeats (no repeatmasker run)
-primspec
Only checks for primate specific repeats (no repeatmasker run)
```

These two previous options check the contamination of rodent repeats in primate and vice versa. As notified below, there is no RepeatMasker run against the library. There is no option to check the contamination of non-mammal species. These options are mainly created by A.R.A Smith to check if the sequence belongs to rodent species or primate species.

6.3. Masking Options

These options change all standard output files: the map file changes according to the options, and the masked file and the summary file change according to the map file.

```
-cutoff [number]
Sets cutoff score for masking repeats when using -lib (default 225)
```

For each repeat hit, a Swith-Watermann score is calculated (Column 1 in Fig. 3). Usually, the hit written in the map file has a Swith-Watermann score higher that a threshold. This value equals 225 by default. Using a local library (option '-lib'), the user can choose this option and change the threshold value. Decreasing the value increases the number the false matches and vice versa. The author stipulates that below 200, the result contains false matches and up to 250 the results give only real matches. The user must know that the low-complexity region in repeats sequences such as non-autonomous elements increase false matches.

```
-nolow /-low
Does not mask low_complexity DNA or simple repeats
```

This option removes the low-complexity DNA matches from all output files.

```
-nolow /-low
Only masks low complex/simple repeats (no interspersed repeats)
```

On the contrary, this option removes all matches excepted to be low-complexity DNA matches.

```
-alu
Only masks Alus (and 7SLRNA, SVA and LTR5)(only for primate DNA)
```

The Alu option removes all repeat from the map file excepted Alu families hits. This options only works with the primate sequence, i.e., when the user tapes the option '-specie' followed by human, primate, monkey …

```
-div [number]

Masks only those repeats < x percent diverged from consensus seq
```

Each line in the map file contains the divergence percentage from consensus sequence (second column in Figs. 2 and 3). This option creates a user's threshold for the maximal value of divergence between the hit and the corresponding consensus sequence in the library (Repbase or custom).

```
- norna

Does not mask small RNA (pseudo) genes
```

By default, RepeatMasker screens the input sequence to look for the small polIII transcribed RNAs (mostly transfer RNA (tRNAs) and small nuclear RNA (snRNAs)) because they are very similar to SINE elements. This option does not mask the identified pseudogenes and small RNA genes.

6.4. Output Options

These options do not change RepeatMasker or the map file runs but adds new output file or the format of the output files.

```
-a(lignments)

Writes alignments in .align output file
```

This option creates a supplementary output file. All alignment hits from the search engine are saved in a '.align' file (Fig. 8). The alignments are in the cross_match/SWAT format, in which mismatches rather than matches are indicated: transitions with an i and transversions with a v. Note it exists some differences between the alignment file and the map file. The map file is produced by ProcessRepeats that the main task is to defragment the original map file and the alignment file is created from the original map file: the difference between them comes from the defragmented hits.

```
-inv

Alignments are presented in the orientation of the repeat (with option -a)
```

This option works only with the previous option. By default, the alignment file keeps the orientation of the query sequence. With this option, the alignment has the same orientation than that of the consensus repeat in the library

```
-x

Returns repetitive regions masked with Xs rather than Ns
```

```
Matrix = 20p35g.matrix
Transitions / transversions = 1.29 (9 / 7)
Gap_init rate = 0.03 (3 / 92), avg. gap size = 1.33 (4 / 3)

   273  14.51 0.00 7.22  chr1_Arabidopsis        2     105 (3499825) C ATREP18        DNA                 (375)   1416    1320

  chr1_Arabidop        2 CCTAAACCCTAAACCCTAAACCCTAAACCTCTGAATCCTTAATCCCTAAA 51
                                                    v - i      v  -
C ATREP18#DNA       1416 CCTAAACCCTAAACCCTAAACCCTAAAGC-CTAAATCCTAAA-CCCTAAA 1369

  chr1_Arabidop       52 TCCCTAAATCTTTAAATCCTACATCCATGAATCCCTAAATACCTAATTCC 101
                         -       i   -   i    v -- v v     v  i i  -   vi
C ATREP18#DNA       1368 -CCCTAAACCTT-AAACCCTAAA--CTTTAATCCATAGACAC-TAAGCCC 1324

  chr1_Arabidop      102 CTAA 105
                         i
C ATREP18#DNA       1323 TTAA 1320


Matrix = Unknown
Transitions / transversions = 0.50 (1 / 2)
Gap_init rate = 0.00 (0 / 42), avg. gap size = 0.0 (0 / 0)

   267   7.89 0.00 0.00  chr1_Arabidopsis     4291    4328 (3495602) C (TA)n          Simple_repeat         (1)    179     142

  chr1_Arabidop     4291 ATATATATATATATATATATATGTGGATATATATATAT 4328
                                               i iv
C (TA)n#Simple_      179 ATATATATATATATATATATATATATATATATATATAT 142
```

Fig. 8. An example of optional alignment file of the query with the matching repeats. The alignments are in the cross_match/SWAT format, whereby mismatches rather than matches are indicated: transitions with an i and transversions with a v.

This option replaces the N character in masked file by X character.

```
-xsmall
returns repetitive regions in lowercase (rest capitals)
rather than masked
```

It does not use the N or X characters for the repeat hit but writes in lowercase for the repeats and in uppercase otherwise.

```
-small
Returns complete .masked sequence in lower case
```

This option writes out the masked sequence in lower case instead of uppercase.

```
-poly
Reports simple repeats that may be polymorphic (in
file.poly)
```

Creates a new output file that contains the list of polymorphic microsatellites. The extension name is '.polyout'. The polymorphic microsatellites correspond to the (X)n satellites in map file. For example, (C)n and (TA)n hit present in Fig. 2 are present in the '.polyout' file.

```
-xm
Creates an additional output file in cross_match format
(for parsing)
-ace
Creates an additional output file in ACeDB format
-gff
Creates an additional Gene Feature Finding format
output
```

The '-xm', '-ace,' and '-gff' options create an additional output file in cross match, ACeDB, and Gene Feature Finding format respectively.

```
-u
Creates an additional annotation file not processed by
ProcessRepeats
```

This option creates an additional '.ori.out' file before the ProcessRepeats run. The difference between this new file and the previous map file mainly comes from the fragmented repeats: the fragmented ones correspond to many lines in this file and only one line after the ProcessRepeats run. The '.ori.out' also contains the repeat hits removed by ProcessRepeats: for example, a small hit included between a fragmented repeat should be removed. Then the user can compare and/or check the defragmented repeat hit in the query sequence.

```
-fixed
Creates an (old style) annotation file with fixed width
columns
```

It creates another map file with fixed width column. Regardless of the repeats query name, it simplifies the creation of generic scripts that can parse this file but this option can shorten a long name and create ambiguous names.

```
-no_id
Leaves out final column with unique ID for each element
(was default)
```

Since September 2000, a column displaying a unique number (ID) for each repeat is printed by default. A fragmented single element has the same number that allows better interpretation of the data. This option removes the ID column.

```
-e(xcln)
Calculates repeat densities (in .tbl) excluding runs of
>=20 N/Xs in the query
```

This option calculates the summary table file (.tbl), excluding the runs of equals or more consecutive N or X characters. This option cans change the proportion of TEs families in sequences containing a lot of N characters (for example draft sequences).

```
-noisy
Prints search engine progress report to screen
(defaults to .stderr file)
```

The option '-noisy' prints a progress report on the user's screen instead of writing a .stderr file.

```
-dir [directory name]
Writes output to this directory (default is query file
directory,
"-dir ." will write to current directory).
```

By default, the output files are written in the directory that contains the query sequence. With this option, RepeatMasker writes the output files in the user directory. Note that the directory is not created by RepeatMasker but it just checks if it exists. Note that during the runs, RepeatMasker creates a temporary directory under the format 'RM XXX.Date', where XXX is the random number and Date is the date the user runs RepeatMasker. This temporary directory is deleted in the end of the run.

6.5. Speed and Search Options

These options change indirectly the options of the search engine tools or the way ProcessRepeat assemble and store the repeats before writing the output files.

```
-e(ngine) [crossmatch|wublast|abblast|ncbi|decypher]
Use an alternate search engine to the default.
```

In the configuration step the user chooses a default search engine and may choose other search engines. This option selects the search engine from the list of configured search engines and use the selected ones for this RepeatMasker run. Note that it is possible to select NCBI BLAST, even if it is not possible to select NCBI BLAST in automatic configuration (see paragraph 3).

```
-pa(rallel) [number]
The number of processors to use in parallel (only works
for batch files or sequences over 50 kb)
```

This option determines the number of simultaneous search engine runs. The limit number of parallel is 16.

```
-s Slow search; 0-5% more sensitive, 2-3 times slower
than default
-q Quick search; 5-10% less sensitive, 2-5 times faster
than default
-qq Rush job; about 10% less sensitive, 4->10 times
faster than default
```

The three previous options (-s, -q, and -qq) change the speed of RepeatMasker runs. The increase of speed depends on the input sequence and the library sequence. The scale of speed increase given by the author was verified in my experience. The speed comes from the parameters given to the search engine. In fact, these three options change the minimal size of the initial word (called seed in BLAST algorithm) used by the search engine: a search engine, like BLAST, looks for an exact word of defined size contained in the library sequence on a sequence query. When this exact word (seed) is found, the algorithm tries to extend the seed by allowing gap and mutations with the query sequence. The size of the seed depends on the search engine and the parameter given by the user. For more explanations, read the Altschul et al. algorithm (1). The more the seed size increases, the less the algorithm finds the seed in the query sequence. For example, the seed size in RepeatMasker script is 9 for normal speed, 8 for slow search, 10 for quick search, and 11 for the rush job option. These values correspond to the default repeats, and they change according to the type of repeats.

As stated by the author, the sensitivity of RepeatMasker decreases according to the speed options. The loss of sensitivity depends on the type of repeats. More precisely, it depends on the repeat age. A transposable element generally accumulates mutations when it becomes older. These mutations decrease the size of the words that are the exact copy with the consensus sequence. As a consequence, an old family of transposable is generally more difficult to find with a larger seed than a younger family.

```
-frag [number]
Maximum sequence length masked without fragmenting
(default 60000,300000 for DeCypher)
```

This option changes the size of batch fragment. The minimal size of the fragment must be twofold longer than the size of the overlap fragment. This corresponds to 4,000 nucleotides. A smaller fragment can improve the detection of transposable elements when the fragment contains large regions of DNA with significantly different GC contents. The maximal size of the fragment depends on the amount of memory used. A larger fragment decreases the error of defragmentation hits by ProcessRepeat. However, if your computer does not have sufficient memory, RepeatMasker will redo the failed search with a longer seed that will decrease the sensitivity of your research (see above).

```
-gc [number]
Use matrices calculated for 'number' percentage
background GC level
-gccalc
RepeatMasker calculates the GC content even for batch
files/small seqs
```

These two parameters change the matrices used by RepeatMasker to identify the transposable elements. Normally, RepeatMasker calculates the average GC content (%) of the query sequence and uses the corresponding matrices (see paragraph 7) to identify the repeats. In some cases, transposable elements have a different GC content that the rest of the genome sequence and the RepeatMasker algorithm uses this difference to detect the repeats. The -gc [number]' option forces RepeatMasker to use a defined GC percent instead of the calculated GC content. The different CG contents that can be applied by RepeatMasker vary from 35 to 53%. The 'gccalc' option calculates the GC content of each fragment instead of the whole query sequence.

```
-nopost
Do not postprocess the results of the run ( i.e. call
ProcessRepeats). NOTE: This option should only be used
when ProcessRepeats will be run manually on the
results.
```

This option skips the ProcessRepeats step. The fragmented hits are not assembled with this option.

References

1. Altschul SF, *et al.* (1997) Gapped Blast and Psi-Blast: a new generation of protein database search programs. J Nucleic Acids Res 25: 3389–402
2. Benson G (1999) Tandem repeats finder: a program to analyze DNA sequences. Nucl Acids Res 27:573–580
3. Wicker T, et al. (2007) A unified classification system for eukaryotic transposable elements. Nat Rev Genet 8:973–82
4. Kapitonov VV, Jurka J (2006) Self-synthesizing DNA transposons in eukaryotes. Proc Natl Acad Sci USA 103:4540–4545
5. Jurka J, et al. (2005) Repbase Update, a database of eukaryotic repetitive elements. Cytogenet Genome Res 110, 462–467
6. Gish W (1996–2009) http://blast.advbiocomp.com
7. Johnson M, et al. (2008) NCBI BLAST: a better web interface. Nucl Acids Res 36:W5-W9

Chapter 3

Roadmap for Annotating Transposable Elements in Eukaryote Genomes

Emmanuelle Permal, Timothée Flutre, and Hadi Quesneville

Abstract

Current high-throughput techniques have made it feasible to sequence even the genomes of non-model organisms. However, the annotation process now represents a bottleneck to genome analysis, especially when dealing with transposable elements (TE). Combined approaches, using both de novo and knowledge-based methods to detect TEs, are likely to produce reasonably comprehensive and sensitive results. This chapter provides a roadmap for researchers involved in genome projects to address this issue. At each step of the TE annotation process, from the identification of TE families to the annotation of TE copies, we outline the tools and good practices to be used.

Key words: Transposable elements, Genome annotation, Sequence analysis, Bioinformatics, Genomics

1. Introduction

Transposable elements (TEs) are mobile genetic elements that shape the eukaryotic genomes in which they are present. They are virtually ubiquitous and make up 20% of a *Drosophila melanogaster* genome (1, 2), 50% of a *Homo sapiens* genome (3), and 85% of a *Zea mays* genome (4). They are classified into two classes depending on their transposition mode: via RNA for class I retrotransposons and via DNA for class II transposons (5). Each class is also subdivided into several orders, superfamilies, and families (6).

Thanks to the development of new sequencing technologies the number of sequenced eukaryotic genomes is constantly increasing. However, the first step of the analysis, i.e., accurate genome annotation, is still a major challenge, particularly with respect to TEs. The correct genome annotation of genes and TEs is indispensable to thorough genome-wide studies. As a result, efficient computational methods have been proposed for TE annotation (7).

Yves Bigot (ed.), *Mobile Genetic Elements: Protocols and Genomic Applications*, Methods in Molecular Biology, vol. 859, DOI 10.1007/978-1-61779-603-6_3,

Given that the pace at which genomes are sequenced which is unlikely to decrease in the coming years, the process of TE annotation has to be standardized.

Here, we establish a clear road map detailing the order in which computational tools (or combinations of such tools) should be used to annotate TE in a whole genome. We distinguish three main steps: (1) identifying TEs by searching some reference sequences (e.g., full length TE sequences) and building consensus sequences from similar sequences, (2) defining and classifying TE families, (3) annotating every TE copy.

2. Materials

Sequencing costs have dropped dramatically and sequences have thus become easier to obtain. However, sequence analysis remains a major bottleneck. Efficient analyses need to be quick and robust to scale up the pace of data production. In this way, the knowledge of the few specialists able to perform genome analysis on a large scale can be exploited so that TE annotations are made available to a large community of scientists.

We have been involved in many genome projects (2, 8–18), and we have assessed the relative benefits of using different programs for TE detection, clustering, and multiple alignments. Our investigations suggest that only combined approaches, using both de novo and knowledge-based TE detection methods, are likely to produce reasonably comprehensive and sensitive results (2, 13, 19). In view of this, the REPET package (19) has been developed. It is composed of two analysis pipelines, TEdenovo and TEannot. These pipelines launch several different prediction programs in parallel and then combine their results to improve the accuracy and exhaustiveness of the detection. However, even with these sophisticated pipelines, manual curation is still needed. Hence, in addition to the automation of all the steps required for TE annotation, these pipelines compute useful data for the manual curation, such as, TE sequence multiple alignments, TE sequence phylogenies, and TE classification evidence. They have been used with success on many different genomes (1, 2, 9, 11–18, 20). To take advantage of all its possibilities, the REPET package (http://urgi.versailles.inra.fr/index.php/urgi/Tools/REPET) requires the following software:

1. Unix-like system, Python, MySQL, and SGE (Sun Grid Engine).
2. BLASTER suite (http://urgi.versailles.inra.fr/index.php/urgi/Projects/URGI-softwares/BLASTER), already included in the REPET package.
3. NCBI-BLAST (http://www.ncbi.nlm.nih.gov/BLAST/download.shtml).

4. WU-BLAST (http://blast.wustl.edu/).
5. RECON (for the version 1.06, http://www.repeatmasker.org/).
6. PILER (http://www.drive5.com/piler/).
7. CENSOR (http://www.girinst.org/censor/download.php).
8. RepeatMasker (http://www.repeatmasker.org/RMDownload.html).
9. Tandem Repeat Finder (http://tandem.bu.edu/trf/trf.download.html).
10. Mreps (http://bioinfo.lifl.fr/mreps/).
11. Shuffle (from squid in hmmer2 package, or esl-shuffle in hmmer3 package http://hmmer.janelia.org/).

    Several other programs can also be useful to analyze the results returned by the pipelines from the REPET package and thus ease the manual curation.

12. PhyML (http://atgc.lirmm.fr/phyml/).
13. JalView (http://www.jalview.org/).
14. MAFFT (http://mafft.cbrc.jp/alignment/software/).
15. TEclass (http://www.compgen.uni-muenster.de/tools/teclass/).

3. Methods

3.1. All-by-All Alignment and Clustering of Interspersed Repeats

The most comprehensive strategy is to identify TEs by self-alignment of the genomic sequences (see Notes 1–3 and 5 for alternative approaches). Analysis is then performed by all-by-all alignment of the genomic sequences (contigs, scaffolds, or pseudo-molecules). Such a genome comparison strategy has been implemented in a pipeline called TEdenovo (Fig. 1). This pipeline is part of the REPET package (19), designed to be used on a computer cluster for fast calculations, and it implements the following steps.

1. BLASTER (8) performs the all-by-all comparison by launching BLAST (21) repeatedly over the genome sequences (see Note 6 for an alternative approach). As the amount of input data is usually high, the computations are intensive. Therefore, stringent parameters are applied: best results are obtained when matches shorter than 100 bp or with identity below 90% or with an E-value above 1e-300 are dismissed (see Note 7). In addition, as most TEs are shorter than 25 kb, segmental duplications can be filtered out by removing longer matches. To speed up the computations, such alignment tools can be launched in parallel on a computer cluster.
2. Once the matches corresponding to the repeats have been obtained, they need to be clustered into groups of similar

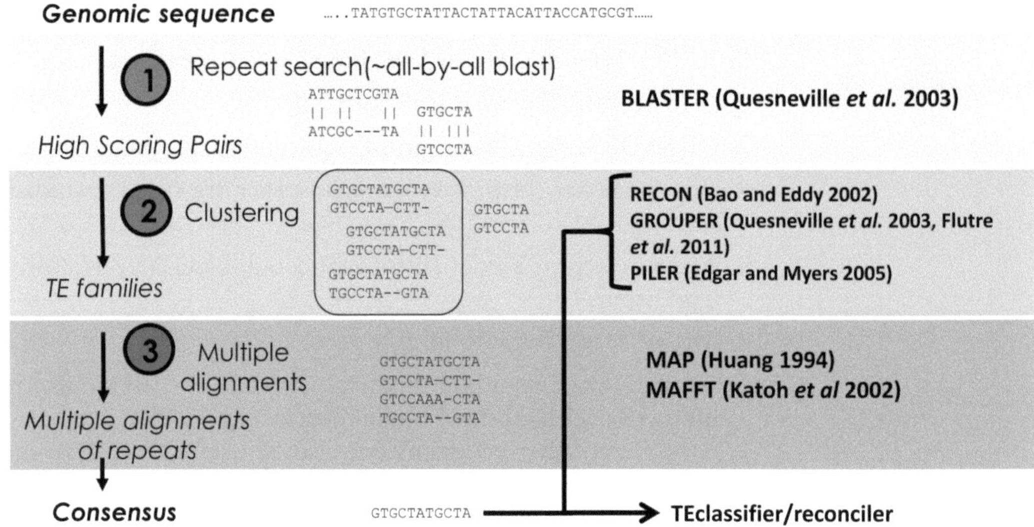

Fig. 1. Workflow of the 3-step de novo TE detection pipeline (19).

sequences. Each cluster would hopefully correspond to copies of a single TE family. However, TEs are divergent interspersed repeats, often nested within each other, making the task difficult. Algorithms have been designed to cluster the identified sequences properly, limiting the artifacts induced by nested and deleted TE copies and non-TE repeats such as segmental duplications. Various tools exist and make different assumptions about (a) the sequence diversity within a TE family, (b) the evolutionary dynamics of TE sequences, (c) nested patterns, and (d) repeat numbers. Three different types of software are recommended to be used simultaneously. Note 8 indicates our observation on their different behavior.

(a) GROUPER (2, 8, 19) starts by connecting fragments belonging to the same copy by dynamic programming, and then applies a single-link clustering algorithm with a 95% coverage constraint between copies of the same cluster. The rationale is to detect copies that have the same length, as they most probably correspond to mobile entities. Indeed, copies diverge rapidly by accumulating deletions leading to copies with different size. Only copies that are almost intact can transpose and then conserve their original, presumably functional, size.

(b) RECON (22) also starts with a single-link clustering step. Then, if a cluster is chimerical because of some nested repeats, it could be subdivided according to its all-by-all genome alignments ends distribution. Indeed, nested repeats exhibit a specific pattern in alignments of sequences obtained in an all-by-all genome comparison; the alignment ends of any one inner repeat are all in the same position.

(c) PILER-DF (23) identifies lists of matches covering a maximal contiguous region, defines them as piles, and then builds clusters of globally alignable piles. The rationale is identical to that used by GROUPER where copies of identical length are sought, but PILER-DF has no specific consideration for insertion–deletion events.

3. Once clusters are defined, a filter is applied to retain only those having at least three members, thus discarding the vast majority of segmental duplications. Finally, for each remaining cluster, a multiple alignment is built from which a consensus sequence is derived. Numerous algorithms are available for this but they must comply with the following criteria: (a) speed, because the number of clusters is usually very high, and (b) ability to handle sequences of different lengths appropriately. MAP (24) and MAFFT (25) comply with these criteria and give good results. From our experience, taking the 20 longest sequences is generally sufficient to build the consensus. The set of consensus sequences obtained then represents a condensed view of all TE families present in the genome being studied.

3.2. Automatic Classification of Transposable Element Sequences

When they amplify, TE copies may nest within each other in complex patterns (1), thereby fragmenting the elements. With time, they accumulate (1) point substitutions, (2) deletions that fragment copies, and (3) insertions that interrupt their sequences (26). These events generate complex remnants of TEs. Various de novo tools use these remnants to try to infer the transposed ancestral sequence.

When starting with a self-alignment of genomic sequences, the optimal strategy would be to use several tools and even combine them. However, whatever the tools are, each de novo approach can encounter difficulties when trying to distinguish true TEs from segmental duplications, multimember gene families, tandem repeats, and satellites. It is therefore strongly recommended to confirm that the predicted sequences can be classified as being TEs (see Note 9). At this stage, automatic analyses still need to be complemented by manual curation. The following steps are also implemented in the TEdenovo pipeline from the REPET package. Alternative tools are mentioned in Note 10.

1. Sequences believed to correspond to TEs can be classified according to their similarity to known TEs as those recorded in databases like RepbaseUpdate (27). Hence, the comparison of TE candidates with a reference data-bank using BLASTN, BLASTX, and TBLASTX can provide hints for classification, as long as the TE candidate has similarities with elements in this kind of reference data-bank.

2. However, for most previously unknown TE sequences obtained via de novo approaches from non-model organisms, classification will require the specific identification of several TE features

(see Chapter 6 for complete description). By searching for structural features, such as terminal repeats, one can identify long terminal repeats specific to class I LTR retrotransposons, terminal inverted repeats specific to class II DNA transposons, and poly-A or SSR-like tails specific to class I non-LTR retrotransposons. Tandem Repeat Finder, and TRsearch and polyAtail from the REPET package, can be used separately or in the frame of the TEdenovo pipeline.

3. It is also judicious to search for matches for TE-specific protein profiles in TE sequences. For example, the presence of a transposase is highly indicative of a class II DNA transposon. Such protein profiles can be obtained from the Pfam database which gathers protein families represented by multiple sequence alignments and hidden Markov models (HMM) (28). These profiles can be used by programs such as HMMER to find matches within the candidate TE sequences.
4. A tool called TEclass (29) implements a support vector machine, using oligomer frequencies, to classify TE candidates. It can be used to classify sequences which are still unclassified at this step.
5. TEclassifier from REPET package (19) allows the removal of redundancy among potential TE sequences, which are a consequence of using several clustering programs. The TEclassifier uses the classification to remove redundant copies (a sequence contained into a longer one), and considers well-classified TE candidate sequences as not removable versus less well-classified sequences. This tool is particularly useful in reconciling several independently obtained TE reference libraries because it guarantees the keeping of well-classified TE candidate sequences.

   Once unknown TE sequences have been classified, some consensus sequences may still be unclassified and there may still be some redundant consensus sequences. Manual curation is crucial because the annotation of TE copies described in the next section depends on the quality of the TE library. One way to curate a library of TE consensus sequences is to gather these sequences into clusters that could represent TE families. Indeed, whereas some TE consensus sequences can represent different structural variants of the same family, other TE consensus sequences may simply be redundant and should therefore be removed (19).

6. A tool like BLASTCLUST in the NCBI-BLAST suite (30) can quickly build such clusters via single-link clustering based on sequence alignment coverage and identity. Eighty percent identity and coverage, as proposed by Wicker et al. (6), gives good results (see Note 11). Typical clusters will contain well-classified consensuses (e.g., class I—LTR—Gypsy element) as well as unclassified consensuses (without structural features,

Fig. 2. Alignment (Jalview (54) screenshot) of de novo TE consensus sequences with Athila, the best matching known TE for these consensuses from Repbase Update. They are represented with some of the features shown: LTRs (*light grey*), ORFs (*dark grey*), and matches with HMM profiles (*black*). The differences between the consensuses obtained by different methods, here RECON (cons1) and GROUPER (cons2, cons3, cons4), are illustrated. Manual curation would remove cons3 as it corresponds to a single LTR with short sequences not present in the Athila family and cons4 as it corresponds to a LTR probably formed from the Athila solo-LTRs of the genome. A good consensus for the family would be a combination of cons1 and cons2.

and no sequence similarity either with known TEs or any TE domain). Computing a multiple sequence alignment (MSA) for each cluster gives a useful view of the relationships between the consensus sequences so that it is possible to assess whether they belong to the same TE family. One of the programs detailed above, MAP, or MAFFT, can be used. It can also be informative to build an MSA with the consensus, and with the genomic sequences from which these consensuses were derived and/or the genomic copies that each of these consensuses can detect. In such cases, we advise first building a single MSA for each consensus with the genomic sequences that it detects and then building a global MSA by aligning these multiple alignments together, for example using the "profile" option of the MUSCLE program (31).

7. Finally, after a visual check of the MSA with the evidence used to assign a classification to the consensus, it is then possible to tag all consensus sequences in the same cluster with the most frequent TE class, order, superfamily and family, if one has been assigned (Fig. 2). The MSA can be also edited by splitting it or deleting sequences to obtain a MSA corresponding to a single TE family. Indeed, in some cases, consensuses are only similar over a small region or have a high sequence divergence. In these cases, the MSA can be split into as many MSAs as there are candidate TE families. In other cases, an insertion appears to be specific to one consensus sequence and may sometimes show evidence (e.g., BLAST hits) for a different TE order. This could highlight a chimeric consensus that can be either removed from the library, if it seems artifactual according to the sequences used to build the consensus (also visible

in the MSA), or used to build a new TE family (if several copies support it) (see Note 12).

8. Phylogenies of TE family copies and/or consensus sequences provide another view of the members within a TE family. This can help the curation if the cluster has many members, or when two or more subfamilies are present. In such cases, subfamilies can be hard to detect by examination of the MSA alone but become evident in a phylogeny when sub-trees appear distinctly and appear to be independent of the other sequence phylogeny. Such phylogenies can be constructed from the MSA with currently available software, including the PhyML program (32). However, as most phylogeny programs do not consider gaps, branch length may be biased when consensus sequences are of very different lengths. Divergence between the sequences can also be a criterion.

3.3. Annotation of Transposable Element Copies

This second phase involves annotating all of the TE copies in the genome, resolving the most complex degenerate or nested structures. This requires a library of reference sequences representing the TE families. In the best case, the library is both exhaustive and non-redundant, i.e., each ancestral TE, autonomous or not, is represented by a single consensus sequence. We usually use the manually curated library built as detailed in the previous section. Then, with these reference sequences, all TE fragments are found on the genome and connected as though they belong to the same TE copies. The TEannot pipeline (2) from the REPET package implements the following steps:

1. Detecting TE fragments. The first step mines the genomic sequences with the TE library via local pairwise alignments. Several tools were designed specifically for this purpose, such as REPEATMASKER (33), CENSOR (34, 35), and BLASTER. Some of these tools incorporate scoring matrices to be used with particular GC percentages, as is the case for isochores in the human genome. All these tools propose a small set of parameter combinations depending on the level of sensitivity required by the user. Although similar, these tools are complementary. We have previously shown that combining these three programs is the best strategy (2).

   (a) RepeatMasker is run with the WU-BLAST search engine with sensitive parameters. It is noteworthy that we found this search engine to be more sensitive and much faster than Cross-match, under sensitive parameters. RepeatMasker is used with parameters "`-cutoff 200 -w -s -gccalc -nolow -no_is.`"

   (b) Censor is used at high sensitivity with parameter "`-s -ns.`"

(c) BLASTER with WU-BLAST as a search engine. It is used with parameters "`-W -S 3.`"

2. The MATCHER program (2, 19) can then be used to assess these multiple results and keep only the best. It is used to combine the results obtained from BLASTER, RepeatMasker, and Censor after normalizing alignment scores to be the hit length multiplied by the identity percentage. The MATCHER program has been developed to map match results onto query sequences by first filtering overlapping hits. When two matches overlap on the genomic (query) sequence, the one with the best alignment score is kept, the other is truncated so that only nonoverlapping regions remain on the match. As a result of this procedure, a match is totally removed only if it is included in a longer one with a best score. All matches that have a length ≤ 20 are eliminated.

3. Whatever parameters are used for the pairwise alignments, some of the matches will be false positives, i.e., a TE reference sequence will match a locus although no TE is present. For protein-coding genes, full-length cDNAs can be used for confirmation; unfortunately, there is no equivalent way of checking for TE annotation. To assess the false-positive risk, an empirical statistical filter such as the one implemented in the TEannot pipeline (REPET package) is useful. The genomic sequences are shuffled and screened with the TE library. The alignments obtained are false positive, then the highest alignment score is used to filter out spurious alignments obtained on the true genome. Only the matches on the true genomic sequences having a higher score are kept. This procedure guarantees that none of the observed matches can be obtained with random sequences.

4. Short simple repeats (SSRs) are short motifs repeated in tandem. Many TE sequences contain SSRs but they are also present in the genome independently. It is therefore necessary to filter out TE matches if they are restricted to SSR that may be contained within the TE consensus. This is achieved by annotating SSRs and then removing TE matches included in SSR annotations. Several efficient programs, for example TRF (36), MREPS (37), and REPEATMASKER, are available for SSR annotation. These three programs are launched in parallel and their results can be subsequently combined in the TEannot pipeline (REPET package) to remove hits only due to SSR within TE consensus sequences (see Note 13 for the case of mini-satellites and satellites).

5. Even when TE fragments have been mapped on the genome, the work is only half-finished. TE copies can be disrupted into several fragments. A complete TE annotation requires retrieval

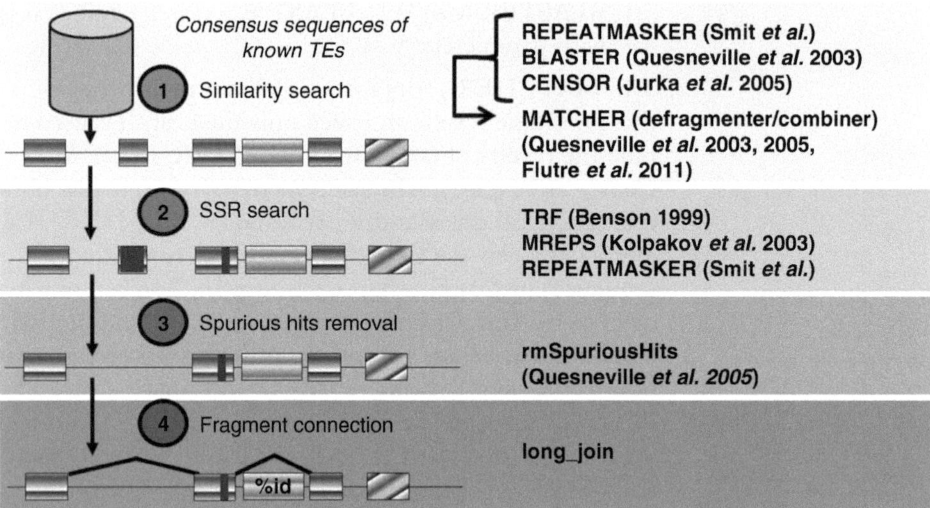

Fig. 3. The four steps of the TEannot pipeline (2, 19).

of all copies, and thus joining fragments which belong to the same copy when it has transposed. Joining TE fragments to reconstruct a TE copy is known as a "chain problem" as it corresponds to finding the best chain of local pairwise alignments. The optimal solution is found via dynamic programming as implemented in MATCHER. Subsequently, an additional procedure implemented in the TEannot pipeline (Fig. 3) called "long join," can be used to account for additional considerations related to TEs biology. Two TE fragments distant from each other but mostly separated by other TE fragments (e.g., at least 95% as in heterochromatin) can be joined as long as the TE fragments between them are younger. The age can be approximated using the percent identity of the matches between the TE reference sequences and the fragments.

Once this phase is finished, a first release of the annotation of TE copies is available. As mentioned above, and if necessary, it is now possible to continue the manual curation of the TE library initiated at the end of the de novo phase. Finally, one possesses a set of TE consensus sequences along with their classification, a set of TE families possibly gathering several TE consensus sequences and a set of coordinates for TE copies.

4. Notes

1. Computing highly repeated words is a fast alternative approach. Software, such as the recent TALLYMER (38) and P-CLOUDS (39), were designed to find repeats quickly in genome sequences

by counting highly frequent words of a given length *k*, called *k*-mers. These programs are very useful for providing a rapid view of the repeated fraction in a given set of genomic sequences, especially with unassembled sequences. However, they do not provide much detail about the presence of TE in these sequences. Their output only corresponds to highly repeated regions without indicating precise TE fragment boundaries or TE family assignment. These methods are quick and simple to use but allow only limited biological interpretations, with no real TE annotation.

2. Other methods also start by counting frequent *k*-mers but then go on to try to define consensus. ReAS (40) applies this approach directly to shotgun reads. For each frequent *k*-mer, a multiple alignment of all short reads containing the *k*-mer is built and then extended iteratively. REPEATSCOUT (41) has a similar approach, but works on assembled sequences. These tools return a library of consensus sequences. Although their results are more biologically relevant than those of previous methods, the consensus sequences are usually too short and correspond to truncated versions of ancestral TEs. Substantial manual inspection and editing is therefore needed to obtain a meaningful list of consensus sequences.
3. Knowledge-based TE detection methods (i.e., based on structure or similarity to distant TEs) have distinct advantages over all-by-all sequence comparison repeat discovery methods. They depend on prior knowledge of TE structure such as Long Terminal Repeats (LTR) or Terminal Inverted Repeats (TIR). The majority of TEs in plant genomes are class I LTR retrotransposons, and many bioinformatic tools have been developed for their detection. They search first for the LTR structure of these retrotransposons and then for other internal features of these elements. LTRharvest (42) is one such tool, recently developed to detect LTR TEs in large genomic sequences. Helitrons can be identified by looking for structured hairpins; this method is used by a tool called HelSearch (43). Class II TEs, with the exception of *Helitrons*, are structurally characterized by TIRs. They can be found by searching for their TIRs; an approach used by a tool called Must (44). This was designed to find MITE elements by searching for TEs containing TIRs adjacent to Direct Repeats (i.e., the Target Site Duplication, also called TSD, see Note 4), within 500 bp. It finds all of the TIRs in the genomic sequence first and then uses sequence alignments to predict MITE candidates. Knowledge-based de novo identification is generally efficient. Many good tools already exist; here we only cite recent ones as an illustration (see refs. 7, 45 for a more exhaustive review). They capitalize on prior knowledge based on the large number of previously

reported TE sequences. Thus, they are more likely to detect *bona fide* TEs, including even those present only as a single copy in the genome. However, these methods are not well suited to detecting new TEs (especially of new types) and have intrinsic ascertainment biases. For example, miniature inverted repeat transposable elements (MITEs) and short interspersed nuclear elements (SINEs) will be underrepresented if we rely on similarity, because these TEs are composed entirely of non-coding sequences. In addition, this approach can only identify well-known TEs with a strong structural signature. Some TEs do not have such characteristics and thus cannot be found using this approach.

4. Many TE families generate a double-strand break when they insert into the DNA sequence. The break is caused by the TE enzymatic machinery that generally cuts the DNA with a shift between the two DNA strands. After insertion, the DNA repair generates a short repeat of few nucleotides (up to 11) at each end; these repeats are called Target Site Duplications (TSDs). TSDs are hallmarks of a transposition event, but they can be difficult to find in old insertions because they are short, and they can be altered by mutations or deletions. In addition, the size of the TSD depends on the family and not all TEs generate TSD upon insertion.

5. The presence of a long indel identified by genome alignments between two closely related species may indicate the presence of a TE. The rest of the genome can then be searched for the presence of this sequence to assess its repetitive nature. This approach has already been used (46) and appears to work well with recent TE insertions. Indeed, only TE insertions that occur after speciation can be detected. Using several alignments with species diverging at different times may lead to more TEs being identified (46) because each alignment would allow detection of TEs inserted at different times. However, it is difficult to correctly align long genomic sequences from increasingly divergent species. This idea could be also used within a genomic sequence using segmental duplications. A long indel in sequence alignments of genomic duplications may also indicate the presence of a TE (47). However, various controls are needed to confirm the TE status of the sequence. For example, TE features such as terminal repeats (e.g., LTR, TIR) or similarity to other TE sequences could be used. This approach only detects TE insertions that occur after the duplication event and may thus be limited to rare events.

6. Exact algorithms can be used. PALS uses "q-Gram filters" which, unlike heuristic algorithms (e.g., BLAST), rapidly and stringently eliminate a large part of the search space from

consideration before the alignment search and guarantee not to eliminate a match-containing region (48).

7. With these parameters, only closely related TE copies will be found. The aim of this step is not to recover all TE copies of a family, but to use those that are well conserved to build a robust consensus sequence (see further). Our experience indicates that stringent alignment parameters are crucial for successful reconstruction of a valid consensus. Even with these stringent criteria, this approach is still more sensitive than other all-by-all alignment methods for identifying repeats. However, it is also the most computer intensive and will miss single-copy TE families because at least two copies are required for detection by self-alignment.
8. The three clustering programs behave differently, according to sequence diversity of TE families. For instance, GROUPER better distinguishes groups of mobile elements differing by their sizes inside a TE family. It also better recovers fragmented copies due to its dynamic programming joining algorithm. However, it produces more redundant results, and only correctly recovers TE families if there are at least three complete copies. RECON is better for TE families with less than three complete copies, being able to reconstruct the complete element from the TE fragments. PILER is fast and very specific. It could be an interesting option on large genomes when time is limited, or if a non-exhaustive search is sufficient.
9. For TE families that are easy to identify, i.e., those with full-length copies that are very similar to each other, all clustering methods will find roughly the same consensus. However, for the other families, which may be numerous, different methods generate different clusters because they rely on different assumptions. Therefore manual curation is required to find a correct set of representative sequences.
10. Other tools classify TE sequences according to their features, usually via a decision tree. The TEclassifier in the REPET package and REPCLASS (49) search both for all of the TE features. In addition, REPCLASS allows TE candidates to be filtered on the basis of the number of their genomic copies.
11. Some authors (6) have suggested an 80-80-80 rule, which considers two sequences as belonging to same TE family if they could be aligned on more than 80 bp, over more than 80% of their length, with over 80% identity. This rule is empirical, but appears to be useful to classify TE sequences in families that are consistent for the following annotation step, i.e., the annotation of their copies. These authors also suggest a nomenclature for naming new TEs that can be followed.

12. We believe that existing automatic approaches still need to be supplemented by expert manual curation. At this step, careful examination is required as some identified families that may appear as artifactual can in fact be unusual TE families. Indeed, well-documented cases show that TE families can appear confusing as they may (a) include cellular genes or parts of genes (e.g., pack-MULEs (50) or *Helitrons* (51)), (b) be restricted to rDNA genes (e.g., the R2 Non-LTR retroelement superfamily (52)), or (c) form telomeres as in *Drosophila* (20). Close examination of non-canonical cases may reveal new and interesting TE families or particular transposition events, e.g., macrotranspositions (53).
13. Satellites are longer motifs, around 100 bp long, also repeated in tandem. Although they are not TEs, they are sometimes difficult to distinguish because they contain parts of TEs. PILER-TA (23) detects pyramids in a self-alignment of the genomic sequences. These pyramids can be used to make a consensus of the satellite unit motif. These consensus sequences can then be aligned on the whole genome to identify all of their occurrences and to distinguish them from TEs.

Acknowledgments

This work was supported in part by grants from the Agence Nationale de la Recherche (Holocentrism project, to HQ [grant number ANR-07-BLAN-0057]) and the Centre National de la Recherche Scientifique—Groupement de Recherche 2157 “Elements Transposables.” TF was supported by a PhD studentship form the Institut National de la Recherche Agronomique. EP was supported by a Post-Doctoral fellowship form the Agence Nationale de la Recherche.

References

1. Bergman CM, et al. (2006) Recurrent insertion and duplication generate networks of transposable element sequences in the Drosophila melanogaster genome. Genome Biol 7:R112
2. Quesneville H, et al. (2005) Combined evidence annotation of transposable elements in genome sequences. PLoS Comput Biol 1: 166–175
3. Lander ES, et al. (2001) Initial sequencing and analysis of the human genome. Nature 409:860–921
4. Schnable PS, et al. (2009) The B73 maize genome: complexity, diversity, and dynamics. Science 326:1112–1115
5. Finnegan DJ (1989) Eukaryotic transposable elements and genome evolution. Trends Genet 5:103–107
6. Wicker T, et al. (2007) A unified classification system for eukaryotic transposable elements. Nat Rev Genet 8:973–982
7. Bergman CM, Quesneville H (2007) Discovering and detecting transposable elements in genome sequences. Brief Bioinform 8:382–392
8. Quesneville H, Nouaud D, Anxolabehere D (2003) Detection of new transposable element families in Drosophila melanogaster and Anopheles gambiae genomes. J Mol Evol 57 Suppl 1:S50-59

9. Cuomo CA, et al. (2007) The Fusarium graminearum genome reveals a link between localized polymorphism and pathogen specialization. Science 317:1400–1402
10. Nene V, et al. (2007) Genome sequence of Aedes aegypti, a major arbovirus vector. Science 316:1718–1723
11. Vitte C, Panaud O, Quesneville H (2007) LTR retrotransposons in rice (Oryza sativa, L.): recent burst amplifications followed by rapid DNA loss. BMC Genomics 8:218
12. Abad P, et al. (2008) Genome sequence of the metazoan plant-parasitic nematode Meloidogyne incognita. Nat Biotechnol 26:909–915
13. Buisine N, Quesneville H, Colot V (2008) Improved detection and annotation of transposable elements in sequenced genomes using multiple reference sequence sets. Genomics 91:467–475
14. Martin F, et al. (2008) The genome of Laccaria bicolor provides insights into mycorrhizal symbiosis. Nature 452:88–92
15. Cock JM, et al. (2010) The Ectocarpus genome and the independent evolution of multicellularity in brown algae. Nature 465:617–621
16. d'Alencon E, et al. (2010) Extensive synteny conservation of holocentric chromosomes in Lepidoptera despite high rates of local genome rearrangements. Proc Natl Acad Sci USA 107:7680–7685
17. Martin F, et al. (2010) Perigord black truffle genome uncovers evolutionary origins and mechanisms of symbiosis. Nature 464: 1033–1038
18. Spanu PD, et al. (2010) Genome expansion and gene loss in powdery mildew fungi reveal tradeoffs in extreme parasitism. Science 330:1543–1546
19. Flutre T, et al. (2011) Considering transposable element diversification in de novo annotation approaches. PLoS One 6:e16526
20. Clark AG, et al. (2007) Evolution of genes and genomes on the Drosophila phylogeny. Nature 450:203–218
21. Altschul SF, et al. (1997) Gapped BLAST and PSI-BLAST: a new generation of protein database search programs. Nucleic Acids Res 25:3389–3402
22. Bao Z, Eddy SR (2002) Automated de novo identification of repeat sequence families in sequenced genomes. Genome Res 12: 1269–1276
23. Edgar RC, Myers EW (2005) PILER: identification and classification of genomic repeats, Bioinformatics 21 Suppl 1:i152-158
24. Huang X (1994) On global sequence alignment. Comput Appl Biosci 10:227–235
25. Katoh K, et al. (2002) MAFFT: a novel method for rapid multiple sequence alignment based on fast Fourier transform. Nucleic Acids Res 30:3059–3066
26. Blumenstiel JP, Hartl DL, Lozovsky ER (2002) Patterns of insertion and deletion in contrasting chromatin domains. Mol Biol Evol 19:2211–2225
27. Jurka J, et al. (2005) Repbase Update, a database of eukaryotic repetitive elements. Cytogenet Genome Res 110:462–467
28. Finn RD, et al. (2010) The Pfam protein families database. Nucleic Acids Res 38:D211-222
29. Abrusan G, et al. (2009) TEclass – a tool for automated classification of unknown eukaryotic transposable elements. Bioinformatics 25: 1329–1330
30. NCBI. NCBI suite
31. Edgar RC (2004) MUSCLE: a multiple sequence alignment method with reduced time and space complexity. BMC Bioinformatics 5:113
32. Guindon S, Gascuel O (2003) A simple, fast, and accurate algorithm to estimate large phylogenies by maximum likelihood. Syst Biol 52:696–704
33. Smit AFA, Hubley R, Green P (1996–2004) RepeatMasker Open-3.0., Institute for Systems Biology
34. Jurka J, et al. (1996) CENSOR – a program for identification and elimination of repetitive elements from DNA sequences. Comput Chem 20:119–121
35. Kohany O, et al. (2006) Annotation, submission and screening of repetitive elements in Repbase: RepbaseSubmitter and Censor. BMC Bioinformatics 7:474
36. Benson G (1999) Tandem repeats finder: a program to analyze DNA sequences. Nucleic Acids Res 27:573–580
37. Kolpakov R, Bana G, Kucherov G (2003) mreps: Efficient and flexible detection of tandem repeats in DNA. Nucleic Acids Res 31:3672–3678
38. Kurtz S, et al. (2008) A new method to compute K-mer frequencies and its application to annotate large repetitive plant genomes. BMC Genomics 9:517
39. Gu W, et al. (2008) Identification of repeat structure in large genomes using repeat probability clouds. Anal Biochem 380:77–83
40. Li R, et al. (2005) ReAS: Recovery of ancestral sequences for transposable elements from the unassembled reads of a whole genome shotgun. PLoS Comput Biol 1:e43
41. Price AL, Jones NC, Pevzner PA (2005) De novo identification of repeat families in large genomes. Bioinformatics 21 Suppl 1:i351-358

42. Ellinghaus D, Kurtz S, Willhoeft U (2008) LTRharvest, an efficient and flexible software for de novo detection of LTR retrotransposons. BMC Bioinformatics 9:18
43. Yang L, Bennetzen JL (2009) Structure-based discovery and description of plant and animal Helitrons. Proc Natl Acad Sci USA 106:12832–12837
44. Chen Y, et al. (2009) MUST: a system for identification of miniature inverted-repeat transposable elements and applications to Anabaena variabilis and Haloquadratum walsbyi. Gene 436:1–7
45. Lerat E (2010) Identifying repeats and transposable elements in sequenced genomes: how to find your way through the dense forest of programs. Heredity 104:520–533
46. Caspi A, Pachter L (2006) Identification of transposable elements using multiple alignments of related genomes. Genome Res 16:260–270
47. Le QH, et al. (2000) Transposon diversity in Arabidopsis thaliana. Proc Natl Acad Sci USA 97:7376–7381
48. Rasmussen K, Stoye J, Myers EW (2006) Efficient q-gram filters for finding all e-matches over a given length. J Comput Biol 13:296–308
49. Feschotte C, et al. (2009) Exploring repetitive DNA landscapes using REPCLASS, a tool that automates the classification of transposable elements in eukaryotic genomes. Genome Biol Evol 1:205–220
50. Jiang N, et al. (2004) Pack-MULE transposable elements mediate gene evolution in plants. Nature 431:569–573
51. Morgante M, et al. (2005) Gene duplication and exon shuffling by helitron-like transposons generate intraspecies diversity in maize. Nat Genet 37:997–1002
52. Eickbush TH, et al. (1997) Evolution of Rl and R2 in the rDNA units of the genus Drosophila. Genetica 100:49–61
53. Gray YH (2000) It takes two transposons to tango: transposable-element-mediated chromosomal rearrangements. Trends Genet 16:461–468
54. Clamp M, et al. (2004) The Jalview Java alignment editor. Bioinformatics 20:426–427

Chapter 4

To Detect and Analyze Sequence Repeats Whatever Be Their Origin

Jacques Nicolas

Abstract

The development of numerous programs for the identification of mobile elements raises the issue of the founding concepts that are shared in their design. This is necessary for at least three reasons. First, the cost of designing, developing, debugging, and maintaining software could present a danger of distracting biologists from their main bioanalysis tasks that require a lot of energy. Some key concepts on exact repeats are always underlying the search for genomic repeats and we recall the most important ones. All along the chapter, we try to select practical tools that may help the design of new identification pipelines. Second, the huge increase of sequence production capacities requires to use the most efficient data structures and algorithms to scale up tools in front of the data deluge. This paper provides an up-to-date glimpse on the art of string indexing and string matching. Third, there exists a growing knowledge on the architecture of mobile elements built from literature and the analysis of results generated by these pipelines. Besides data management which has led to the discovery of new families or new elements of a family, the community has an increasing need in knowledge management tools in order to compare, validate, or simply keep trace of mobile element models. We end the paper with first considerations on what could help the near future of such research on models.

Key words: Repeats, Seeds, String index, Pattern matching, DNA parsing, Grammatical models

1. Introduction

The many types of repeats that occur in genomic sequences have been largely described in the literature and new types are often discovered in newly sequenced species. The management of such a quantity of families rests on dedicated databases (1) or software pipelines, like REPET (see ref. 2 and the chapter inside this volume). Besides the study of these natural repeat families, biologists are routinely concerned with sequence comparison, a task that relies on the search of words common to a given set of sequences and is

Yves Bigot (ed.), *Mobile Genetic Elements: Protocols and Genomic Applications*, Methods in Molecular Biology, vol. 859, DOI 10.1007/978-1-61779-603-6_4, © Springer Science+Business Media, LLC 2012

almost always based on the pre-computation of a repeat index on the sequences. This issue became even more critical with the advent of next-generation sequencing technologies, which is leading each laboratory to access or to produce an increasing quantity of sequence data. Considering that, whole-genome sequencing is becoming an ordinary practice on bacteria and the analysis of the genetic diversity of eukaryotic populations by means of large-scale re-sequencing projects is becoming a general trend. In this context, the presence of repeats causes major assembly issues and requires further algorithmic developments.

On the theoretical side, it is important to try describing all these repeats with a set of precise common concepts in order to better understand their structure and to rationalize the design of search algorithms. Simplified generic models are used to capture important formal properties of biological repeats. Most of them are issued from problems that arose in other domains, such as data compression or Web indexing. Among the corresponding studies, those addressing the fundamental problem of looking for exact repeats prevail. This paper proposes a quick review of concepts at the core of any repeat model in a sequence, mostly focusing on exact repeats. It seems clear to us that any people interested in large-scale study of genomic repeats should have a good understanding of these concepts and we have tried to point all along the chapter at efficient tools that could help turning theory into practice.

2. Working on Exact Repeats

Exact repeats are words with several identical occurrences that are possibly overlapping. The search for approximate repeats is always based on the search for exact repeats that reflect the presence of the repeated structure and serve as anchor points during the exploration. Exact repeats have been extensively studied, starting from simple *k-grams or k-mers*, which are just words of fixed size *k*. A fundamental issue is to limit the number of representatives that are necessary in order to describe all "interesting" repeats. The notion of *maximality* is quite natural in this respect but not so trivial to define properly. For instance, given the string *GTTCGTTTCTTA*, the single letter *T* is repeated seven times in the string, making it the exact repeat with the maximal number of occurrences. Frequency alone is, however, unlikely to be useful and one has to add a criterion on the length of interesting repeats. Most of the people working on sequences are familiar with statistics that help to distinguish repeats with an unexpected number of occurrences with respect to their length (see, e.g., (3, 4)). Counting unexpected words is not easy because strings have combinatorial properties that have to be taken into account (see, for instance, Subheading 2.3), a problem that becomes even harder in strings like genomic sequences that are structured and contain many particular patterns. The first step in any analysis is, thus, to be able to explore the set of repeats without being hampered by combinatorial effects. In other terms, in genomics like in other texts, *the context* of word appearance matters and repeats are in some way made of letters appearing in a common context. In the string instance before, *GT* is a repeat made of co-occurring repeats *G* and *T* that are mutual contexts of their occurrences. In this example in fact, every occurrence of *G* is followed by an occurrence of *T*. The fact that *G* appears with the right context *T* with probability 1 renders repeat *GT* strictly more interesting than repeat *G*. The next section investigates such properties in more detail.

2.1. A Bestiary of Simple Repeats

The simplest idea of maximality is to look for each position at longest repeated substrings ending at this position (longest repeated suffixes) and present elsewhere in the string (very useful for plagiarism detection!). For instance, in the previous string at ending position 6, GT is the longest repeated suffix (it ends at 6 and has another occurrence ending at 2). Considering the occurrence of repeat T at this position is suboptimal in the sense that a longer repeat ends at the same position, conveying potentially more information. Some authors have designed very efficient algorithms for this task (5), and the visualization software FORRepeats has been developed on this basis (6). Note that every time GT occurs in the preceding string GTT also occurs. So "longest" means "longest up to a certain position". In order to select only the string GTT, the concept of maximal repeat is necessary. A *maximal repeat* is a word that cannot be extended to the left or right without decreasing its number of occurrences in the sequence. Maximal repeats (MRs) have very nice properties since they contain longest repeats, are never more numerous than the number of sequence characters, and contain all other repeats. Moreover, these repeats can be used as basic blocks to compute error-prone repeats and they have a well-defined mathematical structure of inclusion: the intersection of two MR remains an MR. As a consequence of this last property, small words are generally maximal repeats: in string GTTCG TTTCTTA, since GTT and TTC are maximal repeats, then TT is for instance also an MR. Some authors select for this reason *supermaximal* repeats, the set of MR that are not included in another MR. Note, however, that this does not guarantee anymore to cover the whole set of repetitions, e.g. the last occurrence of TT in our example is lost if only supermaximal repeats are retained. Another more interesting variation with respect to applications consists of fixing a lower bound for the number of occurrences of a repeat. A *multi-repeat* of multiplicity *k* is a repeat with at least *k* occurrences. Maximal multi-repeats are naturally defined as multi-repeats with at least two occurrences surrounded by two different pairs of characters.

Multi-repeats are particularly useful in the context of multiple sequences. In this case, it may be useful to define both a multiplicity per sequence and a multiplicity over the set of sequences (quorum). In case of multiple sequences, it is sometimes possible to distinguish query sequences and target sequences. New types of maximal repeats are then relevant. *Maximal substring matches* are words present in both the query and the target with at least one occurrence in the query and one in the target surrounded by different pairs of characters. If these two occurrences are the sole occurrences of the word in the target and the query, it is called a maximal unique match, and if the word is the whole query, it is called a complete match.

Repeats in general may occur everywhere in the string, possibly at overlapping positions. Genomic sequences offer sometimes more constrained distributions. The occurrences of tandem repeats have to be consecutive in genomic sequences contrary to interspersed repeats. Technically, the term used to denote consecutive copies is *repetition*. The study of repetitions in sequences constitutes a foundation of stringology, a field concerned with the combinatorial study of strings and initiated at the beginning of the previous century by Thue (see, e.g., (7)). Sequences have subsequences that form periodic signals in the same way as numerical functions and this serves as the basis for very efficient pattern matching algorithms. Thus, the sequence ACAAACAAAC has period 4 (for every position P, the character at P is the same as the character at $P+4$) and is a repetition made of the word ACAA. Maximal repetitions or runs may be defined in the same way as maximal repeats and corresponds to an abstract notion of tandem repeat purely generated by amplification and with no mutation: it denotes repetitions that cannot be extended to the left or right without losing their periodicity—in the example, AAA is a run of A starting at positions 3 and 7, but AA is not since it has period 1 like AAA. Once again, maximality does not guarantee to get the longest runs. For instance, the sequence ACAACACAACAG starts with the run ACAACA of period 3 but contains a longer run, namely, ACAACACAACA of period 5. Maximal repetitions are representative of all repetitions and contain branching tandem repeats, a concept used in Vmatch software defined as squares *ww* which are not followed by the first letter of word *w*.

2.2. Suffix Array: The Leading Data Structure for Tools Working on Repeats in a Sequence

Genomics has been one of the driving forces together with Web searching in the development of new index structures on sequences. Conversely, virtually all new developments for genome-scale sequence analysis should be concerned with progress made in sequence data structure elaboration. Since various tasks have to be fulfilled on genomic sequences that may reach several giga base pairs, it is of utmost importance to achieve fast sequence pretreatments that render such tasks independent of the length of the sequence. In practice, this requires the development of indexing algorithms with a linear or sub-linear behaviour in time and space. A *suffix tree* is a sequence structure made of a hierarchical dictionary of the subsequences starting at each position, each level of the tree corresponding to different possibilities of subsequence's beginning and each leaf to a subsequence's position. For instance, the suffix tree of ATATAC is tree (A(T(1, 3), C(5)), C(6), T(2, 4)). Suffix trees have long been the most efficient way to index sequence and allow the search of any word in a sequence in a time proportional to its length. However, since there is a big performance gap between main memory and disc storage and current computers have only several gigabytes of main memory, even the strict linear bound on algorithms may be insufficient for some usages—the most efficient suffix tree implementation requires 12 bytes per nucleic acid.

In the current state of the art, the best structure is the *suffix array*, a simple list of starting positions in a sequence sorted according to the ascending lexicographical order of each corresponding subsequence. For instance, the suffix array of sequence GTTCGTTTCTTA is (12, 4, 9, 1, 5, 11, 3, 8, 10, 2, 7, 6) in accordance with the lexicographical order of suffixes A, CGTTTCTTA, CTTA, GTTCGTTTCTTA, GTTTCTTA, TA, TCGTTTCTTA, TCTTA, TTA, TTCGTTTCTTA, TTCTTA, TTTCTTA. Note how repeats are naturally clustered in contiguous positions in the array: for instance, the last four elements of the array are positions of repeats TT. Proposed in 1990 by Manber and Myers (8), the structure has been made largely available via a number of studies and implementations (9) since the work of Abouelhoda, Kurtz, and Ohlebusch (10) showing that suffix arrays can replace suffix trees in every aspect. In practice, some additional pre-computed tables are necessary towards this aim—based on explicit recording of prefixes common to two consecutive entries in the array—a minor change with respect to the main structure. Suffix arrays are closely related to the so-called Burrows–Wheeler transform (BWt[1]), a reversible permutation of sequence characters that tends to generate runs in the transformed sequence. This nice property has been first used to design efficient lossless data compressors, such as bzip2. It is now increasingly used in genomics. We mention here an efficient tool, BWtrs (11), able to look at all runs, i.e. all maximal exact tandem repeats at eukaryotic genome-wide scale. This tool improves on previous ones either by its memory consumption or the fact that it is not limited in small length repeat motif like microsatellites, nor to specific alphabet. BWtrs accepts a GenBank or user-defined sequence in the FASTA format and four input parameters: the minimum and maximum motif size of tandem repeats, their minimum total length, and the minimum number of units (repeat ratio).

[1] If SA denotes the suffix array of sequence s, then BWt$[i]$, the ith letter of the BWt, is $s[\mathrm{SA}[i]-1 \bmod |s|]$. In our example, Bwt corresponds to string *TTTACTTTCGTG*.

As for suffix trees, suffix array can be built with a linear time and space complexity. The best implementation available to date is probably SAIS, which has been conceived by Nong (SAIS, (12)) and further enhanced by Mori. It is based on a clear divide-and-conquer recursive approach and uses only 5 bytes per character. In practical applications, it may be useful to get a more complete package, including the management of auxiliary tables, allowing multiple sequences and to make profit of multiple CPU computers via multi-threading. The Bielefeld University proposes an open-source program, mkESA, which answers all these constraints (13). Its output is compatible with mkvtree, another widely used implementation included in the package Vmatch ((10), see Subheading 4.2).

The story of sequence data structures does not end here. Research studies are now focusing on compressing the BWT while keeping interesting average computing time. Given the current increase of main memory capacity on computers, further compression is only useful for very large strings and it is not completely clear which impact it will have on the analysis of genomic sequences. We anticipate that the added value will come from structures taking into account basic operations needed in this context: dynamic structure adjustment for genome re-sequencing projects, management of internal mutation or insertion/deletion (indel) in repeats, and management of palindromic secondary structures. For instance, Schnattinger et al. (14) have recently presented a promising technique tested on the search for miRNA genes, using a variation on wavelet trees allowing bidirectional search and thus adapted to palindromic repeats. A wavelet tree is a natural companion of BWt originating from the field of data compression, coding sequences without loss through a tree of bit strings.

2.3. Introducing Don't Care Positions in Exact Repeats

A striking characteristic of genomic sequences is that possible variations occur in a non-uniform way along the positions. This reflects the existence of various biases, such as the GC content, or of many repeats, and more generally this is a consequence of the fact that genomic sequences are highly constrained sequences. Since all comparison methods are based on the search of exact repeats that are then extended (the seeds), this has important consequences on the sensitivity/specificity that can be achieved with respect to the length of the detected repeats. Getting high sensitivity requires to limit the size of seeds as much as possible while being specific requires both a lower bound size for these seeds (in practice, Blast is used with a default seed length $k=11$ on DNA sequences) and to filter low-complexity repeats from sequences before any search in order to get meaningful statistics of scores. A program like Blast is particularly confused by the existence of interspersed repeats. A number of tools have been developed using *k-mer* seeds as Blast does for the identification of genomic repeats (e.g. Repeatscout (15) or ReAS (16)). For large genomes, authors recommend to use $k=12$ and more generally, for a genome of size n, $k \cong \log 4(n)$ gives the best results (Table 1).

Table 1
Main software tools cited in the text

Name	Purpose	Site	Code or Web	References
SAIS	Suffix array sequence indexing	http://sites.google.com/site/yuta256/sais	Code	(12)
mkESA	Extended suffix array sequence indexing	http://bibiserv.techfak.uni-bielefeld.de/mkesa/	Code	(13)
BWtrs	Exact tandem repeats genome-scale identification	http://bwtools.polsl.pl/BWtrs/input.jsp	Web site	(11)
Bidirectional	Palindromic repeat search	http://www.uni-ulm.de/en/in/institute-of-theoretical-computer-science/research/seqana.html	Source Code	(14)
YASS	Similarity search with subset seeds in DNA sequences	http://bioinfo.lifl.fr/yass/	Code and Web server	(17)
Idera	Subset seed design for more sensitive similarity search	http://bioinfo.lifl.fr/yass/iedera.php	Code	(18)
Plast	Parallel Blast on protein data banks using subset seeds	http://www.irisa.fr/symbiose/projects/plast/	Code	(19)
Last	Similarity search with adaptive seeds in DNA sequences	http://last.cbrc.jp/	Code and Web server	(20)

(continued)

Table 1
(continued)

Name	Purpose	Site	Code or Web	References
Gepard	Genome-scale zoomable dotplots	http://www.helmholtz-muenchen.de/en/mips/services/analysis-tools/gepard/	Code and Java Webstart	(21)
Pygram	Hierarchical maximal repeat display along genomic sequences	http://pygram.genouest.org	Code and Web site	(22)
TandemGraph	Hierarchical tandem repeat display along genomic sequences	http://tandem.sci.brooklyn.cuny.edu/ https://github.com/ramin32/tandemgraph	Code and Web site	(23)
Circos	Synthetic circular visualization of comparative genomics data	http://circos.ca/	Code and Web site	(24)
ModuleOrganizer	Displaying the module organization of a family of transposable elements	http://moduleorganizer.genouest.org	Code and Web site	(25)
Vmatch	Large-scale sequence matching	http://www.vmatch.de/	Web site	(10)
Logol		http://webapps.genouest.org/LogolDesigner/	Logol	(26)

Since a few years, a number of improvements have been carried out on Blast-like algorithms in order to choose better seeds with increased sensitivity for a same computational cost. The idea is to introduce don't care positions—jokers—along the seed (*spaced seeds*) and more generally some constraints on the allowed matches at each position (*subset seeds*). Considering simple combinatorial effects on words helps to understand the interest of spaced seeds in the reduction of dependencies. Given, for instance, an occurrence of word TTTT at some position, the probability to get at least another occurrence of the same word at the next three positions is 0.25, whereas it is 0.0625 for ATAT and 0 for the word ATTT. Spaced seeds and multiple spaced seeds (optimal set of seeds instead of a single one) have been used in the tool PatternHunter (27) and recent versions of Blast. The YASS software designed by Noe and Kucherov (17) is going a step further extending Blast with the use of subset seeds. YASS has proved to be useful for the pairwise alignment of mobile elements (28, 29). Seeds may be specified using a seed motif built over a three-letter alphabet #, @, and –, where # stands for a nucleotide match, – for a don't care symbol, and @ for a match or a transition –A/G or C/T–, transitions taking precedence over transversions, in both coding and non-coding regions. For instance, using the pattern ##@–##, the sequence CGCCCG will be a hit on sequence ACGTACGT on position 2 since it can be aligned with a transition T/C and a substitution A/C. Note that no score is used here. The weight of a pattern, defined as the number of # plus half the number of @, is the main characteristic of seed selectivity/sensitivity, together with the seed model itself. For instance, the pattern ######### is the Blast motif used with parameter $k=9$, meaning that nine consecutive characters have to be present in order to start an alignment. Its weight is 9 and sensitivity is 0.453, estimated with respect to alignments of size 50 generated with a simple Bernouilli sequence model. If one is interested in other patterns of weight 9, of size at most 14 and containing 2 w and 8 #, i.e. allowing 2 transitions and requiring 8 matching positions, the pattern ##@–#@#––#–### can been shown to have far greater sensitivity (0.625) and to be optimal with respect to the seed model. Designing an optimal set of seeds depends on the application at hand and programs do exist to generate them. Iedera (18) can be used as a pre-processing step to YASS for this purpose. It allows a number of parameters to be taken into account for seed selection and can be applied to DNA or protein sequences (30). At first sight, it seems hard to apply seed models on proteins since seeds have to be very short ($w=3$ for Blast) and hits have to consider the similarity of aligned sequences instead of just a match. However, subset seeds provide a good way to manage similarity without the need for score; and in practice, for sufficiently long matches, it is possible to improve the search sensitivity through the design of appropriate seeds. Idera has been used, for instance, in the software Plast, a parallel alignment search tool dedicated to the comparison of large protein banks that runs three to five times faster than the NCBI-BLAST software on this task (19).

A recent approach has further demonstrated the interest of studying seeds, introducing a new idea of adaptive seed. Among other experiments, Kielbasa et al. (20) show an impressive comparison of chimpanzee and human Y chromosomes. This genomic region is known to be hard to compare due to a very low information content correlated to its repeat richness and a number of rearrangements. In 2010, Hughes et al. succeeded in sequencing the male-specific region of the chimpanzee Y chromosome and published a surprising paper in Nature (31) claiming that >30% of its content share no alignment with the human corresponding region. This corresponds to an unexpectedly high level of divergence since this level reduces to less than 2% for the rest of the genome. Authors used ClustalW with default parameters in their analysis. Kielbasa et al. show by using adaptive seed-based comparison that the level of divergence is in fact less than 14%. It remains a high level, but this implies that their method recovers more than 15% of possible alignments. *Adaptive seeds* are, for any target sequence and starting position in a query, the shortest seeds at this position such that the matching sequence occurs at most *f* times in the target, *f* being a fixed parameter threshold. The idea is to explore a large set of seeds (the seed length is not fixed a priori) and to quickly select the most promising ones on the basis of their level of specificity. Adaptive seeds can be computed very directly using a suffix array data structure and, moreover, they are compatible with spaced and fixed seeds, leading to adaptive spaced seeds and adaptive subset seeds. In the human/chimpanzee Y chromosome study, repeats are made of microsatellites that have to be masked (detection of tandem repeats) and other classes, such as LINEs and SINEs, that can be safely handled by adaptive seeds in order to increase the sensitivity of the search. Kielbasa et al. have developed downloadable software, Last (20), and a restricted version that can be run on a Web server.

We end this section with some remarks on the way to extend exact seeds to get approximate matches taking into account possible variations. Consider first the case where only mutations occur inside repeat copies. Seeds are extended to the left while the number of mutations remains acceptable and no other seed is reached. The resulting left-extended seeds are then extended to the right while the total number of mutations remains acceptable, possibly integrating other seeds. This way, all maximal matches are reported. If insertions and deletions are allowed, it is necessary to maintain a set of extension possibilities instead of a single value at each step. Note that the matching algorithm remains linear with respect to sequence length or number of seeds but becomes cubic with respect to the number of errors. Under natural constraints on the cost of indel with respect to match/mismatch[2] and with the assumption

[2] Namely, Cost(Indel) = Cost(Mismatch) − Cost(Match)/2.

that the total number of differences remains small, it is possible to design an improved greedy algorithm, that has a quadratic instead of cubic complexity.

3. Displaying Repeats

3.1. Dotplots

Various kinds of tools displaying repeats at genome level have been proposed in the past, one of the oldest and still in use being dotplots. Dotplot is a quadratic representation of repeats, crossing in a 2D array a sequence either with itself or another sequence. The interest of this type of representation have been greatly enhanced since a few years by integrating suffix array indexing possibilities. Gepard (21) can switch from a classical window-based calculation for small dotplots to a suffix array-based computation looking at all repeats of a given length for large-scale dotplots, something that was out of reach of programs like Dotter. Gepard is compatible with the Vmatch package (10) and offers a command-line mode or a Java Webstart run mode. Annotated sequences can also be uploaded from the PEDANT database. Figure 1 displays a dotplot of a CRISPR (clustered regularly interspaced short palindromic repeats) region in the bacterial genome of *H. ochraceum*, a repeat-rich region, where repeats form a skeleton of regularly spaced words flanking foreign genomic material.

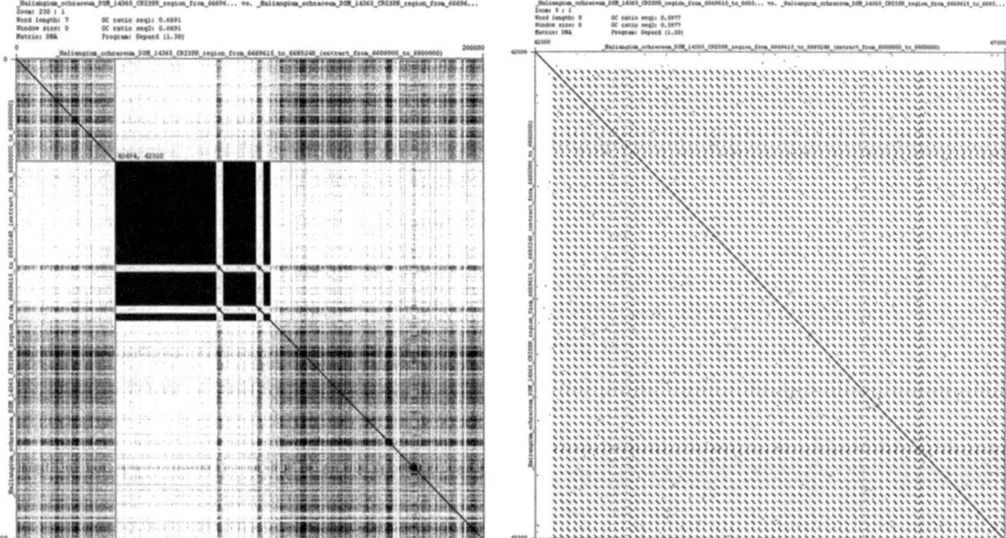

Fig. 1. Dotplots produced by Gepard. The *right dotplot* has been obtained by zooming on a repeat-rich CRISPR region, the *upper left black square* of the left dotplot.

3.2. Landscapes and Pygrams

Dotplots present the disadvantages of a quadratic representation: apart from the computational complexity, it entails an interpretation complexity that may hinder the understanding of structures made of complex repeat arrangements. We have proposed to take maximal repeat as a basis for the visualization of all repeats. Pygram (22) introduces the pyramid diagram, an abstract representation of the organization of repeated structures in genomic sequences. A pyramid diagram is a hierarchical representation of repeats along the sequence that makes use of the fact that intersections of MR are MR. Choosing a different colour for each MR and displaying the smallest on top of the largest repeats, this property ensures that no maximal repeat is masked by others. Technically, a Pygram for a genome sequence S is a bi-dimensional plot, where S and all its maximal repeats are mapped along the x-axis. Given an x-axis-magnifying factor zx and a y-axis-magnifying factor zy, mapping is defined as follows: the ith nucleotide of S is located at position $(i/zx,0)$, and the MR of size m located at position i within S corresponds to an isosceles triangle (a pyramid) of height dm/zy and base $[i/zx, (i+m)/zx]$ on the x-axis—$d=1$ for an MR on the normal strand or -1 for an MR on the reverse strand. The pygrams may be considered as a rational reconstruction of landscapes—defined by Clift, Stormo et al. in 1986 and extended in 1998—fully characterizing the structure that is displayed without requiring the computation of intermediate repeats. Pygram introduces various practical improvements over landscapes (two-strand display, zoom lenses) and offers several additional features, including frequency visualization and multi-genome repeat analysis. Most important, Pygram visualization is closely associated with a query system designed to locate repeats that share specific properties. When combined, the query system and visual interface provide an efficient repeat browser that is useful for discovering unexpected structures in genomes.

Figure 2 provides the pygram of the CRISPR structure of Fig. 1. The zoomed region not only shows the regular spacing of the repeat unit skeleton, but it also points to the absence of repeat in the spacer region, a clear signature of the presence of extra-genomic material.

Pygrams have been the basis of the visualization software TandemGraph that has been used for tandem repeat display along the human genome and is proposed in the database TRedD (23). Pygrams are also used in the CRISPR database Crispi (32).

3.3. The Modulome/ Mobilome

The complex repeat architecture characteristic of every mobile element family is an interesting source of knowledge on the way they have evolved but is hardly depicted by the previous display modes. Two converging facts contribute to this difficulty. First, the variations inside a given family lead to a complex multiple alignment since starting from identical copies of an ancestor

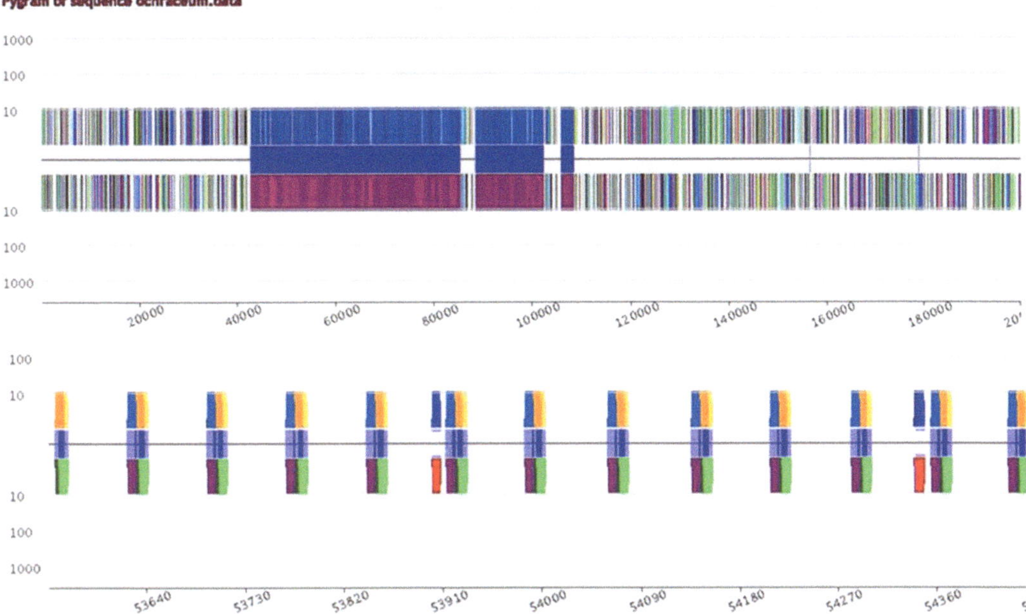

Fig. 2. Pygrams of the same genomic regions than in Fig. 1.

sequence, mutation, deletions, or large insertions cumulate over time to produce very divergent copies. This is particularly observable in the case of non-autonomous elements. These copies exhibit generally a modular structure, where modules may appear in various orderings. Second, graphical tools on multiple alignments often consider them as multidimensional data that have to be projected into a 2D space like dotplots. For tools like VISTA (33) or GATA (34), a reference sequence has to be chosen so as to ensure a linear number of projections with respect to the number of sequences in the family. A generic software allowing beautiful and sophisticated graphics for all genomic comparison tasks has been designed by Krzywinski: Circos (24) uses a concentric circular layout, a practical way to display multiple information together with relationships between pairs of positions with connecting ribbons, which encode the position, size, and orientation of related genomic elements. Circos can produce either bitmap images or high-quality vector images. It offers many options and graphical representations (line, histogram and scatter plots, heat maps, connectors …) with a major concern of producing accurate, discernable, and flexible display. For these reasons, it is a valuable asset in the development of new comparison tools with high-quality outputs. Unfortunately, this tool is lacking a true dedicated programming language and a complete user manual, but many tutorials are available on the Web and show in details the wide range of Circos display modes. Figure 3 (see also the book's cover) highlights

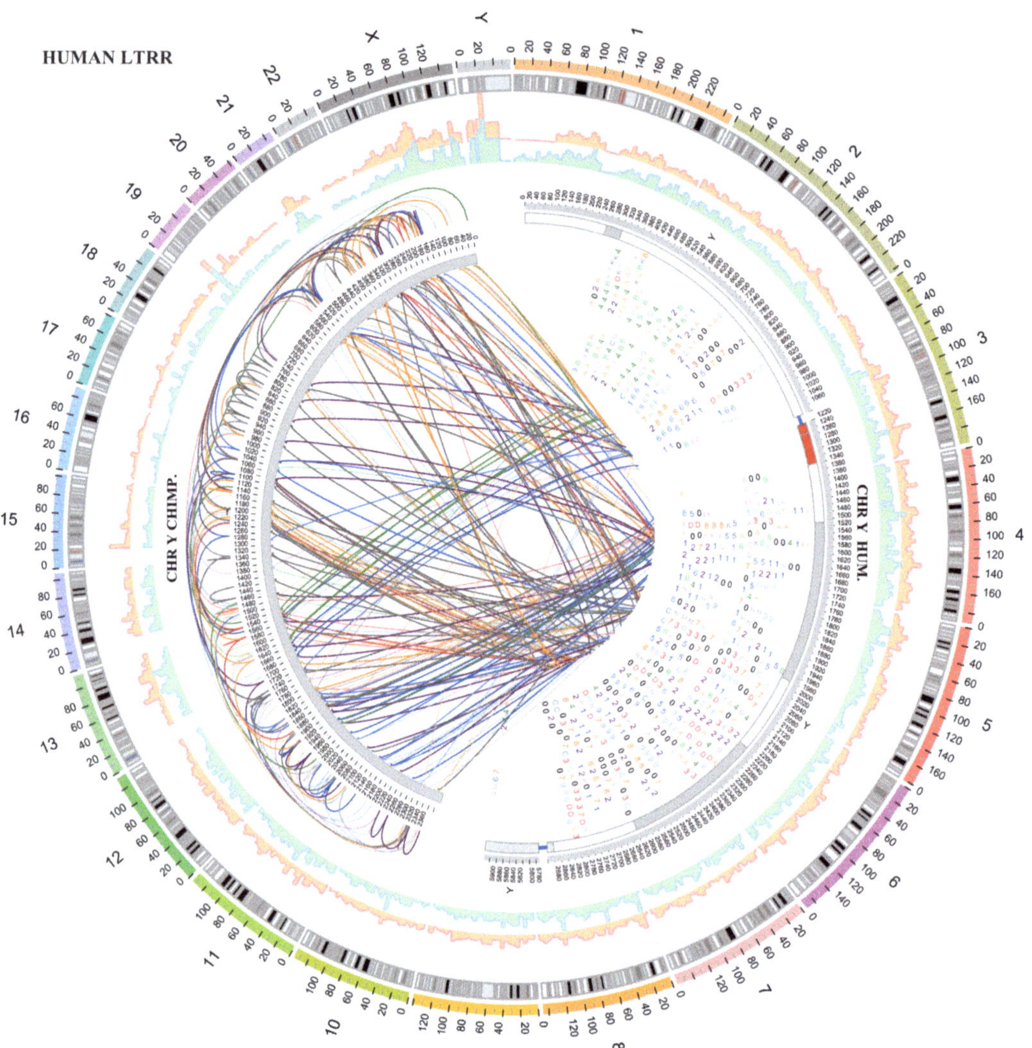

Fig. 3. Distribution of LTR retrotransposons (LTRRs) in the human genome. Data have been extracted from the most recent version of UCSC genome tables: assembly Feb. 2009 GRCh37/hg19 for the human genome and assembly Oct. 2010 CGSC 2.1.3/panTro3 for the chimpanzee genome. The figure shows the distribution of LTRRs all along the human genome. Chromosomes are displayed in the *outer circles* using an adapted luminance-corrected version of the UCSC genome browser colour convention. Histograms are shown for the *direct and reverse strand* in the *orange and green circles* just under the circle showing cytogenetic bands. Sexual chromosomes have numerous LTRRs and their list is given in the *inner right circles* for chromosome Y. Note that chromosome Y has been cut at two points, where no LTRR occurs in the corresponding regions. Each character is coding for a particular subfamily and displayed using an associated colour. For instance, character 0 = HERVK = {HERVK-int, HERVKC4-int, HERVK3-int, HERVK9-int, HERVK11-int, HERVK13-int, HERVK14-int, HERVK14C-int, HERVK22-int, …} is coloured in *black*. The *left inner circles* are dedicated to the chimpanzee Y chromosome. This part of the figure shows results of a comparative analysis between human and chimpanzee LTRR for the direct strand (*most inner circle*) of the Y chromosome. *Links* are coloured with the same palette than linked occurrences. Intrachromosomal links are also given for the chimpanzee Y chromosome.

some of the possibilities of Circos on LTR retrotransposon data. For casual users, it is highly recommended to use predefined configuration files and Perl script utilities, such as tableviewer—dedicated to the graphical display of data tables. To our knowledge,

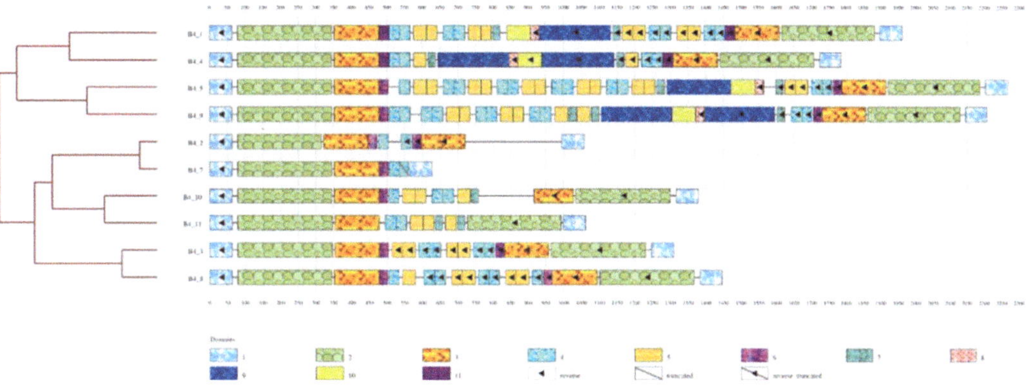

Fig. 4. Module organization of Foldback4 in the *D. melanogaster* genome.

there exists to date no specific Circos environment for the study of mobile element repeats, apart from a small tool, Circoletto (35), combining Blast with Circos.

We have proposed a tool dedicated to the visualization of families of mobile elements, ModuleOrganizer (25), an extension of DomainOrganizer (36) that represents a given family of TE sequences as an assembly of elementary blocks called modules. A module is a flexible motif present in at least two sequences of a family of transposable elements and built on a succession of maximal repeats. Flexibility is founded on two simple criteria that delimit the possible spacers between consecutive repeats: the length of the spacer should not be greater than the linked part and the distance between spacers should not exceed a certain threshold. We have proposed to create such modules by a recursive assembly method working on the set of maximal repeats present in the family sequences and added the possibility to detect palindromic modules or truncated modules. Each sequence may then be summarized by a vector of counters associated to module occurrences. The method results in a hierarchical graphical view of sequences segmented into modules, a representation that allows the exploration of transformations that have occurred between them (see Fig. 4 for the analysis of a non-autonomous family).

4. Perspective: Modelling Repeats

The common approach for the analysis of particular families is the development of dedicated pipelines assembling on-the-shelf tools and tuning their parameters to get the best cover of known repeats while excluding repeats associated to other families. Among recent workflows characteristic of the variety of available programs, one can cite REPET ((2) and chapter inside this volume) and REPCLASS (37), some general pipelines for the annotation of transposable elements; DAGGPAWS (38), a workflow established to annotate transposable elements in plants or MITE-Hunter (39), a pipeline that focuses on small class II non-autonomous transposable elements, such as MITE. These are valuable resources that reflect an increasing knowledge on the way mobile elements are generated and evolve over time.

4.1. What Is the Next Step?

From one side, pipelines are costly to develop and even more to maintain. Most of them use (possibly large) parameter sets that may be hard to tune for a new data set because it is not direct to link input parameters with the desired result. Once empirical identification methods have been designed that produce good results, a high added value can be produced by people trying to rationalize them and to formalize the type of repeats that have been characterized. This can also help to solve a major issue on the fact that users have to test and compare the results of an increasing number of programs. This can also help to solve a minor issue on the fact that elements in few copy number (e.g. up to 3) are generally ignored since automatic methods have to rely on multiple occurrences of repeats for their detection.

From the other size, there exist expert centres that have acquired through large-scale scans of genomes and observation of multiple instances of repeat families a good grasp of some of their major features. How to transform such expertise into scientific hypotheses that can be tested and validated on an increasing volume of genomic sequences?

We are convinced that major improvements on these questions can only be achieved by combining two types of developments: the development of generic, efficient solvers for a variety of well-defined string matching problems and the development of generic languages for modelling the complex architecture of repeats naturally occurring in genomic sequences. The rest of this section elaborates further on these two key ideas.

4.2. Vmatch, a General Framework for the Search of Similarities in Genomes

The field of word algorithmics and stringology has clearly matured since a few years in the sense that practical programs have been developed for a broader audience. It is now possible to develop efficient pipelines with reduced code acting on top of generic softwares computing well-defined string selections. Vmatch (10) is a package resulting from continued efforts for years in the field of indexing and pattern matching for genomic sequences (a previous version was called Reputer) and is maintained since 2003 by Kurtz. It offers a flexible framework where one can proceed to a very large variety of queries. Vmatch is free for academic research and can be obtained after downloading a license agreement form. It proposes a command language whose complete possibilities are described through associated user manuals (40).

The first step in using Vmatch is to build an efficient index of the sequences to be analyzed. This is the role of command mkvtree, which accepts all common sequence input formats—it even accepts gzipped files—and can be further specified with many options for the construction of auxiliary search tables or the specification of the alphabet used for sequences. The alphabet may be any user-defined alphabet not larger than 250 printable symbols. It is specified by a file storing a series of lines of equivalent letters—useful for protein sequences—and a last line of wild card characters. Another command, mkdna6idx, generates a six-frame translation index of DNA sequences.

Once the index is built, the command vmatch allows a number of different matching tasks, including Blast-like operations. It is possible to extract almost all types of fundamental repeats that have been described in Subheading 1: maximal repeats, supermaximal repeats, branching tandem repeats, maximal substring matches, reverse complemented, matches of length k, and approximate matches with errors .... Furthermore, there exists a rich palette of possible post-processing treatments on the set of solutions, either with predefined options or with user-defined C code. Basically, solutions may be sorted, filtered from sequences for masking purpose, clustered together according to pairwise similarities or positions, or chained together in order to obtain maximally covering subsets of matches.

4.3. First Steps in Modelling

Instead of developing brand new pipelines for an in-depth treatment of each mobile element family, it is tempting to go a step further and directly describe their characteristics in a suitable language that is then compiled in order to generate an operational searching procedure. A purely structural approach has some advantages with respect to more procedural approaches. Unlike de novo repeat discovery methods, structure-based methods rely on detecting specific models of transposable elements (TEs) architecture, rather than just the expected results of the transposition process (i.e. dispersed repeats with similar boundaries). Potentially, they can detect low copy number families, have high specificity to detect TE repeats, and can provide a preliminary structural classification of the newly identified TE. In contrast to homology-based methods, structure-based methods are less biased by similarity to the set of known elements.

Funding the analysis on models is a tendency that is particularly observable on well-known families of mobile elements, such as LTR retrotransposons (LTRRs) for which a global architecture can be easily described (Fig. 5). The SMaRTFinder platform, for instance, (41) has been developed to conduct efficient searches in DNA for structured sets of motifs, including those shared among

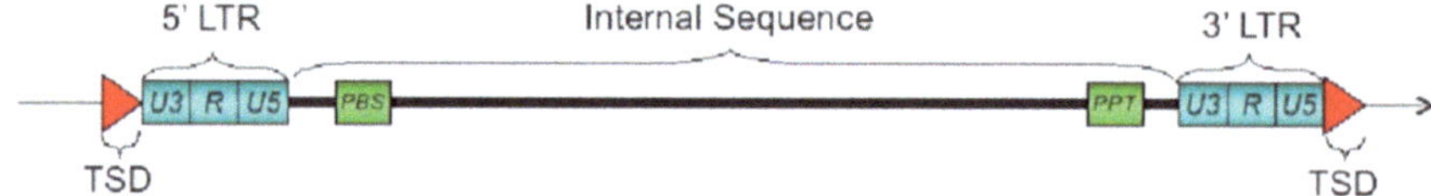

Fig. 5. Abstract structure of an LTR retrotransposon.

LTRRs. A structured motif is an ordered set of motifs and a list of intervals specifying the distances between motifs. In the case of LTRR elements, these motifs can be LTR end motifs, the PBS or PPT, or the DNA sequence of a highly conserved domain in an ORF. This generalized approach first starts by locating instances of individual motifs (using suffix trees in this case) and then solves a constraint satisfaction problem by constructing a graph with motif instances as nodes and edges between nodes which satisfy order and distance constraints. Zhang and Zaki have proposed an improvement of this method with SMOTIF (42). More recently, LTRharvest (43) allows a flexible incorporation of knowledge on LTR transposon structural features.

Using formal languages, it is possible to define at a more abstract level such structures. Formal languages are a framework introducing models as a set or rewriting rules acting on a starting axiom. The set of rules is called a *grammar*. For instance, a protein in a bacterial genome may be roughly recognized using the following grammar:

Axiom → Start Codons Stop

Start → NotC TG

Stop → TAG| TGA | TAA

Codons → Codon

Codons → Codon Codons

Codon → Na Na Na

NotC → A | G | T

Na → A | C| G | T

In such model, the left part of a rule rewrites into its right part. Any genomic sequence that can be generated by a finite application of such rules, starting from the axiom rule, is accepted as a putative ORF by the model. Conversely, it is possible to check (i.e. to parse) a given sequence by applying the rules from right to left. Among general bioinformatics tools for genomic sequences modelling, a major contribution in this framework is due to Searls. He was the first to supervise developments allowing users to conceive grammars representing their biological models, and parse real genomic sequences with them (44, 45). One of the key ideas of Searls is to try to find a balance between the well-founded framework of algebraic languages, a particular class of languages that offer a good expressivity/efficiency trade-off, and the necessity to describe easily basic biological mechanisms, such as copy (direct or reverse) that is at the core of genome evolution. He has proposed to introduce in grammars a new kind of object for this purpose, the string variable, which can represent any substring and be subject to various constraints and transformations. The resulting formalism is called String Variable Grammars (SVG). From the point of view of expressivity on biological sequences, this allows to take into account hierarchical aspects of life and presence of copies. For instance, in the case of LTRR, the top-level rule of the grammar could be represented by the following expression—it is given for illustration purpose only and does not pretend to be fully realistic:

LTRR→DR:(2..6), «tg», (U5,R,U3):(80..750), «ca»,
(1..100), pbs, (1..100), gag, (1 000..15 000), ppt, (1..100),
«tg», (U5:80%, R:90%, U3:80%), «ca», DR.

In this expression, DR, U5, R, and U3 are string variables. Its meaning is: "The sequence is surrounded by two exact copies of a direct repeat (DR) of size between 2 and 6. The LTRs are starting by nucleotides 'tg' and ending by nucleotides 'ca' and are made of three parts named U5, R, and U3 with a global length between 80 and 750. The right LTR is an approximate copy of the left one. The central part (R) is the most preserved—because of the hybridization between both R during duplication—with a 90% minimum identity level, whereas U3 and U5 need only to have 80% level identity. The central part of the sequence must contain at constrained distances a primer binding site (pbs), a group-specific antigen (gag), and poly purine tract (ppt), which are described by other grammatical rules".

The community is still lacking efficient parsers for the previous expression (Genlang, a parser developed by Searls has not been maintained and is no more available). Although it has not been tested on a large-scale mobile elements modelling task, we mention the existence of Logol, an ongoing project developed in our team towards this goal. The Logol Software Suite—a major update of the former program Stan (46)—is a set of software composed of a Logol language interpreter and pattern search tool (LogolMatch), a graphical Web-based editor, and a result analyzer. It is still a beta-release and its access is provided for research purpose only. The software is designed to run on a single computer (Linux), with one or several CPU, or on a cluster. Additionally, Logol Designer is an online graphical tool to create some Logol grammar templates. It provides a drag-and-drop component interface to build the template. LogolMatch takes as input a sequence—DNA, RNA, or protein—and a grammar file and compile them using Vmatch and a Sicstus Prolog interpreter. Result files contain the matches on the sequence(s) with all required details.

Logol is a highly descriptive language (26) dedicated to the representation and search of complex models on biological sequences. Models use constrained string variables (supporting overlaps, substitution, and distance errors) that can be subject to various transformations (e.g. inverse complement), gaps, and repetitions of a pattern along the sequence, negation, and alternatives to define different possibilities. As in every formal grammar, components can be grouped in a view to get a high-level representation of a subset of components.

To sum up, the analysis of TE is clearly facing exciting new challenges that are going beyond the routine procedure of sequence data gathering, Blast comparison, and production of descriptive statistics. We have briefly presented principles and tools that allow to efficiently compare and display sequence *structures* at genome scale. The word "structure" that usually refers to 2D or 3D spatial characteristics of proteins has now to be applied on DNA sequences to denote the issue of understanding the complex architecture of modules that make genomic families. Such structures are closely related to the family-associated mechanisms and are, thus, providing invaluable hints for a better understanding of their role. Mobile elements seem to offer a particularly interesting field of research with respect to this structural analysis. From one side, they are generally hard to analyze using standard local comparison procedures due to the existence of embedded and degenerated structures. On the other side, there exist general transposition mechanisms that constrain the architecture of mobile elements and a growing knowledge on this architecture built from literature and the results generated by analysis pipelines. Moreover, a same sequence generally contains ancient and recent copies of a given element and this is a unique opportunity with respect to the understanding of the conservation of structures during the evolution of sequences.

The future of bioinformatics will lie in a gradual transition from data management to knowledge management tools, i.e. a transition towards more explicit models that can be confronted, refined, and validated on the large base of whole-genome studies. As in other contexts, the development of new tools is a matter of supply and demand. Producing and keeping a precise trace of mobile element models, whatever the modelling language used for this purpose and even if tools are still lacking, would be a valuable contribution to the advances in this domain.

Acknowledgments

This work was supported in part by a grant from the Agence Nationale de la Recherche [project Modulome ANR-05-MMSA-0010-01].

References

1. Jurka J, et al. (2005) Repbase Update, a database of eukaryotic repetitive elements. Cyt Gen Res. 110:462–467
2. Flutre T., et al. (2011) Considering transposable element diversification in de novo annotation approaches. PLoS ONE. 6:1
3. Reinert G, Schbath S, Waterman MS (2005) Probabilistic and Statistical Properties of Finite Words in Finite Sequences. J Berstel and D Perrin (eds.). In Applied Combinatorics on Words. Cambridge University Press
4. Ussery D, Wassenaar T, Borini S (2009) Word Frequencies and Repeats. Computing for Comparative Microbial Genomics: Bioinformatics for Microbiologists. Computational Biology. s.l.: Springer. 2009, Chapters 7 and 8, pp. 111–150
5. Lefebvre A, Lecroq T, Alexandre J (2003) An improved algorithm for finding longest repeats with a modified factor oracle. Journal of Automata, Languages and Combinatorics 8:347–658
6. Lefebvre A, et al. (2003) FORRepeats: detects repeats on entire chromosomes and between genomes. Bioinformatics 19:319–326
7. Crochemore M, Ilie L, Rytter W (2009) Repetitions in strings: algorithms and combinatorics. Theoret Comput Sci 410(50):5227–5235
8. Manber U, Myers G (1990) Suffix arrays: A new method for on-line string searches. In Proceedings of the 1st ACM-SIAM Symposium on Discrete Algorithms. Ed. Edited Dana Randall, pp. 319–327
9. Puglisi SJ, Smyth WF, Turpin AH (2007) A taxonomy of suffix array construction algorithms. ACM Comput. Surv 39:1–31
10. Abouelhoda MI, Kurtz S, Ohlebusch E (2004) Replacing suffix trees with enhanced suffix arrays. J Disc Algo 4:53–86
11. Pokrzywa R, Polanski A (2010) BWtrs: A tool for searching for tandem repeats in DNA sequences based on the Burrows-Wheeler transform. Genomics 96:316–321
12. Nong G, Zhang S, Chan W. (2009) Linear Suffix Array Construction by Almost Pure Induced-Sorting, Proceedings of 19th IEEE Data Compression Conference (IEEE DCC). Mar. 2009, Snowbird, UT, USA, pp. 193–202
13. Homann R, et al. (2009) mkESA: enhanced suffix array construction tool. Bioinformatics. 25:1084–1085
14. Schnattinger T, Ohlebusch E, Gog S (2010) Bidirectional search in a string with wavelet trees. In Proceedings of the 21st annual conference on Combinatorial pattern matching (CPM'10). Amihood Amir and Laxmi Parida (Eds.). Springer-Verlag. pp. 40–50
15. Price AL, Jones NC, Pevzner PA (2005) De novo identification of repeat families in large genomes. Proceedings of the 13th Annual International conference on Intelligent Systems for Molecular Biology (ISMB-05). Detroit, Michigan
16. Li R, et al. (2005) ReAS: Recovery of ancestral sequences for transposable elements from the unassembled reads of a whole genome shotgun. PLoS Comput 1:4
17. Noe L, Kucherov G (2005) YASS: enhancing the sensitivity of DNA similarity search. Nucl Acids Res 33: 540-W543
18. Kucherov G, Noe L, Roytberg M (2006) A unifying framework for seed sensitivity and its application to subset seeds. J. Bioinf Comp Biol 4:553–569
19. Nguyen VH, Lavenier D (2009) PLAST: parallel local alignment search tool for database comparison BMC Bioinformatics 10:329
20. Kiełbasa SM, et al. (2011) Adaptive seeds tame genomic sequence comparison. Genome Res 21:487–493
21. Krumsiek J, et al. (2007) A rapid and sensitive tool for creating dotplots on genome scale. Bioinformatics 23:1026–1028
22. Durand P, et al. (2006) Browsing repeats in genomes: Pygram and an application to non-coding region analysis. BMC Bioinformatics 7:477
23. Sokol D, Atagun F (2010) TRedD: A database for tandem repeats over the edit distance. Database: article ID baq003
24. Krzywinski M, et al. (2009) Circos: an information aesthetic for comparative genomics. Gen Res 19:1639–1645
25. Tempel S, et al. (2010) ModuleOrganizer: detecting modules in families of transposable elements. BMC Bioinformatics 11:474
26. Belleannée C, Nicolas J (2007) Logol: Modelling evolving sequence families through a dedicated constrained string language. Inria Research report RR-6350:19
27. Li M, et al. (2004) Highly sensitive and fast homology search. J Bioinform Comput Biol 2:417–439
28. Weber MJ (2006) Mammalian Small Nucleolar RNAs Are Mobile Genetic Elements PLoS Genet 2:e205
29. Grzebelus D, et al. (2007) Diversity and structure of PIF/Harbinger-like elements in the genome of Medicago truncatula. BMC Genomics 8:409
30. Roytberg M, et al. (2009) On Subset Seeds for Protein Alignment. IEEE/ACM Transactions on Computational Biology and Bioinformatics. 6:483–494

31. Hughes JF, et al. (2010) Chimpanzee and human Y chromosomes are remarkably divergent in structure gene content. Nature 463: 536–539
32. Rousseau C, et al. (2009) CRISPI: a CRISPR interactive database. Bioinformatics 25: 3317–3318.
33. Brudno M, et al. (2007) Multiple whole genome alignments and novel biomedical applications at the VISTA portal. Nucl Acids Res 35:W669-W674
34. Nix DA, Eisen MB (2005) GATA: a graphic alignment tool for comparative sequence analysis. BMC Bioinformatics 6:9
35. Darzentas N (2010) Circoletto: visualizing sequence similarity with Circos. Bioinformatics 26:2620–2621
36. Tempel S, et al. (2006) Domain organization within repeated DNA sequences: application to the study of a family of transposable elements. Bioinformatics. 22:1948–1954
37. Feschotte C, et al. (2009) Exploring repetitive DNA landscapes using REPCLASS, a tool that automates the classification of transposable elements in eukaryotic genomes. Gen Biol Evol 1:205–220
38. Estill JC, Bennetzen JL (2009) The DAWGPAWS pipeline for the annotation of genes and transposable elements in plant genomes. Plant Met 5:8
39. Han Y, Wessler SR (2010) MITE-Hunter: a program for discovering miniature inverted-repeat transposable elements from genomic sequences. Nucl Acids Res 38:e199
40. Kurtz S (2011) The Vmatch large scale sequence analysis software. A Manual. Unpublished report. Center for Bioinformatics Univ. of Hamburg, http://www.vmatch.de/virtman.pdf; + 2 other manuals “Chaining pairwise matches using the program chain2dim. Manual” and “Clustering Matches using the program matchcluster. Manual”
41. Morgante M, et al. (2005) A Structured motifs search. J Comput Biol. 12:1065–1082.
42. Zhang Y, Zaki MJ (2006) SMOTIF: efficient structured pattern and profile motif search. Algorithms Mol Biol 21:1–22
43. Ellinghaus D, Kurtz S, Willhoeft U (2008) LTRharvest, an efficient and flexible software for de novo detection of LTR retrotransposons. BMC Bioinformatics 9:18
44. Searls DB (1993) String variable grammar: a logic grammar formalism for the biological language of DNA. J Logic Program 24:73–102
45. Searls DB (2002) The language of genes. Nature 420:211–217
46. Nicolas J et al. (2005) Suffix-tree analyser (STAN): looking for nucleotidic and peptidic patterns in chromosomes. Bioinformatics 21:4408–4410

Chapter 5

Exploring Bacterial Insertion Sequences with ISfinder: Objectives, Uses, and Future Developments

P. Siguier, A. Varani, J. Perochon, and M. Chandler

Abstract

We describe here the use of the ISfinder database and its associated software. ISfinder was conceived initially as a comprehensive database for prokaryotic insertion sequences (ISs). It now includes software for visualising complete and partial IS copies in whole genomes (ISbrowser) and for high-quality genome annotation (ISsaga).

Key words: Bacterial insertion sequence, Genome annotation

1. Introduction

ISfinder (http://www-is.biotoul.fr (1)) is a dedicated database for bacterial insertion sequences (ISs). These mobile genetic elements (MGEs) have played a preponderant role in shaping prokaryotic genomes by contributing massively to horizontal gene transfer (HGT) and genetic rearrangements. Comparative analyses of multiple bacterial genomes have revealed that some bacterial species possess extremely plastic genomes resulting from MGE activity (2). For example, in 20 completely sequenced strains of *Escherichia coli*, the large differences in size of the chromosome range from 4.6 to 5.5 Mbp due largely to MGEs (3). This "optional" genomic repertoire composed has been called the "mobilome" (mobile genome). The importance of MGEs is underlined by the observation that transposases, the enzymes which catalyse their movement, are by far the most numerous and ubiquitous genes in nature (4).

ISs are among the simplest autonomous MGEs and a key component of many of the more complex transposable elements (the compound transposons) found both in bacterial and archaeal

Yves Bigot (ed.), *Mobile Genetic Elements: Protocols and Genomic Applications*, Methods in Molecular Biology, vol. 859,
DOI 10.1007/978-1-61779-603-6_5,

genomes. There are few, if any, major differences between prokaryotic ISs and those found in eukaryotes (5). They are small genetically compact transposable DNA segments of between 0.7 and 3.5 kbp and generally include only a Tpase-encoding gene. Many carry short (<40 bp) imperfect terminal inverted repeats (IRs) at their ends. One of these repeats, defined by convention as the left IR, contains the Tpase gene promoter. Many ISs generate small (between 2 and 14 bp) duplications (DR) of the target DNA flanking the point of insertion and whose length is specific for each IS family. ISs are grouped into more than 25 different families on the basis of various shared features (e.g. related transposases, strong conservation of the catalytic site, conservation of organisation, and similar IRs; (6)). Major features of these families are listed in Table 1.

In spite of their numerical and functional importance, IS elements are generally poorly annotated. A large number of bacterial and archaeal genomes in the public databases are seriously compromised in this way. Even more serious is that incompletely and inaccurately annotated genomes are often used to draw (probably erroneous) conclusions about genome evolution by analysis of IS distribution, acquisition, and loss. The scale of the potential problem should not be underestimated: there are over 1,524 complete and 6,158 ongoing archaeal and bacterial genomes and 307 metagenomes (GOLD Genome Database, April 2011). Often, only the transposase reading frame is annotated (and frequently as "integrase" or "unknown function"). Even if this is intact, it does not indicate whether the IS is active since, for transposition to occur, the transposase must act on the IS DNA ends. If these are not intact, the IS will be inactive. Correct IS identification and annotation at the DNA level (IRs, DRs, and correct ends), and family identification, are therefore essential. Moreover, genomes often contain many partial IS copies: versions with truncated transposases (which do not appear in open reading frame annotations), miniature inverted repeat transposable elements (MITEs, which carry both IS ends but are devoid of internal reading frames), mobile cassettes (MICs, in which the transposase is substituted by another unrelated gene), and solo IRs (isolated copies of the ends, equivalent to isolated LTRs in eukaryotes). More recently, IS copies with additional passenger genes but with no transposition functions have been identified ((7); Fig. 1).

Some of these generally unannotated genetic objects, such as the MITES and MICs, can be activated by a cognate transposase from an intact related IS in the same cell, and all provide a substrate for genome rearrangements by homologous recombination. Moreover, in addition to their impact on genome structure and evolution, their presence also reflects scars of previous recombination events which have shaped the genome and their detailed analysis can provide information central to understanding the relationship between different genomes of the same or more distant species.

Table 1
IS families

Families	Groups	Size range	DR (bp)	Ends	IRs	Nb ORF	Tpase type
IS*1*		740–1,180	8–9	GGnnnTG	Y	2	DDE
	Only 1 ORF	800–1,200	0–9			1	
	IS*Mhu11*	900–4,600	0–10			2	
IS1595	IS*Pna2*	1,000–1,150	8	**GG**Cnn**TG**	Y	1	DDE
	+pass	1,500–2,600	8	C**G**CTC**T**T		1 or +	
	IS*H4*	1,000	8	**GG**Ggc**tg**		1	
	IS*1016*	700–745	7–9	CcTGA**T**T		1	
	IS*1595*	900–1,100	8	nnn**G**cnTA**T**C		1	
	IS*Sod11*	1,000–1,100	8	**gg**nna**t**TAT		1	
	IS*Nwi1*	1,080–1,200	8	C**G**Gnn**T**T		1	
	+pass	1,750–4,750	8			1 or +	
	IS*Nha5*	3,450–7,900	8			1	
IS*3*	IS*150*	1,200–1,600	3–4	TG	Y	2	DDE
	IS*407*	1,100–1,400	4	TG			
	IS*51*	1,000–1,400	3–4	TG			
	IS*3*	1,150–1,750	3–4	TGa/g			
	IS*2*	1,300–1,400	5	TG			
IS*481*		950–1,300	4–15	TGT	Y	1	
IS*4*	IS*10*	1,200–1,350	9	CT	Y	1	DDE
	IS*50*	1,350–1,550	8–9	C		1 or +	
	IS*Pepr1*	1,500–1,600	7–8	-T-AA			
	IS*4*	1,400–1,600	10–13	-AAT			
	IS*4Sa*	1,150–1,750	8–10	CA			
	IS*H8*	1,400–1,800	10	CAT			
	IS*231*	1,450–5,400	10–12	CAT			

(continued)

Table 1 (continued)

Families	Groups	Size range	DR (bp)	Ends	IRs	Nb ORF	Tpase type
IS*701*		1,400–1,550	4		Y	1	DDE
IS*H3*		1,225–1,500	4–5	C-GT	Y	1	DDE
IS*1634*		1,500–2,000	5–6	C	Y	1	DDE
IS*5*	IS*903*	950–1,150	9	GG	Y	1	DDE
	IS*L2*	850–1,200	2–3	-GC		1	
	IS*H1*	900–1,150	8	Ga/g		1	
	IS*5*	1,000–1,500	4	GAa/g		1	
	IS*1031*	850–1,050	3	Ga/g		1	
	IS*427*	800–1,000	2–4			2	
IS*1182*		1,350–1,950			Y	1	DDE
IS*6*		700–900	8	GG	Y	1	DDE
IS*21*		1,750–2,600	4–8	TG	Y	2	DDE
IS*30*		1,000–1,700	2–3		Y	1	DDE
IS*66*		2,000–3,000	8–9	GTAA	Y	3	DDE
	IS*Bst12*	1,350–1,900	8–9	GTAA	Y	1	DDE
IS*91*		1,500–2,000	0		N	1	Y2
IS*110*	IS*1111*	1,200–1,550	0		N	1	DDE
IS*200*/IS*605*	IS*200*	600–750	0		N	1	Y1
	IS*608*	1,300–2,000				2	Y1
	IS*1341*	1,200–1,500				1	

IS*607*	1,700–2,500	0		N	2	Serine
IS*256*	1,200–1,500	8–9	Ga/g	Y	1	DDE
IS*630*	1,000–1,400	2		Y	1 or 2	DDE
IS*982*	1,000	3–9	AC	Y	1	DDE
IS*1380*	1,550–2,000	4–5	CC	Y	1	DDE
IS*As1*	1,200–1,500	8–10	CAGGG	Y	1	DDE
IS*L3*	1,300–2,300	8	GG	Y	1	DDE
Tn*3*	>3,000	0	GGGG	Y	>1	DDE
IS*Azo13*	1,250–2,200	0–4	Ga/g	Y	1	

Size range in base pairs (bp) represents the typical range of each group. N, no; less frequently observed lengths are included in parentheses; Ends, typical nucleotide sequences at the very ends of the element. Presence (Y) or absence (N) of terminal inverted repeats is indicated. DDE represents the common acidic triad presumed to be part of the active site of the transposase

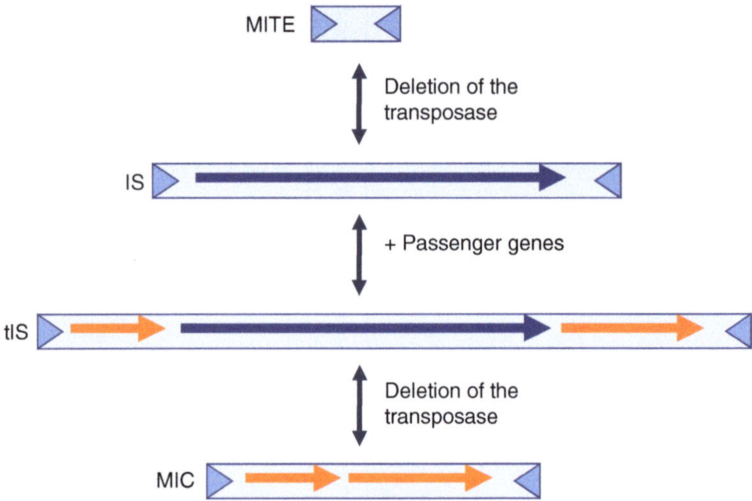

Fig. 1. Relationship between ISs and their derivatives. The terminal inverted repeats (IRs) are shown as *light blue triangles*, the transposase gene is shown as a *blue arrow*, and passenger genes by *an orange arrow*.

Correct identification and annotation of these simple MGEs are surprisingly complex and available annotation programmes do not provide an authoritative IS annotation.

ISfinder was initially designed to assign IS names providing a focal point for a coherent nomenclature. It is also the international reference centre and repository for IS sequences. Each new IS is indexed with descriptives (see below). ISfinder is also used to continuously monitor public databases for new IS groups and families, provide comprehensive groupings or families, and supply some insight into their phylogenies. The site contains extensive background information on ISs and transposons in general.

ISfinder is composed of several modules: the database per se which includes an online BLAST (8) facility against the entire bank and will be supplemented in the future with additional integrated online tools (e.g. alignment capability, PsiBLAST, and HMM profiles); ISbrowser, which provides an overview of IS distribution in individual genomes and includes a graphic module; and ISsaga (*I*nsertion *S*equence *s*emi-*a*utomatic *g*enome *a*nnotation), a pipeline developed for semi-automatic IS annotation of individual genomes.

2. Using the ISfinder Web Site

2.1. ISfinder

ISfinder is limited to ISs and does not yet deal with other MGEs. As of May 2011, the database included nearly 4,000 different expertly annotated ISs. ISfinder is divided into several sections.

The first includes extensive background information on ISs and transposons in general (*Information\General information*) and is based on two broad reviews (6, 8) enhanced by supplementary material (a large bibliography and figures). Information concerning IS regulation, structure, catalytic mechanism, and target site specificity is presented in subsections. This section also provides detailed descriptions of the characteristics of different IS families (*Information\IS families*) based on their genetic organisation, similarities between their transposases and the relationship of their IRs (6), catalytic properties of their transposases (*The DDE motif*), and more detailed individual family characteristics (*Family information\Major features of prokaryote IS families*), including a general section (*Occurrence, Variety, and Systematics*).

This grouping is a constantly evolving aid to classification and management of the high number and variety of ISs which are being identified in the various genome sequencing projects.

Another role of ISfinder is to assign IS names and provide a focal point for a coherent nomenclature. A second section (*Information\Nomenclature and Attribution requests*) contains an explanation of IS nomenclature and suggestions for nomenclature for different bacterial species (*IS Nomenclature*). It also includes the original blocks of IS numbers assigned to individuals, groups, or institutions (9); a listing of these original ISs and the last known address of the attribution (*Reserved blocks of IS numbers previously attributed*); IS names attributed by ISfinder or which have appeared in the literature in a form which did not correspond to the recommended nomenclature (*List of IS names currently attributed*); and an online form for registering new ISs (*Attribution requests*). If the protein sequence is more than 98% similar and/or the DNA sequence is more than 95% identity to an entry in ISfinder, the IS is considered as isoform and does not require a separate attribution.

The ISfinder database is also the repository for IS sequences. Each IS is indexed together with information, such as its DNA sequence, potential *orfs*, and sequence of the ends and target sites, its origin, distribution, and family attribution, together with a bibliography where available. Information on each IS is stored as an individual file in the MySQL database, with links to the NCBI Taxonomy Browser and to the relevant public database file through its accession number. ISfinder can be readily searched (*Using the database\Search*) using the online search tool available to extract different types of information (an entire file with all information concerning the IS of interest; the original bacterial host; a table of all ISs found in the search together with various characteristics, such as Synonyms, Isoforms, Family, Origin, Accession Number, Length, IR, DR; other bacterial species which also contain the IS; a short bibliography on the IS; general comments). A more advanced search (*Extensive search*) allows a larger set of output layouts. Online tools are gradually being added. At present, an online

BLAST facility (*Using the database\Analysis\BLAST*) against the entire bank is available.

Following the attribution of an IS name, the sequence should be deposited in the database (*Using the database\Submission*). This enriches the base and helps subsequent users. It also enables us to maintain an overview to prevent multiple names being attributed to a single IS. An online form is available and minimally involves completing the boxes marked "*". Following online submission, each submitted sequence file is verified before inclusion in the public database. Confidential sequence information is retained in a secure database prior to being released (following accord) into the public domain.

ISfinder also includes a section of bacterial genomes composed of two modules, ISbrowser and ISsaga.

2.2. ISbrowser

ISbrowser (10) provides information about the IS content of sequenced prokaryotic genomes. It includes only those genomes which have been expertly annotated and verified by ISfinder annotators. At present, it contains more than 40 expertly annotated genomes (119 replicons) listed alphabetically (*Genomes\ISbrowser*). Existing genomes are also regularly updated when new types of IS appear in the ISfinder database. This process will be greatly improved and accelerated by the semi-automatic annotation tool, ISsaga (see below).

The major feature of ISbrowser is a visualisation tool: a circular graphic representation of each genome (CGview (11)) showing the positions and orientations of IS sequences and their family attributions. Individual complete and partial ISs are distinguished by a colour code. Additional details concerning a given IS can be obtained simply by clicking on each individual example. The ISbrowser suite also includes sets of tables which provide a more detailed picture of the IS content and permit the user to visualise individual multi- or single-copy ISs on the genome; determine the content in user-defined sub-regions of the genome; obtain alignments of multicopy ISs (the ends—IRs, the entire DNA, and amino acid sequences); and obtain information on the IS family through a link to the ISfinder information section.

2.3. ISsaga

The ISsaga module (12) was written to assist accurate IS annotation of whole genomes using ISfinder standards. ISsaga provides comprehensive computational tools and methods for rapid, high-quality IS annotation. It is integrated as a module into ISfinder and was designed specifically for use with the ISfinder database. ISsaga is a semi-automatic system. However, although some results are generated automatically they must be validated by the user. The procedure is described in the online ISsaga manual. The user must identify any new IS elements not already present in ISfinder using the toolbox provided.

The semi-automatic annotation system works at two levels: protein and nucleotide. Examples of a completed genome annotation and a genome "in progress" performed using ISsaga can be found on the Web site. Pre-annotated GenBank files are recommended, but FASTA nucleotide files are accepted as are FASTA protein files if a corresponding FASTA nucleotide file is also included. GenBank format is preferred because it provides information concerning the neihbouring environment of the IS. Up to ten replicons (chromosomes and plasmids) can be analysed concurrently in a single project.

ISsaga identifies IS-associated *orfs* using an *orf* identification module which also attributes them to one of the ISfinder-defined IS families. BLASTX analysis is then performed to ensure that all potentially relevant *orfs* have been identified. A prediction of the number of full or partial IS copies and of IS families is then generated. This creates an annotation table (*Annotation tab\"Annotation table"*) which is completed during the annotation process. The initial table includes the *orfs* identified; their family attribution; similarity with ISs in ISfinder; genome coordinates; and fields concerning the subsequent nucleotide annotation. Orfs annotated as such in the source GenBank file but not belonging to an existing ISfinder-defined family are labelled as "putative new family".

ISsaga will not automatically identify ISs if they are very different to those in the database, e.g. ISs using different chemistries to the classical DDE transposases (13) will not be found unless a copy is included in ISfinder. Contributions from the community obtained from direct identification of ISs from individual transposition events (e.g. insertional mutation of cloned genes) are important in improving IS identification and extending the accuracy of annotation. The probability that ISs will not be identified will decrease with the increasing use of ISsaga to supplement the ISfinder database.

The nucleotide annotation module automatically identifies ISs present in ISfinder generating a list of ISs present in the genome (*Semi-automatic tab\"List Annotated IS(s)"*) and a report including details of each individual copy. Once validated by the user, these are automatically added to the annotation table. ISfinder attributes a block of names (one for each new IS) on request. New ISs should be submitted to ISfinder for verification (*Validation tab\"Submit IS to ISfinder"*). They are then included in ISfinder (either in the public or private sections, as initially chosen by the user when opening the project), added to the list of ISs present in the genome, and, after validation of this IS report, included in the annotation table.

Prokaryotic genomes often carry intercalated IS clusters in which one IS is interrupted by insertion of additional ISs. ISsaga also includes a tool in the annotation report to resolve such structures and to reconstruct the associated ISs.

A series of graphic representations of the annotation status can be generated, including a pie chart and histograms as well as a circular representation of the IS distribution using an integrated CGView tool (11). This is dynamic and, together with a summary table, provides a continuous snapshot of annotation progress. This can be compared directly with the results obtained from the automatic prediction.

At the end of the annotation process, the identified IS(s) and the annotation result can be retrieved in a spreadsheet format or as a new GenBank file which can then be used in other applications for further analysis.

Often, accurate annotation may not be necessary, but an estimate of the number of ISs (both complete and partial copies) and the number of different IS families in a given genome would be useful. A prediction can be automatically generated in the initial step after loading the genome file. There are a number of rules, defined from our present experience with IS identification, which automatically remove many of the major annotation ambiguities encountered due to the diversity and complexity of ISs (e.g. the presence of more than one *orf* in an IS, overlapping reading frames, programmed translational frameshifting, etc.). Although we have included filters to eliminate some of these, we have voluntarily set the filters at a level which retains a small fraction. This ensures that we do not eliminate real but distantly related IS-associated *orfs*. Although many false positives are removed from the predictor results, they are included in the final annotation table to enable individual examination and manual deletion or validation in the final annotation.

In spite of its limitations, the predictor is the most reliable available software for automatic IS prediction and its reliability will evolve with time and experience.

3. Discussion

At present, we possess only a very partial understanding of the diversity, distribution, and activity of ISs and their derivatives within prokaryotic genomes. One major problem in assessing IS impact lies in the accurate identification of both their transposases and the DNA components. At present, we have full IS annotations of only a small fraction of available prokaryote genomes. The numerical importance of transposases in nature (14), and therefore, presumably, the genetic objects on which they function, makes their correct annotation imperative. However, although ISs are arguably the simplest autonomous transposable elements, their diversity and complexity probably exclude the development of an entirely automatic annotation procedure. To resolve this problem, we have

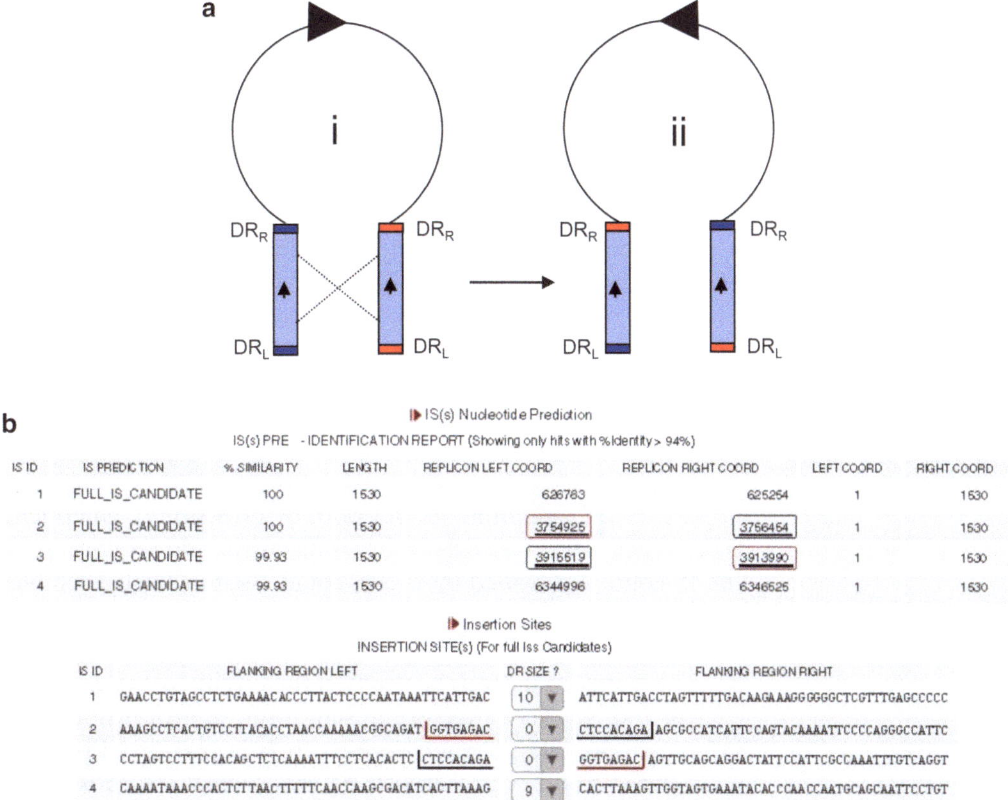

IS(s) Nucleotide Prediction

IS(s) PRE - IDENTIFICATION REPORT (Showing only hits with %Identity > 94%)

IS ID	IS PREDICTION	% SIMILARITY	LENGTH	REPLICON LEFT COORD	REPLICON RIGHT COORD	LEFT COORD	RIGHT COORD
1	FULL_IS_CANDIDATE	100	1530	626783	625254	1	1530
2	FULL_IS_CANDIDATE	100	1530	3754925	3756454	1	1530
3	FULL_IS_CANDIDATE	99.93	1530	3915619	3913990	1	1530
4	FULL_IS_CANDIDATE	99.93	1530	6344996	6346525	1	1530

Insertion Sites

INSERTION SITE(s) (For full Iss Candidates)

IS ID	FLANKING REGION LEFT	DR SIZE ?	FLANKING REGION RIGHT
1	GAACCTGTAGCCTCTGAAAACACCCTTACTCCCCAATAAATTCATTGAC	10	ATTCATTGACCTAGTTTTTGACAAGAAAGGGGGGCTCGTTTGAGCCCCC
2	AAAGCCTCACTGTCCTTACACCTAACCAAAAACGGCAGAT GGTGAGAC	0	CTCCACAGA AGCGCCATCATTCCAGTACAAAATTCCCCAGGGCCATTC
3	CCTAGTCCTTTCCACAGCTCTCAAAATTTCCTCACACTC CTCCACAGA	0	GGTGAGAC AGTTGCAGCAGGACTATTCCATTCGCCAAATTTGTCAGGT
4	CAAAATAAACCCACTCTTAACTTTTTCAACCAAGCGACATCACTTAAAG	9	CACTTAAAGTTGGTAGTGAAATACACCCAACCAATGCAGCAATTCCTGT

Fig. 2. Detecting recombination events. Panel (**a**) The figure shows two ISs (*light blue boxes*) in inverse orientation (*black arrows*) together with two different direct target repeats (DRs) flanking the left DR_L and right DR_R ends (*red and dark blue*). Recombination between the ISs exchanges one DR of each pair and inverts the intervening genome sequence. This outcome can also be generated by transposition via a replicative transposition cointegrate route. (**b**) The upper section shows the genome coordinates of each IS copy as obtained from the individual ISsaga IS report. Copies 2 and 3 are located at some distance from each other on the genome. The lower section shows the flanking 49 base pairs and the corresponding DRs. DR_L of copy 2 (*red*) is present as DR_R of copy 3 (*red*), whereas DR_R of copy 2 (*black*) is present as DR_L of copy 3 (*black*).

developed ISsaga, a suite of annotation tools which functions with the ISfinder database. Enrichment of ISfinder with additional ISs (and their derivatives) found in the course of genome annotations will not only facilitate further accurate automatic annotation, but in turn will also provide a full picture of IS diversity and impact.

ISsaga has several features to examine IS impact within and between genomes. It includes the genome context (i.e. flanking genes each annotated IS) allowing identification of IS-induced gene disruption and rearrangements. For example, inspection of the list of IS target sites and the flanking regions can identify one DR copy associated with one IS and the second associated with another in this list indicating either recombination between the two ISs (Fig. 2) or, alternatively, the point of insertion of a composite transposon.

The analysis also provides evidence of IS-mediated synteny interruption between closely related strains (e.g. 15). Inspection of flanking genes or gene fragments can also identify a variety of local genomic modifications: insertion hot spots relating to target specificity; intercalated or tandem IS insertions; and IS-driven flanking gene expression (e.g. formation of hybrid promoters) (see ref. 8, as well as genes interrupted by the insertion).

Additional functions are being implemented to further improve ISfinder. These include the ability to submit multi-Fasta sequences containing multiple contigs from a single genome. This assists in genome assembly by exploiting flanking IS DRs rather than masking ISs. Rapid comparison of sets of closely related strains relies increasingly on mapping contigs to a common scaffold. ISsaga will be developed to provide strong support for assembly of the scaffold for synteny studies. In addition, these developments will provide support for metagenomic studies both by provision of an enriched database and direct analysis of raw metagenomic sequence data. The IS context provided by ISsaga could assist in small assemblies, but more importantly it provides identification tags for ISs whose distribution is limited and which may be used to determine some of the genera and even species present in the original sample.

ISs and IS derivatives represent only a proportion of all prokaryotic mobile genetic elements. It is hoped that ISsaga will be extended to other prokaryotic MGEs, such as transposons (see http://www.ucl.ac.uk/eastman/tn/) and integrative conjugative elements (ICEs (16)), and linked to other databases, such as the general database for mobile genetic elements, ACLAME (http://aclame.ulb.ac.be/), and more specialised databases available for integrons and integron cassettes ((17); INTEGRALL: http://integrall.bio.ua.pt/) and inteins (http://www.neb.com/neb/inteins.html; http://bioinfo.weizmann.ac.il/~pietro/inteins/) as well as that for eukaryotic transposons (REPBASE: http://www.girinst.org/repbase/).

References

1. Siguier P, et al. (2006) ISfinder: the reference centre for bacterial insertion sequences. Nucleic Acids Res 34:D32–36
2. Dobrindt U, et al. (2004) Genomic islands in pathogenic and environmental microorganisms. Nat Rev Microbiol 2:414–424
3. Binnewies TT, et al. (2006) Ten years of bacterial genome sequencing: comparative-genomics-based discoveries. Funct Integr Genomics 6:165–185
4. Altschul SF, et al. (1990) Basic local alignment search tool. J Mol Biol 215:403–410
5. Hickman AB, Chandler M, Dyda F (2010) Integrating prokaryotes and eukaryotes: DNA transposases in light of structure. Crit Rev Biochem Mol Biol 45:50–69
6. Chandler M, Mahillon J (2002) Insertion sequences revisited. In Craig NL, Craigie R, Gellert M, Lambowitz A (eds) Mobile DNA. ASM press, Washington DC
7. Siguier P, Gagnevin L, Chandler M (2009) The new IS1595 family, its relation to IS1 and the frontier between insertion sequences and transposons. Res Microbiol 160:232–241

8. Mahillon J, Chandler M (1998) Insertion sequences. Microbiol Mol Biol Rev 62:725–774
9. Campbell A, Berg DE, Botstein D, et al. (1979). Nomenclature of transposable elements in prokaryotes. Gene 5:197–206
10. Kichenaradja P, et al. (2010). ISbrowser: an extension of ISfinder for visualizing insertion sequences in prokaryotic genomes. Nucleic Acids Res 38:D62-68
11. Stothard P, Wishart DS (2005). Circular genome visualization and exploration using CGView. Bioinformatics 21:537–539
12. Varani AM, et al. (2011) ISsaga is an ensemble of web-based methods for high throughput identification and semi-automatic annotation of insertion sequences in prokaryotic genomes. Genome Biol 12:R30
13. Curcio MJ, Derbyshire KM (2003). The outs and ins of transposition: from Mu to Kangaroo. Nat Rev Mol Cell Biol 4:865–877
14. Aziz RK, Breitbart M, Edwards RA (2010). Transposases are the most abundant, most ubiquitous genes in nature. Nucleic Acids Res 38:4207–4217
15. Zerillo MM, et al. (2008) Characterization of new IS elements and studies of their dispersion in two subspecies of Leifsonia xyli. BMC Microbiol 8:127
16. Burrus V, Waldor MK (2004) Shaping bacterial genomes with integrative and conjugative elements. Res Microbiol 155: 376–386
17. Mazel D (2006) Integrons: agents of bacterial evolution. Nat Rev Microbiol 4:608–620

Chapter 6

Methods and Software in NGS for TE Analysis

Cristian Chaparro and Francois Sabot

Abstract

The recent development of next-generation sequencing (NGS) technologies allowed various authors to imagine, test, and validate new approaches for TE analysis, in their nature, type, activity, or quantity. In this chapter, we describe briefly the technologies used, then the various approaches and methods used already, and finally some potential new methods. In contrast to the more molecular chapters of the book, the approaches described here are purely bioinformatics, and have a set of NGS data as a starting point. Moreover, as these analyses are quite recent in the field, most of them were only performed once, and we cannot be sure that they could be reused in other species or context than the original one. However, there are a lot of interesting approaches and results that NGS can provide in the TE field.

Key words: NGS, Repeats, Interspersion, Transposon, Retrotransposon

Abbreviations

BAC	Bacterial artificial chromosome
LTR	Long terminal repeat
NGS	Next-generation sequencing
siRNA	Small interfering RNA
SNP	Single-nucleotide polymorphism
TE	Transposable element
VS	Structural variant

1. Introduction to NGS Technologies

Currently, most of the transposable element (TE) analyses using next-generation sequencing (NGS) are performed with 454 or Illumina data. The other NGS technology available at the time of

Yves Bigot (ed.), *Mobile Genetic Elements: Protocols and Genomic Applications*, Methods in Molecular Biology, vol. 859, DOI 10.1007/978-1-61779-603-6_6, © Springer Science+Business Media, LLC 2012

writing, SOLiD, is mainly used in metagenomics, and is not used in TE analysis. These last years, the companies developing these technologies have concentrated at making them more efficient and in making the runs cheaper, thus lowering the cost per base. With the apparition of IonTorrent (now owned by Life Technologies) on the market with its bench sequencer, the race to produce small versions of the 454 (GS Junior) and Illumina (MySeq Personal Sequencing System) machines, affordable by more scientists, has begun. While these smaller versions have smaller throughputs, they present an undeniable interest to the mass of laboratories that cannot afford the bigger versions. One other technology is entering the field, the SMAR-T technology from Pacific Biosciences, which promises long reads in the order of several thousand base pairs.

1.1. The Pyrosequencing/454

The 454 technology is based on the detection of the released phosphate during DNA synthesis. The system will detect the A, then the T, then the G, and then the C, sequentially. If three A's are following, the machine will detect a threefold increase in signal and provide a three A information for peak calling. The current technology allows up to one million reads of 300–500 bases each (Titanium kit) per run. Thus, we obtain a quite large array of long reads and 400 Mb per run in less than 2 days after DNA (or RNA) extraction to results. The main limit with this technology is that for homopolymers longer than four bases (e.g., more than four As in a row) the detection system reaches the limits and cannot decipher the correct number of bases. The company released a smaller more affordable version of their machine (GS Junior), which still produces 400-bp reads and can produce 35 Mbases per run in 10 h. This technology, being the oldest of NGS, is fully mature and no more technical advances are expected.

1.2. The Solexa/ Illumina

The second NGS technology used in TE studies is the Illumina system. This system relies on the detection of fluorescent dyes released after the incorporation of the tagged bases. It will detect one base at a time, thus avoiding the homopolymers' effect from 454. The current machines (HiSeq2000) provide a minimum of 50 millions of clusters per lane (8 lanes per flow cell), in which 100 high-quality bases are read from each border (pair-end sequencing). Thus, around 10 Gbases of sequences per lane are expected in around 8 days (the sequencing lasts for 6 days) with a total of 180–200 Gbytes of raw data per run. The main limit is the size of the reads that are produced, currently two times 100 bases and two times 150 bases expected for the end of 2011. The Illumina company continues the development of this technology, increasing read length, read quality, and cluster density, and as mentioned above, they also started recently to propose a smaller bench version of their machine, the MySeq, providing 3.2 million clusters per run, representing from 120 Mbases (for 35-bases reads) to 1 Gbases (for 150-bases paired-end reads).

1.3. The Upcoming Technologies

Currently, other technologies are being developed and perfected and will provide new possibilities for TE analysis. These are the IonTorrent/PGM (Personal Genome Machine) and Pacific-Biosciences/SMAR-T technologies. The first, based on a variant of pyrosequencing, aims at providing cheap systems with a limited amount of reads (100,000 of 100–300 bases, for $500). The second technology is geared toward de novo sequencing and the production of long reads (currently, 3 kb, up to 40 kb announced). While these machines do not provide enough throughput to analyze whole large eukaryotic genomes for the moment, an array of new analyses will become available for individual laboratories, from variant detection to sequencing of BAC pools. These next years will then see the emergence of new methods and ideas for TE analysis.

2. Approaches

2.1. Recurrence of Sequences and Recurrence of Insertions

One straightforward possibility of using NGS is to detect TEs using their "More-Than-One-Copy" property, as was performed by Wicker et al. (1). For that, authors used a 0.1× (10% of the genome, which represents a low coverage) sequencing of the barley genome. Using this set of Illumina data, they counted the number of occurrences of specific k-mers. Those k-mers were small sequences of 20 bases, and for each sequence they can report then a score representing its frequency in the genome. Plotting this frequency on the sequence of barley BACs, i.e., reporting the corresponding k-mers' frequency at each position (1–20, 2–21, etc.), they generated graphs representing the repeated level of the underlying sequence. The higher the plots, the more repeated the region is. The automatic identification was compared to the manually curated annotations available for those BACs, and more than 90% of the TE annotations were common between the two methods. Thus, even without any previous knowledge, it is possible to perform repeats annotation on a genome and even to detect LTR retrotransposons (their specific features of two quite identical LTR for each copy provide a "Batman ears" effect on the graph; Fig. 1). The software used here is *Tallymer* (2), on a currently standard bioinformatics personal machine. The counts were plotted logarithmically to allow better viewing. To perform such analysis, only short reads are needed, as no identification is possible with this method. The original paper was performed with 36-bases single-end reads from an Illumina GA I sequencer. Such lengths are still available, or more generally a 50-bases single-end set of sequences. The only prerequisite is to have clean data, i.e., without any adapters or primer or vector contamination. Other tools, such as *JellyFish* (3), are also able to perform fast and powerful k-mers counting, reducing the

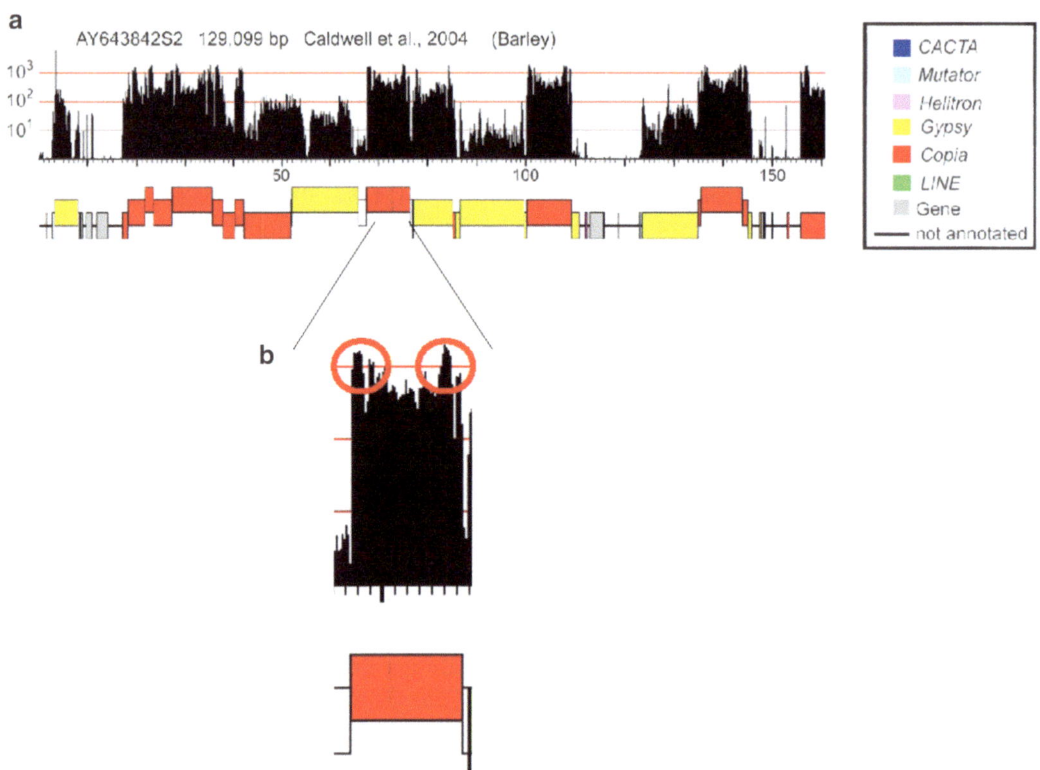

Fig. 1. Automatic TE detection through k-mers count. Images came from Wicker et al. (1). (**a**) Automatic annotation (k-mers logarithmic count) is compared to manual curated annotation. (**b**) "Batman ears" effect produced by the LTRs surrounding a *Copia* element.

time and memory necessary for doing these kind of analyses. However, this method is limited to the possibility of detecting repeated regions without being able to detect the precise boundary of elements or identify them.

2.2. Identification of New Elements Using the Nonassembled Reads in De Novo Sequencing

The amount of data produced by NGS technologies, which makes sequencing a genome a matter of days, led the community to the development of software that would assemble whole genomes. The most important problem that is hindering the assembly of genomes, to at least the same quality of the more expensive and time-consuming Sanger-sequenced genomes, is that due to the short reads the repeated regions cannot be spanned by the assembling algorithms. The repeated nature of these regions produces dead ends or many possible paths in DeBruijn graphs, commonly used in NGS assembly algorithms (Fig. 2). Gnerre et al. (4) have approached this issue in their *Allpaths-LG* software by producing libraries that span different lengths, and thus have shown that they can achieve the same finishing quality as Sanger-assembled genomes. They rely on very high coverage, and they have partly

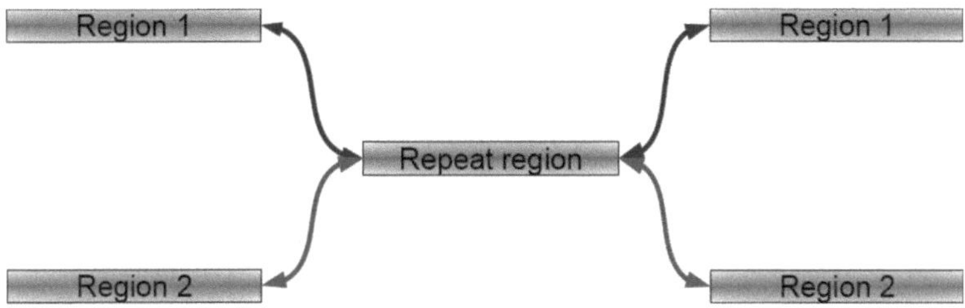

Fig. 2. Simplified schema of a DeBruijn graph depicting the problem of repeated sequences. Repeat regions can be depicted as the crossing of several paths through the genome. When extending region 1, eventually the extended sequence will enter the repeat region. At the end of the repeat region, the graph will have multiple possible paths to follow, in this simplified case region 1 and region 2. As the correct path cannot be established, the sequence will be reverse complemented and the algorithm will start tracking back reaching another impasse at the other end of the repeat region.

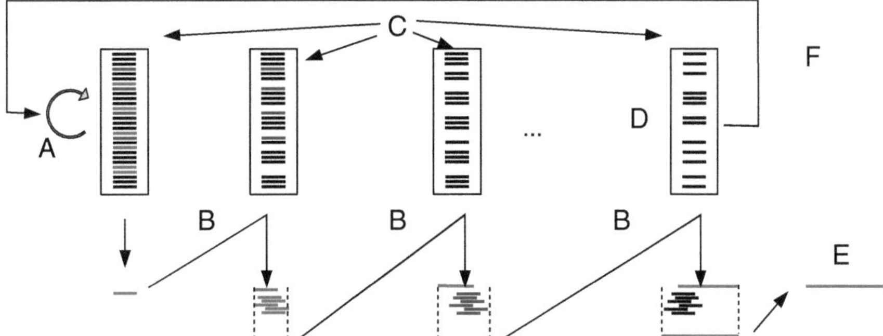

Fig. 3. Search for repeated elements in nonassembled reads. (**a**) Blast the reads database against itself in order to select the most abundant read. (**b**) Blast of the extended read to the reads database. (**c**) The reads that match the extension are extracted from the nonassembled reads database. (**d**) When the extension cannot continue, reverse complement the sequence and continue extending the other end. (**e**) The identified reference sequence for the repeat element is manually curated.

solved the repetitive sequence problem by collapsing read pairs if there is enough evidence that they overlap or span a gap. Still, over 60% of the unaligned reads pertains to transposable elements.

One can already suspect that most of the nonassembled sequences in such cases belong to TE sequences. So, new methods were developed to identify new TEs in the nonassembled portion of sequences. It consists on taking the most represented reads from the pool of nonassembled sequences, and then these sequences are extended by blasting one end against the remaining nonassembled reads. In this case, the best matching sequence that extends the repeated sequence is kept and all similar sequences are deleted from the database. This extension process is repeated until the sequence cannot be extended any more. The next step extends the other end of the sequence until no more sequences can be added or an overlap with the already assembled sequence is detected (Fig. 3). Afterward, the whole process is repeated and the remaining pool of sequences are searched for other repeated elements.

This method was used in Cocoa (5), where the *Gaucho* LTR retrotransposon was identified. Generally, homemade customized scripts are used for such purpose. The sequences identified with this method are considered as reference sequences and serve as a representative sequence of a family as it is almost impossible to identify a true genomic copy through this method.

2.3. Transcriptome and Expressed TEs

In a more classical way, the NGS can be used in transcriptomic analyses to detect expressed TEs. The method is more reliable and reproducible than microarray detection (6), and also provides more information, as all expressed sequences can be detected, even the currently unknown. The main limit in such analysis is the same as in microarray, i.e., identifying the stages or conditions where the TEs are expressed. However, the multiplexing capacities of NGS technologies provide an increased potential of identifying expressed TEs.

2.4. Improved Annotation of TEs Using siRNA

Small interfering RNAs (siRNAs) are supposed to be generated in order to block the expression of TEs (7). Thus, if a set of siRNAs are sequenced and mapped upon their reference genome, one can determine genomic loci that present an important production of siRNA. One could infer that such loci might pertain to repeated sequences. Thus, if a region is identified by an important set of siRNA but is not annotated yet as containing repeats, it will become a good candidate place to identify a new type of TE.

2.5. Identification of Active TEs Through Structural Variant Detections

The main interesting application of NGS technology is the detection of actively transposing elements by structural variant detection. While transcriptomics can be used to identify transcribed elements, it is not possible to know if these transcribed elements are capable of completing the transposition cycle. Using a whole-genome sequence as a template, it is possible to detect structural variants (SV) in this genome. The idea is to detect an anomaly in the mapping of a pair in two far away positions or in two chromosomes, or even with only one mate mapped (Fig. 4). The method, originally used on human genome, is extensively employed today in tumor cell studies, and recently in the TE field. In Asian rice, this method was used to show that in a single regeneration event at least 34 new insertions appeared in the close vicinity of genes (8). In *Arabidopsis*, this method was successfully used to detect the neoinsertions of the *Evade* elements (9). Recently, Fiston-Lavier et al. (10) released a program called *T-Lex* allowing the detection of specific copies in a sample compared to a reference. The software relies on the detection of NGS reads that map to the border between the repeated element and its surrounding genomic region. The strength of this software is that it will return an estimate of the population frequency for each annotated transposable element. Such method is very useful and powerful for population genomics of TEs and

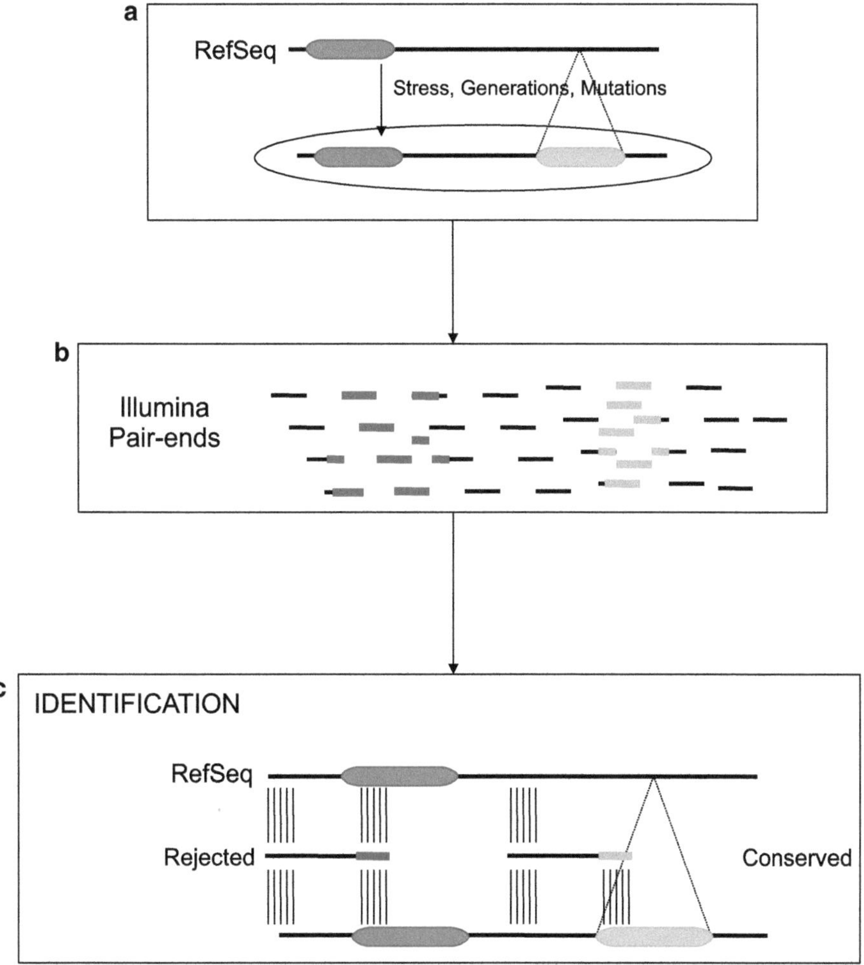

Fig. 4. Detection of structural variants. (**a**) A genome for which a reference (RefSeq) is available is resequenced after stress/generations/mutations. (**b**) The sequencing is performed using (Illumina or 454) pair-ends fragments. (**c**) Pairs are mapped on the RefSeq, and mapping anomalies are detected.

for studying the potential impact of specific insertions upon a specific phenotype.

This method relies on the initial knowledge of annotated TE insertions and on the existence of the reference genome. Only a good coverage of the genomic sequence is needed. One drawback is that this approach is not efficient in identifying insertions of transposable elements into other repeated or duplicated sequences as the flanking regions are not unique.

In the same way, markers based on TEs can be quickly generated using this method, with a bulk of DNAs from various individuals, and in detecting all the SV in the mix. Those SV positions can then be transformed in retrotransposon-based insertion polymorphism (RBIP)-like markers, suitable for genetic screening or positional cloning.

This is generally performed using pair-ended reads, sometimes using mate paired. This can also in theory be performed using longer single reads provided that a very stringent analysis is performed, as anomalies in mapping will probably result in a "non-mapped" state of the overlapping read.

3. Methods

3.1. Mapping with a Direct Reference

If one possesses the corresponding genome (or transcriptome) from the studied organism, the main used method (and the easiest) is the mapping. It consists in positioning the reads at the most probable position upon the reference. The software generally first calculates all the potential occurrences of each read, and then provides the best one in an outfile. If the data are pair-ended reads, the output will take into account the best reciprocal position to map the pair. There are various algorithms, methods, and programs to map short reads. The most used so far (and the most robust) are *Eland* (Illumina software), *BWA* (11), *Bowtie* (12), *MAQ* ((13); less used by now), and *SOAP2* (14). The classical tools, such as *BLAST*, *BLAT*, and *MegaBLAST*, are far too slow to ensure a good analysis. The parameters depend on the heterogeneity and type of organism: Is it homozygous or heterozygous? Is the species really variant in terms of SNP? Generally, one can assume that having a mapping allowing three mismatches in the short sequence (up to 75 bases, 4 for longer reads) will ensure a good mapping in terms of sensibility, even at the detriment of sensitivity (less false positive, but more false negative). The main limits of mappers are that, in the case of TEs, they cannot decipher between the various possibilities for each pair and provide a random location over all the possible ones. However, mapping is highly efficient and, coupled with sufficient posttreatment, can effectively be used to detect and identify SV.

In the case of a transcriptome as reference or as data set, either the same tools mentioned above or more specialized software, like *Exonerate* (15), *CuffLinks* (16), or *TopHat* (17), can be used. Those last will take into account the exonic nature of data to try to predict intron splicing sites in order to improve the mapping.

Mapping is not a highly time-consuming method these days, and it can be performed by the previously cited tools on normal desktop/laptop computers in a decent period of time (less than 1 h for a 10× coverage of a 400 Mbases genome, such as rice).

3.2. Mapping Without a Direct Reference

If no reference from the same species is available, the mapping can be performed however upon a more distant one (but not too far). The same tools can be used, relaxing even more the mapping tolerance (e.g., allowing more SNP/indels or larger gaps in the mapping). Another tool, *LAST* (18), originally dedicated to perform alignments on large sequences (as genomes), was enhanced to allow the so-called "xeno-mapping", i.e., mapping upon a reference which is not the one from which the data originated. Tested on *Drosophila* genomes (*melanogaster* data upon *simulans* genome), this approach ensures a better mapping. The program in fact increases the mapping using the provided information of theoretical genetic distance between the two organisms.

3.3. Assembly of TEs

Classical assembly softwares can be used, such as *Newbler* (454 software), *Velvet* (19), *Abyss* (20), or *Trans-Abyss* (for transcriptomics data (21)). However, these softwares were not designed to reconstruct repeated regions. It is possible to use part of the program to be able to gather all related reads, and then use homemade scripts to "reassemble" a generic element.

This method is particularly time- and processor-consuming, and cannot be achieved for a classical eukaryotic genome upon a normal computer, but must be run on clusters or grid.

A specific tool for that purpose was developed by DeBarry et al. (22), *AAARF*, and efficiently tested on maize genome.

3.4. Count of TEs

Using NGS data and a set of clean sequences (representative true copies or genome-wide consensus), it is possible to calculate the copy number variation of TEs between individuals or closely related species. For that, a mapping and a coverage calculation upon this reference will theoretically be enough. However, technical biases from DNA extractions, library preparation, and cluster sequencing may provide biased results. Thus, such analysis must be performed using a reiterative experimental procedure to ensure strong and statistically valid results.

References

1. Wicker T, et al. (2008) Low-pass shotgun sequencing of the barley genome facilitates rapid identification of genes, conserved non-coding sequences and novel repeats. BMC Genomics 9:518
2. Kurtz S, et al. (2008) A new method to compute K-mer frequencies and its application to annotate large repetitive plant genomes. BMC Genomics 9:517
3. Marçais G, Kingsford C (2011) A fast, lock-free approach for efficient parallel counting of occurrences of k-mers. Bioinformatics 27: 764–770
4. Gnerre S, et al. (2011) High-quality draft assemblies of mammalian genomes from massively parallel sequence data. Proc Natl Acad Sci USA 108:1513–1518
5. Argout X, et al. (2010) The genome of Theobroma cacao. Nat Genet 43:101–108
6. Marioni JC, et al. (2008) RNA-seq: An assessment of technical reproducibility and comparison with gene expression arrays. Genome Res 18:1509–1517
7. Blumenstiel JP (2011) Evolutionary dynamics of transposable elements in a small RNA world. Trends in Genetics 27:23–31
8. Sabot F, et al. (2011) Transpositional landscape of rice genome revealed by Paired-End Mapping of high-throughput resequencing data. Plant J doi: 10.1111/j.1365-313X.2011.04492.x. [Epub ahead of print]
9. Mirouze M, et al. (2009) Selective epigenetic control of retrotransposition in Arabidopsis. Nature 461:427–430

10. Fiston-Lavier A, et al. (2010) T-lex: a program for fast and accurate assessment of transposable element presence using next-generation sequencing data. Nucleic Acids Res 39:e36
11. Li H, Durbin R (2009) Fast and accurate short read alignment with Burrows–Wheeler transform. Bioinformatics 25:1754–1760
12. Langmead B, et al. (2009) Ultrafast and memory-efficient alignment of short DNA sequences to the human genome. Genome Biol 10:R25
13. Li H, Ruan J, Durbin R (2008) Mapping short DNA sequencing reads and calling variants using mapping quality scores. Genome Res 18:1851–1858
14. Li R, et al. (2009) SOAP2: an improved ultrafast tool for short read alignment. Bioinformatics 25:1966–1967
15. Slater G, Birney E (2005) Automated generation of heuristics for biological sequence comparison. BMC Bioinformatics 6:31
16. Trapnell C, et al. (2010) Transcript assembly and quantification by RNA-Seq reveals unannotated transcripts and isoform switching during cell differentiation. Nat Biotech 28: 511–515
17. Trapnell C, Pachter L, Salzberg SL (2009) TopHat: discovering splice junctions with RNA-Seq. Bioinformatics 25:1105–1111
18. Frith MC, Wan R, Horton P (2010) Incorporating sequence quality data into alignment improves DNA read mapping. Nucl. Acids Res 38:e100
19. Zerbino, D.R., Birney, E., 2008. Velvet: Algorithms for de novo short read assembly using de Bruijn graphs. Genome Research 18, 821–829
20. Simpson JT, et al. (2009) ABySS: A parallel assembler for short read sequence data. Genome Res 19:1117–1123
21. Robertson G, et al. (2010) De novo assembly and analysis of RNA-seq data. Nat Meth 7:909–912
22. DeBarry J, Liu R, Bennetzen J (2008) Discovery and assembly of repeat family pseudomolecules from sparse genomic sequence data using the Assisted Automated Assembler of Repeat Families (AAARF) algorithm. BMC Bioinformatics 9:235

Chapter 7

The Application of LTR Retrotransposons as Molecular Markers in Plants

Alan H. Schulman, Andrew J. Flavell, Etienne Paux, and T.H. Noel Ellis

Abstract

Retrotransposons are a major agent of genome evolution. Various molecular marker systems have been developed that exploit the ubiquitous nature of these genetic elements and their property of stable integration into dispersed chromosomal loci that are polymorphic within species. The key methods, SSAP, IRAP, REMAP, RBIP, and ISBP, all detect the sites at which the retrotransposon DNA, which is conserved between families of elements, is integrated into the genome. Marker systems exploiting these methods can be easily developed and inexpensively deployed in the absence of extensive genome sequence data. They offer access to the dynamic and polymorphic, nongenic portion of the genome and thereby complement methods, such as gene-derived SNPs, that target primarily the genic fraction.

Key words: Retrotransposon, Molecular marker, SSAP, IRAP, REMAP, RBIP, ISBP

1. Introduction

Markers, entities which are heritable as simple Mendelian traits and are easy to score, have long been important in studies of inheritance and variability, in the construction of linkage maps, and in the diagnosis of individuals or lines carrying certain linked genes. Phenotypic and biochemical (enzyme) markers tend to have the disadvantages of a low degree of polymorphism, limiting their ability to be mapped in crosses; relatively few loci, limiting the density of maps which can be produced; and environmentally variable expression, complicating scoring and the determination of genotype. These marker types have, therefore, been largely superseded by DNA-based molecular markers. A DNA molecular marker is in essence a nucleotide sequence corresponding to a particular physical location in the genome. Its sequence needs to be polymorphic

Yves Bigot (ed.), *Mobile Genetic Elements: Protocols and Genomic Applications*, Methods in Molecular Biology, vol. 859,
DOI 10.1007/978-1-61779-603-6_7, © Springer Science+Business Media, LLC 2012

in the individuals under analysis to allow its pattern of inheritance to be easily followed. Molecular marker methods may be defined by the kind of DNA variation or polymorphism they detect and the way in which the polymorphism is detected or visualized. Some methods generate "fingerprints," distinctive patterns of DNA fragments which are typically resolved by electrophoresis in agarose or acrylamide gels and detected by staining or labeling. Other methods detect polymorphisms on solid supports, such as filters, microarrays, or immobilized beads. Polymorphisms can also be detected in silico by analysis of sequencing data.

1.1. Molecular Markers

Restriction fragment length polymorphism (RFLP) was the first DNA-based molecular marker technique and an outgrowth of the development of gene cloning and filter hybridization methods in the 1970s. The polymorphisms it exploits are for the presence or absence of restriction sites in genomic sequences for which a cloned hybridization probe exists. Originally, RFLP analysis required Southern blotting and hybridization (1). The RFLP method is still used to generate widely shared "anchor" markers, which are those used by many researchers to combine segregation data from different experiments onto recombinational maps, though its laboriousness and lack of alleles and loci have led to the adoption of conserved orthologous sequence (COS) markers derived from sequencing projects for this purpose (2). The advent of the polymerase chain reaction (PCR) made possible the detection of variation in randomly amplified polymorphic DNAs (RAPDs) (3). The RAPDs are indeed rapid, being independent of the need for sequence data, but they suffer from low polymorphic information content (PIC), poor correlation with other marker data, and problems in reproducibility due to the low annealing temperatures in the reactions.

Around 1990, methods, which detect variability in the number of simple sequence repeats (SSRs) in microsatellites (4) or which measure variability in the occurrence of two microsatellites close to one another (5), were developed for plants. In the mid-1990s, the Amplified Fragment Length Polymorphism (AFLP®) method was introduced. The AFLP® approach is a conceptual hybrid between RFLP and the PCR methods because, whereas the method is PCR based, its polymorphism is derived from variations in restriction site occurrence or digestibility (6). The AFLP® products were initially resolved for scoring by polyacrylamide gel electrophoresis, but CRoPS® represents an updated version using high-throughput sequencing to collect data (7). The diversity array technology (DArT), introduced about 10 years after AFLP® (8), detects variability in the presence of amplified DNA fragments that are produced by methods similar to that for AFLP® but detected on a solid support by hybridization.

During the mid 1990s, large-scale projects aimed at gene discovery by sequencing segments of mRNAs (expressed sequence tags, ESTs) began to appear (6). As the projects expanded to include more than one accession or variety for a given species, systematic variations at individual nucleotide positions (single-nucleotide polymorphisms, SNPs) became apparent. The utility of SNPs as molecular markers has grown with the power and affordability of sequencing. In addition to sequencing itself, many high-throughput commercial platforms have been developed for SNP genotyping (9).

The polymorphisms detected by the foregoing methods for generating molecular markers are primarily those of small sequence variations. The RFLP, AFLP®, and to some extent DArT methods detect polymorphisms in restriction sites, typically comprising four to six base pairs, whereas SNP methods focus on single nucleotides. As the activity of some restriction endonucleases is also dependent on the DNA methylation state at the recognition site, these methods can also detect differences in DNA methylation that may segregate genetically (10); this feature may have advantages in some circumstances. Although insertions or deletions within a restriction fragment would also generate an RFLP or AFLP® polymorphism, the resolution limits of gel electrophoresis restrict insertions that can be scored to several kilobases in length. Polymorphisms in RAPDs primarily affect the ability of the 9 or 10 nt primers to anneal efficiently under the reaction conditions of particular experiments. Microsatellite alleles are generated by the gain or loss of repeat units of only a few base pairs. The foregoing changes are, furthermore, bidirectional in the sense that further mutations can restore a restriction site or primer binding site. This bidirectionality or homoplasy reduces the usefulness of these marker systems in resolving phylogenies and pedigrees.

An alternative approach is to exploit large physical changes in a genome to visualize genetic diversity. The loci scored by the method should be spread throughout the genome at high frequency, enabling dense and well-distributed recombinational maps to be generated. Such a method should be universal in its application, with low investment required for marker development in any particular species. Generation of the marker pattern should be robust and reproducible, and detection should be inexpensive and technically straightforward. Retrotransposons, described below, meet many of these requirements and have been recently developed as molecular marker systems. After providing an introduction to retrotransposons as biological phenomena, the main marker techniques currently applied to retrotransposons are described in detail.

1.2. LTR Retrotransposons

Retrotransposons or Class I transposable elements (TEs) are an abundant class of mobile DNA that has little in common with the Class II transposable elements, the DNA transposons, such as *Ac*, *En/Spm*, or *Mutator*, or with MITEs, such as *Stowaway* (11, 12).

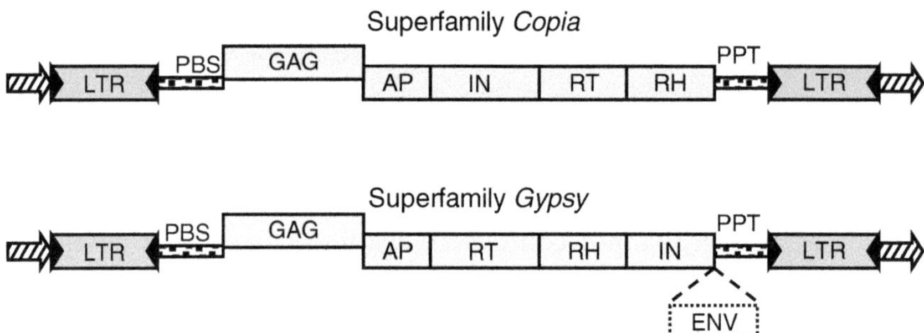

Fig. 1. Organization of long terminal repeat (LTR) retrotransposons. The elements can be classified into two major groups, the *Copia* and *Gypsy* superfamilies, named after the type members of *Drosophila melanogaster*. The elements are flanked by 5-bp direct repeats in the host DNA (*hatched arrows*), formed by the integration of the element. The retrotransposons consist of two LTRs which contain short inverted repeats at their edges (*dark triangles*) and bound the internal synthesized as a polyprotein and contains the GAG domain which encodes the protein forming the capsid of the virus-like particle, the aspartic proteinase (AP), which cleaves the polyprotein into functional units, the integrase (IN), which inserts the cDNA copy into the genome, and the reverse transcriptase (RT) and RNase H (RH), which synthesize the cDNA from the RNA transcript. The GAG may be expressed in some elements in a different reading frame, and is shown shifted upward to reflect this. *Gypsy* elements characteristically differ from *Copia* elements in the position of the IN domain. Some retrotransposons contain, as do retroviruses, an envelope (ENV) domain generally expressed in a separate reading frame.

Unlike the DNA transposons, the retrotransposons do not excise during their invasion of new loci in the genome but instead enter new loci as copies of the mother element, which remains fixed in the genome. Retrotransposons fall into clearly separated Orders, which include the long terminal repeat (LTR)-containing elements and those lacking LTRs, the LINES (long interspersed elements), and SINEs (short interspersed elements) (12).

Retrotransposons share many similarities with the retroviruses in their organization (Fig. 1), the gene products they encode, and in the steps of their life cycles (12, 13). Both retrotransposons (14) and retroviruses (15) propagate through cycles of successive transcription, reverse transcription, and genomic integration. The extant retroviruses are characterized by their possession of an *envelope* (*env*) domain encoding a glycoprotein which is necessary for infective passage from cell to cell through the plasma membrane. The extensive similarities between retrotransposons and retroviruses suggest that present-day retroviruses are derived from superfamily *Gypsy* retrotransposons by gain of the *env* domain; *env*-containing superfamily *Gypsy* elements are in fact widespread in the plants (16). The *gypsy* family of elements from *Drosophila melanogaster*, which gave the name to the superfamily, is in fact transitional between retroviruses and retrotransposons and can be infective under experimental conditions (17). The related defective elements in humans are called human endogenous retroviruses (HERVs) rather than retrotransposons (18).

1.3. Retrotransposons and the Genome

Retrotransposon transcripts each have the formal potential to be reintegrated into the genome as cDNA copies, which can then serve as further sources of transcripts leading to integrating cDNA copies. The newly integrated retrotransposon copies can be inherited if they are present in cells ultimately giving rise to gametes. In view of the many somatic cell divisions that take place prior to the differentiation of germ cells in plants, it is not totally surprising that retrotransposons have succeeded in becoming major genomic components. In plants with large genomes, retrotransposons are the major class of repetitive DNA and can comprise 40–90% of the genome as a whole (19). Independent of genome size, both the *Copia* and *Gypsy* superfamilies are ubiquitous throughout the Plant Kingdom (20–22). The major families of retrotransposons are, with a few exceptions, dispersed throughout the chromosomes in the plant species examined (23–25). In the cases examined, retrotransposon copy number increases, aside from polyploidization, appear to have been a major factor in genome size growth in the plants (26–29). Conversely, loss of retrotransposons through the recombinational production of solo LTRs and accumulation of deletions helps keep small genomes small (30–33).

1.4. Retrotransposons as Molecular Markers

The ubiquitous nature of retrotransposons and their activity in creating genomic diversity by stably integrating large DNA segments into dispersed chromosomal loci makes these elements ideal for development as molecular markers. Integration sites shared between germplasm accessions are highly likely to have been present in their last common ancestor. Therefore, retrotransposon insertional polymorphisms can help establish pedigrees and phylogenies as well as serve as biodiversity indicators (34–37).

Since 1990, several molecular marker methods based on retrotransposons have been introduced (15, 38–44), and are presented in detail below (Fig. 2). All rely on the principle that a joint is formed, during retrotransposon integration, between genomic DNA and the retrotransposon. These joints may be detected by amplification between a primer corresponding to the retrotransposon and a primer matching a nearby motif in the genome. The methods have been named according to the particular motif that provides the second priming site. The sequence-specific amplified polymorphism (SSAP) method (Figs. 2a and 3), the first retrotransposon-based method to be described, amplifies products between a retrotransposon integration site and a restriction site to which an adapter has been ligated. In the inter-retrotransposon amplified polymorphism (IRAP, Figs. 2b and 4) and inter-Primer Binding Site (iPBS) methods (Figs. 2d and 6), segments between two nearby retrotransposons or LTRs are amplified. The retrotransposon-microsatellite amplified polymorphism (REMAP, Figs. 2c and 5) technique detects retrotransposons integrated near a microsatellite or stretch of SSRs. The retrotransposon-based

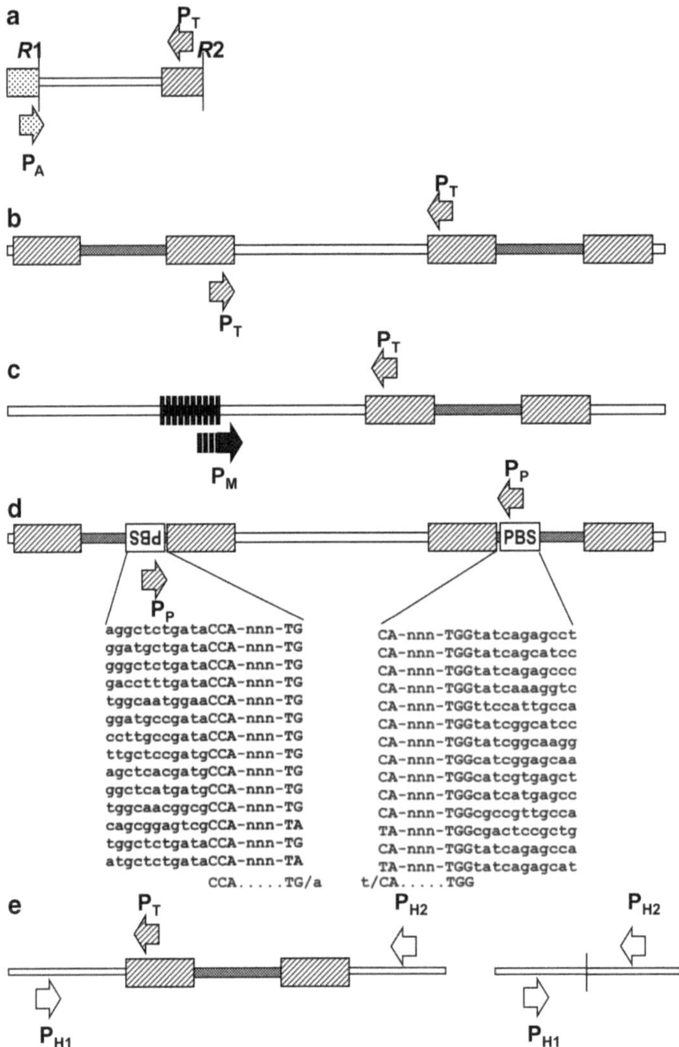

Fig. 2. Marker methods based on long terminal repeat (LTR) retrotransposons. Shared features of retrotransposons for all methods are diagrammed as LTR (*hatched box*); internal domain (*gray bar*); and genomic intervening sequence (*white bar*). (**a**) *SSAP* Sequence-specific amplified polymorphism. The DNA template is digested by two restriction enzymes (*R*1, *R*2), an adapter ligated (*stippled box*), and fragments sharing both a retrotransposon region and restriction site *R*1 are amplified by PCR with adapter primers (P_A) and retrotransposon primers (P_T). (**b**) *IRAP* Inter-retrotransposon amplified polymorphism. Regions of the genome flanked by two retrotransposons are amplified by PCR using either two identical or two different retrotransposon primers (P_T). (**c**) *REMAP* Retrotransposon-microsatellite amplified polymorphism. Regions of the genome flanked by a microsatellite domain (*vertical bars, left*) and a retrotransposon are amplified by PCR using primers containing simple sequence repeats with 3′ anchor nucleotides (P_M) and retrotransposon primers (P_T). (**d**) *iPBS* inter-Primer Binding Site polymorphism. PCR is carried out between primers (P_P) matching the PBS domains. The two retrotransposons must be oriented in opposition and either sufficiently closed for amplification (*shown*) or nested. The PCR product contains both LTRs and PBS sequences, together with the genomic sequence between the LTRs. Below the diagram, the sequence of a set of PBS domains, the 0–5 base spacer and the universal 5′ TG of LTRs, is shown. (**e**) *RBIP* Retrotransposon-based insertional polymorphism. Individual sites at which are polymorphic for retrotransposon insertion can be detected by PCR in both allelic states, full (*left*) and empty (*right*). To detect the presence of the retrotransposon, primers specific to the host DNA that is on one side of the integrated element (P_{H1}) are used together with a retrotransposon primer (P_T). To detect the empty site, primers to the two host flanks are combined (P_{H1}, P_{H2}).

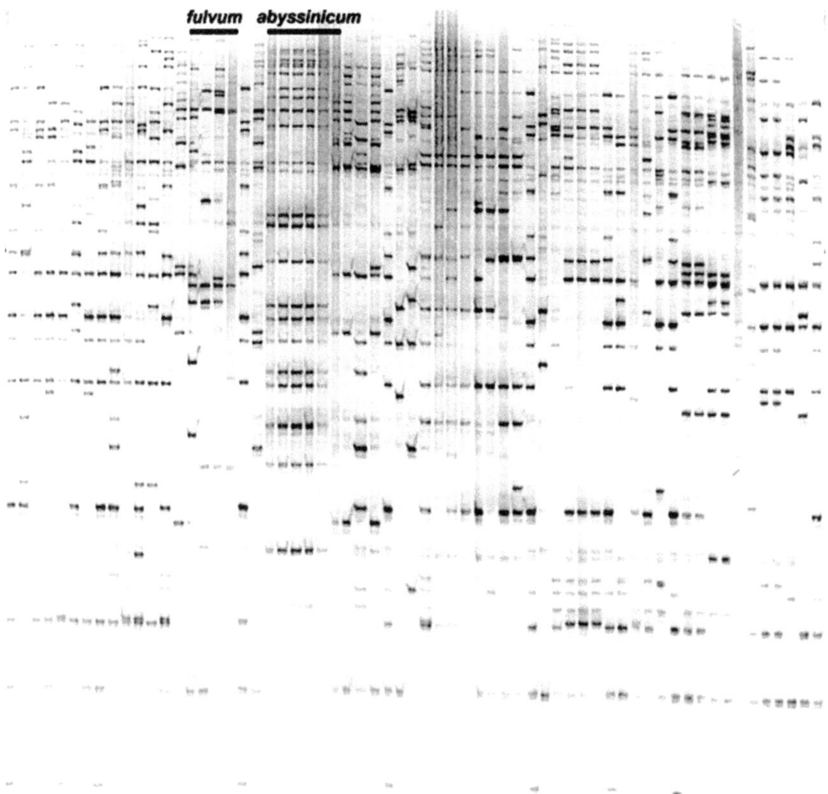

Fig. 3. SSAP analysis. The figure shows an autoradiograph of a sequencing gel resolving SSAP products. Products were generated from a set of *Pisum* accessions (*lanes*) using a ^{33}P-labeled PCR primer specific to the PPT of the *Pisum* retrotransposon *PDR*1 and a primer, with selective bases TT, matching a *Taq*I restriction site adapter. The first set of lanes are *P. sativum* accessions, the set labeled *fulvum* is *P. fulvum*, and the set labeled *abyssinicum*, *P. abyssinicum*. The other *lanes* contain accessions of various *Pisum* species. From ref. 45, with permission.

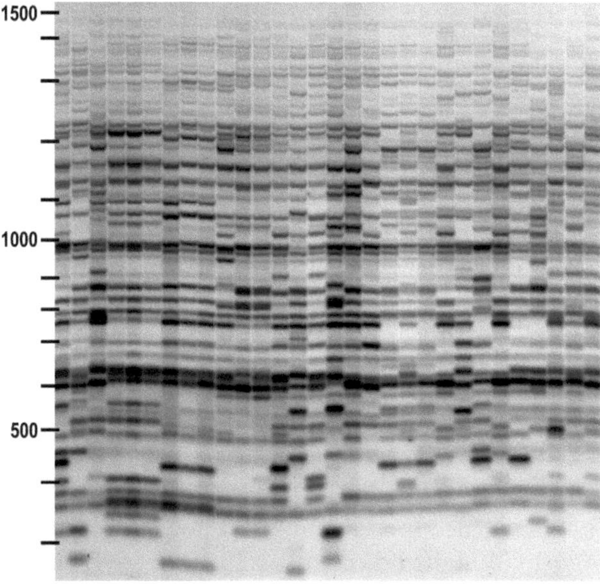

Fig. 4. IRAP analysis. IRAP amplification products from Hordeum spontaneum (wild barley) accessions (46, 107) with a *BARE1* primer is displayed. The gel has been ethidium bromide stained and the fluorescence detected with UV light; a negative image is shown. A 100-bp ladder is shown on the *left.*

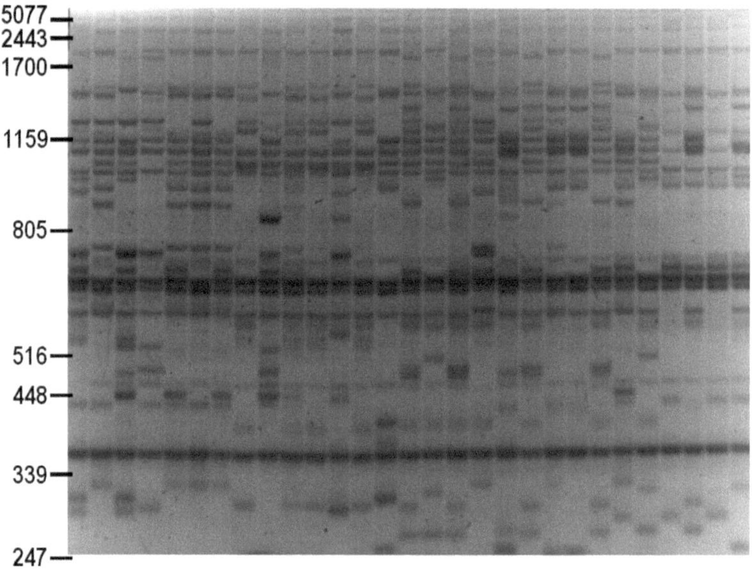

Fig. 5. REMAP analysis. A gel is shown of REMAP amplification products from *Hordeum spontaneum* using *BARE1* primers. The 26 genotypes shown (*gel lanes*) can be distinguished by their *BARE1* insertion patterns. The REMAP system is useful for population studies as well as for cultivar distinction. The banding pattern has been detected as in Fig. 4. Size markers in bp derive from a bacteriophage λ *Pst*I digest. From ref. 48, with permission.

amplified polymorphism (RBIP, Figs. 2e and 7) and insertion site-based polymorphism (ISBP) systems, in contrast to the others, detect a given locus in both alternative states, empty and occupied by a retrotransposon, by using both flanking primers and a retrotransposon primer or primers overlapping the retrotransposon joint with the flanking DNA.

Although these methods are presented here as examples with primers specific to a particular family of retrotransposons, it is important to note that retrotransposon marker methods are generic. Any organism, in which retrotransposons are dispersed components of the genome and in which they have been active over a timescale relevant to the question being asked, can be examined with retrotransposon markers. Direct comparisons of retrotransposon marker methods with AFLP® indicate that the retrotransposon markers are some 25% more polymorphic (45, 49). In principle, retrotransposon-based or retrovirus-based molecular markers could prove highly useful in animals, including mammals and birds.

1.5. Sequence-Specific Amplified Polymorphism

Sequence-specific amplified polymorphism (SSAP) was described by Waugh and coworkers in 1997 (44), but has several origins and embodiments (45, 50–52). The SSAP method can be considered to be a modification of AFLP® (53) or as a variant of anchored PCR (50). The method described by Waugh and colleagues (39, 42–44) has many similarities to AFLP®, especially in that two different enzymes are used to generate the template for the specific primer PCR and that selective bases are used in the adapter primer.

Two implementations of SSAP (Fig. 2a) are described below. The first (Fig. 3) was designed for use with a retrotransposon found in relatively few copies. In this procedure, it is important to maximize the sequence complexity of the template for the specific primer amplification, so a single-enzyme digestion is used (45). As with the method described for *BARE1* (44), the adapter primer is selective. This is a matter of convenience, and nonselective primers could be substituted where the enzyme used for digestion has a larger recognition sequence or if the copy number were lower. In general, LTR ends are convenient for the design of SSAP primers. However, in the case of *PDR1*, the LTR is exceptionally short at 156 bp (54); so a GC-rich primer could be designed corresponding to the polypurine tract (PPT) which is found internal to the 3′ LTR in retrotransposons. The second implementation is for *BARE1* in barley, based on the published method (43, 44). For *BARE1* and other high-copy-number families, the number of selective bases may be increased compared to the first version of the protocol. Furthermore, *BARE1* and most other retrotransposons have long LTRs, necessitating an anchor primer in the LTR near to the external terminus.

Several features of the first protocol are specific to *PDR1*, but could be used with other retrotransposons of similar structure and copy number. The main feature of the procedure that should be modified for other situations is location of the sequence-specific primer (55). The choice of this primer is critical, and can be modified according to need. For example, internal primer sites have been exploited to describe structural variation within retrotransposons (56), and the primers can be applied to defined sequences other than the LTR or PPT.

1.6. Inter-retrotransposon Amplified Polymorphism

IRAP (Figs. 2b and 4) detects two retrotransposons or LTRs sufficiently close to one another in the genome to permit PCR amplification of the intervening region. Unlike AFLP® or SSAP, the method requires only intact genomic DNA as the template and PCR reagents and apparatus for amplification. There are no restriction enzyme digestion or adapter ligation steps. The amplification products are generally resolved by electrophoresis in wide-resolution agarose gels, but if labeled primers are used sequencing gel systems may be employed. The amplified fragments range from under 100 bp to over several kilobases, with the minimum size depending on the placement of the PCR priming sites with respect to the ends of the retrotransposon.

The IRAP method (40, 57) has found applications in gene mapping in barley (53) and its wild relatives (Fig. 4, (48)), in wheat and its relatives (58, 59), as well as in a wide range of other species (60–69). Even given a large genome and a highly prevalent retrotransposon family, one would not expect the IRAP method to produce very many resolvable PCR products. Taking the *BARE1* elements in barley as an example, the genome is approximately $\sim 4.7 \times 10^9$ bp in size (70), and the retrotransposon family is present in $\sim 1.5 \times 10^4$ full-length copies in addition to 1.7×10^5 solo LTRs (28). The full-length *BARE1* is 8,932 bp and the LTRs are 1,809 bp, comprising a total of 4.4×10^8 bp in the genome and leaving 4.3×10^9 bp of the genome *not BARE1*. Were insertions to be random within the genome, they would be expected to follow a Poisson distribution. If the total of 1.85×10^5 intact *BARE1* elements and solo LTRs were equidistantly dispersed within the remaining non-*BARE* part of the genome, they would be situated on average roughly 23 kb apart, with most insertions too far from another for PCR and beyond the resolution of conventional agarose gel electrophoresis.

The IRAP method, however, does produce a range of subkilobase fragments (Fig. 4), in part because barley (32, 71) and at least other large genomes (72) are organized into gene-rich islands surrounded by seas of repetitive DNA. The retrotransposons, which comprise large portions of the repeat seas, tend to be nested, one inserted into another (32, 68, 73). The IRAP amplification products can derive, therefore, variously both from nearby solo LTRs and full-length elements interspersed with nonretrotransposon DNA and from nested retrotransposons. The example given below is for the *BARE1* element in barley. However, the method is widely applicable, as illustrated by the citations above.

1.7. Retrotransposon-Microsatellite Amplified Polymorphism

REMAP (Figs. 2c and 5) is conceptually similar to IRAP, but differs in that it detects polymorphisms in the presence of retrotransposons or LTR derivatives sufficiently near SSRs, often referred to as microsatellites, to allow PCR amplification. Microsatellites are ubiquitous features of eukaryotic genomes, and have served directly to generate molecular markers in many plants (5, 74–76). For this reason, it was of interest to determine if retrotransposons were associated with microsatellites in the genome, and to what extent such associational polymorphism could serve as molecular markers. We found (32) that indeed, for *BARE1* in barley, retrotransposon insertions near microsatellites are considerably polymorphic. This was confirmed by others (77).

The REMAP method combines outward-facing LTR primers of the sort used in IRAP with SSR primers containing a set of repeats and one or more nonrepeat nucleotides at the 3′ end to serve as an anchor. The anchor is necessary to provide specificity to the PCR amplification; otherwise, the repetitive structure of the primer might cause it to anneal in multiple positions in any given microsatellite. Both IRAP and REMAP consist of PCR carried out on undigested template DNA and resolve the products on agarose gels. Following the initial publication of the technique by us (57) and almost simultaneously by others (78) under the guise "*copia-SSR*," REMAP found wide application in studies of genome evolution and gene mapping (48, 58, 60, 62, 64, 79–83). The implementation below is for *BARE1* and is useful in a range of cereals and grasses, but the method is generic and widely applicable.

1.8. Inter- Primer Binding Site Polymorphism

1.8.1. Overview

The factor limiting development of molecular marker systems based on LTR retrotransposons for new plant species, as discussed below under prospects (Subheading 5), is availability of retrotransposon sequences. The iPBS method (30) overcomes this limitation. It exploits the use of a small set of cellular tRNAs by almost all retroviruses and LTR retrotransposons for priming reverse transcription during their replication (84–87). A few LTR retrotransposons, such as the *Tf1/sushi* group of fungi and vertebrates and *Fourf* in maize, are exceptions to this rule and self-prime cDNA synthesis (85, 88).

For all other retrotransposons and retroviruses, the PBS domains match a limited set of tRNAs: $tRNA^{iMet}$, $tRNA^{Lys}$, $tRNA^{Pro}$, $tRNA^{Trp}$, $tRNA^{Asn}$, $tRNA^{Ser}$, $tRNA^{Arg}$, $tRNA^{Phe}$, $tRNA^{Leu}$, or $tRNA^{Gln}$. For the plant retrotransposons, $tRNA^{iMet}$ predominates as the primer. In retrotransposons, the PBS is complementary to 8–18 nt at the 3′ end of the tRNA serving as primer (84–87). With these features in mind, a set of universal PBS primers were designed (38). The PBS primers are then used in a manner akin to IRAP primers (Subheading 1.6), singly or in pairs. The iPBS reaction yields products spanning between retrotransposons in opposite orientation in close enough proximity to be amplified. The iPBS method favors strongly against the amplification of cellular tRNAs. In the genome, the greatest proportion of sequences matching tRNAs are retroelements. The rice genome, for example, which is relatively depauperate of retrotransposons, contains 737 tRNA genes, but 53,302 retrotransposons (46). Furthermore, the iPBS primers contain a discriminatory CCA trinucleotide at their 3′ termini, which is complementary to the 5′ TGG motif in PBS sites but which is not found in eukaryotic tRNA genes.

1.8.2. Applications of iPBS

Due to the nearly universal nature of iPBS primers, they can be used with unsequenced genomes, in the absence of additional information, to obtain LTR sequences of retrotransposons that are either prevalent or tend to cluster or nest in the genome. The LTR sequence can then be combined with the PBS sequence to clone entire elements. The method can be used not only with autonomous retrotransposons, but also with LARDs and TRIMs, which lack intact open reading frames but contain the PBS domain. The method has been applied across the Plant Kingdom for this purpose (89).

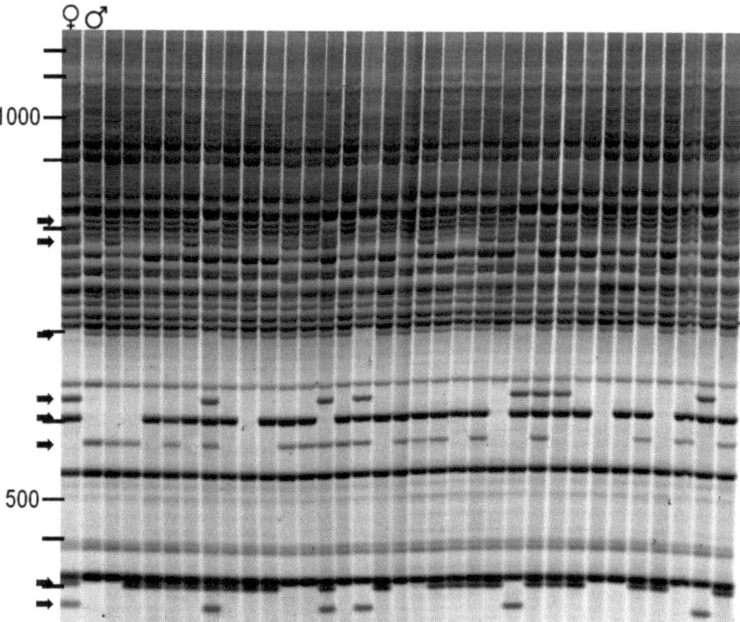

Fig. 6. iPBS analysis. iPBS amplification products, separated and visualized as in Fig. 1, are shown for two mapping parents of barley (*left*), cv Rolfi and CI-9819, respectively (47), and population of doubled-haploid (DH) lines derived from the cross. Bands polymorphic in the parents and segregating in the offspring are indicated by *arrows*. A 100-bp ladder is shown on the *left*.

The most common application of iPBS is for detecting polymorphism (Fig. 6), for which it has been applied to samples from angiosperms, gymnosperms, and lower plants to chickens, pigs, and cattle; in animals, it amplifies endogenous retroviruses (ERVs) (30). The method is effective both for large genomes and for small genomes, such as *Brachypodium distachyon* (30). Both single PBS primers and two different PBS primers in combination are effective and the products, from 100- to 5,000-bp long, can be scored on standard agarose gels stained with ethidium bromide. The method works well for mapping studies (Fig. 6).

The iPBS method can also be applied in silico to identify retrotransposons in sequence databases (38). To carry this out, PBS sequences are first identified, and then LTR segments identified by the presence within 0–4 nt from the 5′ end of the PBS of the universal 5′ … CA 3′ terminus of retrotransposon LTRs. One then clusters the LTR sequences and the adjacent PBS motifs and identifies complete left LTR sequences by the presence of paired terminal inverted repeats (TIRs) matching the 5′ TG … CA 3′ consensus for retrotransposons, the CA being adjacent to the PBS. In the next step, one identifies matching right LTRs with the same TIRs and assembles the intervening internal domain. Finally, one constructs consensus sequences for each pair of LTRs and intervening internal domain. Searches with this technique against the rice, grape, Arabidopsis, and *Drosophila melanogaster* genomes identified many previously unannotated retroelements (38).

1.8.3. Comparison of iPBS to Other Marker Methods

The iPBS amplification products behave as dominant markers, as do those of the IRAP, REMAP, SSAP, and other anonymous PCR-based molecular marker systems. A meta-analysis for three barley varieties was made for SSAP, IRAP, and REMAP data collected earlier (90) with that for iPBS (38). Between 5 and 25% of the iPBS bands were polymorphic; each iPBS primer visualized on average 23.3 bands. The IRAP technique for these three barley varieties demonstrated polymorphism between 10 and 60%, depending on the chosen primer. One major difference is that iPBS is not specific to a given retrotransposon family, whereas SSAP, REMAP, and IRAP, which prime from LTRs, can be. The iPBS method, thus, trades off the ability to tailor the level and evolutionary window of the detected polymorphism possible with these other methods for the simplicity of universal primers.

1.9. Retrotransposon-Based Insertional Polymorphism

1.9.1. Overview

RBIP (Figs. 2d and 7) is in essence the simple PCR-based detection of retrotransposon insertions using PCR between primers flanking the insertion site and primers from the insertion itself. A complementary reaction using primers from the surrounding DNA alone detects the unoccupied site (Fig. 7a). Because retrotransposon

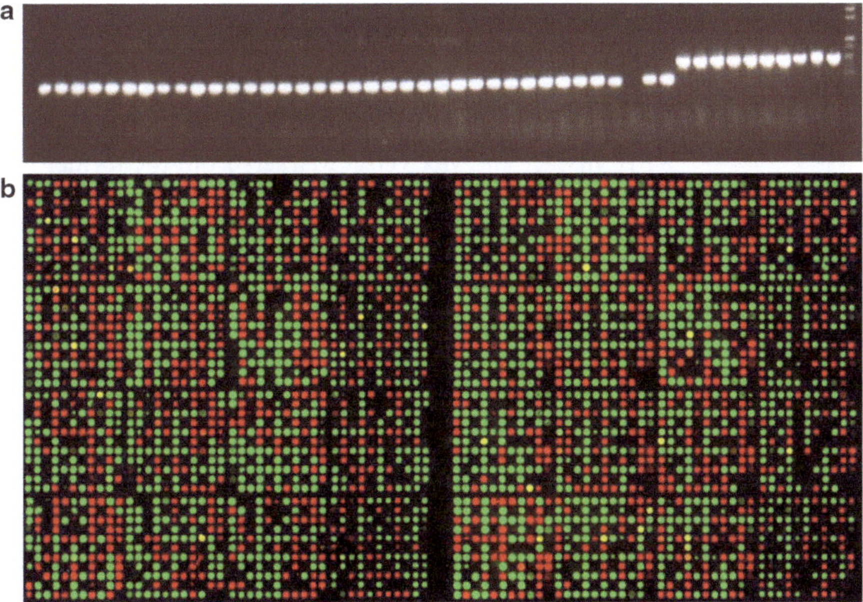

Fig. 7. RBIP analysis. (**a**) Agarose gel electrophoresis products of RBIP PCRs containing two host-specific primers and a retrotransposon-specific primer. Only one of the two possible products is produced per sample if the latter is homozygous for the locus, and the size indicates the product and hence the state of each locus. (**b**) TAM detection of 3,029 samples at a single pea RBIP locus.

insertions are generally thousands of bases in length, the "unoccupied-site PCR" produces no product from an occupied site. The particular feature of RBIP that distinguishes it from other retrotransposon-based marker methods described in this chapter is that it is a single-locus codominant technique.

RBIP is a robust technique. For low numbers of samples, the products are detected by normal agarose gel electrophoresis (Fig. 7b). Both reactions are carried out in the same tube and the size of the PCR product indicates which allele (occupied or unoccupied) has been amplified. The technical problems with this basic RBIP method are all associated with the acquisition of the sequence information for the flanking primers. This is closely analogous to the collection of new flanking sequences for microsatellite or SSR markers. Sequence data for new RBIP markers may be obtained from sequence analysis of genomic clones. Alternatively, SSAP markers (or other multilocus retrotransposon-based markers) can be converted into RBIP markers (see Subheading 1.9.2 below).

The basic RBIP method can be automated by adopting a dot-based assay (Fig. 7) to replace gel electrophoresis (90). Originally, the products were dotted onto nylon membrane and probed with a locus-specific probe (91), but this method has been superceded by a microarray-based fluorescence approach (90) which can handle far more samples simultaneously. Array-based scoring avoids a size-separation step and is scalable up to thousands of DNA samples by robotic spotting. In this case, production of the raw marker data (fluorescent hybridization signals) is independent of sample number; thus, data capture and processing can be automated using the technology developed for scoring microarrays (92).

1.9.2. Converting Other Retrotransposon Markers into RBIP Markers

In principle, a marker from any of the systems discussed above (SSAP, IRAP, REMAP) can be converted into a corresponding RBIP marker and vice versa. Markers from the former set of techniques are very easy to obtain and they can be rapidly prescreened for their potential informativeness before investing in the effort of developing a corresponding RBIP marker. An SSAP electrophoresis band represents one side of the insertion. It is relatively easy to cut out these bands from a gel, amplify the fragments by PCR, and sequence them to obtain the sequence of one side of the insert. This information is sufficient to allow the development of ISBP markers (see Subheading 1.10 below) but is insufficient to allow the detection of the unoccupied site, and this is a disadvantage because a strength of the RBIP technique lies in the very high accuracy of a double (or codominant)-assay method. A description of standard methods for obtaining the sequence corresponding to the other side of the insertion is given in Note 1.

1.9.3. RBIP Compared to the Other Retrotransposon-Based Marker Systems

Retrotransposon-based SSAP, REMAP, and IRAP are well suited to deal with hundreds of markers in tens to hundreds of samples. RBIP is more useful for fewer markers in thousands of samples because it can, in principle, be completely automated. The RBIP method is also very well suited to phylogeny and biodiversity assessments because it is a codominant marker system and retrotransposon insertions are quite stable, with a known ground state, namely, absence of the insertion (91). A strategy akin to RBIP was used successfully to determine the distant phylogenetic relationships between whales and ungulates (35).

1.10. Insertion Site -Based Polymorphism

1.10.1. Overview

Insertion site -based polymorphism (ISBP) is an RBIP-derived method that exploits knowledge of the sequence flanking a TE to design one primer in the TE and the other in one flanking sequence (93). By using only two primers for one side of an insertion site, it overcomes the limitation of RBIP markers, which require long genomic sequences to define both flanks. Therefore, ISBPs can easily be derived both from large BAC sequences as well as from short sequences, such as BAC-end or Roche 454 GS FLX whole-genome shotgun sequences.

Because TEs are often nested in large and highly repetitive genomes where they display unique insertion sites, ISBPs represent unique single-locus molecular markers that have the advantage of being genome-specific for polyploid species. In addition, because of the high methylation level in TEs that leads to an increase in mutation frequency at deaminated sites, ISBPs display a higher rate of SNPs than genic regions (41). As a consequence, ISBPs can be scored for various types of polymorphism, including the presence or absence of a TE, insertional size polymorphism, and SNPs. The basic detection technique is classical PCR amplification followed by agarose gel electrophoresis, as for IRAP, REMAP, and iPBS. Using electrophoresis, ISBPs can be scored for TE insertional and size polymorphism (93). However, the throughput and resolution are consequently limited. In addition, insertional polymorphisms can be difficult to distinguish from PCR failure.

To circumvent the limitations of gel electrophoresis, alternative techniques can also be used to score for polymorphism without prior knowledge of the presence of SNPs. These include melting curve analysis (MCA), high-resolution melt (HRM) analysis, and temperature-gradient capillary electrophoresis (TGCE) (41). Finally, by comparing sequences from different individuals, ISBPs can also be mined for sequence polymorphism and subsequently converted into ISBP-SNPs. These ISBP-SNPs can in turn be scored with a wide range of technologies, including the SNaPshot® Multiplex System, Illumina BeadArray technology, and KASPar (41).

1.10.2. Applications

ISBPs were first described and mainly used in wheat but can be implemented in virtually all large (i.e., TE-rich) genomes. In addition to wheat, the potential of ISBP markers has already been demonstrated in other species, including barley (Dave Laurie, personal communication) and rye (94). In wheat, ISBPs have been successfully used for genetic and physical mapping as well as for radiation hybrid mapping (95). They have been shown to meet the five main requirements for their utilization in marker-assisted selection (41). Finally, since TEs are key actors of genome evolution, ISBPs can also be used for phylogenetic and evolutionary studies (Paux et al., unpublished data).

2. Materials

2.1. SSAP for PDR1 in Pea

1. RL buffer: 10 mM Tris–acetate, pH 7.5, 10 mM Mg acetate, 50 mM K acetate, 5 mM dithiothreitol, 5 ng/μL bovine serum albumin.
2. Primers: For PDR1, the PPT primer is 5′ATTCACCAGCT TGAGGGGAG.
3. Stop solution: 0.25% w/v bromophenol blue and xylene cyanol in 98% formamide, 10 mM EDTA, pH 8.0.
4. Resolution of the SSAP products: Acrylamide gel solution of 4.5% for the casting of polyacrylamide gels, either homemade according to standard protocols or commercially prepared.

2.2. SSAP for BARE-1 in Barley

1. RL buffer: As in Subheading 2.1.
2. Preparation of adaptors: These should *not* be phosphorylated when synthesized or subsequently treated with kinase.

*Mse*I	25 μg	5′GACGATGAGTCCTGAG
	25 μg	3′TACTCAGGACTCAT

Make up to 100 μL with water, incubate at 65°C for 10 min, then place on ice, and add 1 μL 1 M MgOAc. Bring to 37°C for 10 min, and then 25°C for 10 min; place on ice (store at −20°C).

*Pst*I	25 μg	5′CTCGTAGACTGCGTACATGCA
	25 μg	3′CATCTGACGCATGT

Treat as for *Mse*I adaptors.

3. Preparation of primers:

BARE-1 primer	5′CTAGGGCATAATTCCAACAA
*Mse*I primer	5′GATGAGTCCTGAGTAA
*Pst*I primer	5′GACTGCGTACATGCAG

Selective primers are derived from the basic nonselective *Mse*I and *Pst*I primers, above referred to as M(0) and P(0), respectively.

The selective *Mse*I primers are M(C); M(AC); M(ACA). The selective *Pst*I primers are: P(C); P(CG); P(CGA).

4. T0.1E: 10 mM Tris–HCl, pH 7.5, 0.1 mM EDTA.

2.3. IRAP for *BARE1* in Cereals

A protocol for IRAP with *BARE1* is described below and shown for *H. spontaneum* (Fig. 7.4), but the procedure is largely same for other retrotransposons and plants.

1. Preparation of template DNA: DNA prepared by most standard methods or commercial kits is suitable. Inhibitors of PCRs, such as polyphenols (54) or other pigments, that may be present in the template preparation interfere with PCR in IRAP as well.
2. Primers: Primers are made in unphosphorylated, unlabeled form. For separation on sequencing systems, fluorescein or Cy5-labeled primers may be used, but the reaction conditions should be reoptimized as these dyes affect primer annealing to the template.

   Direct *BARE*-1 primer: 5′ CTA CAT CAA CCG CGT TTA TT.

   This corresponds to the LTR at nt 1,993–2,012 of accession Z17327, situated 105–124 nt from the right 3′ end of the LTR.

   Inverse *BARE*-1 primer: 5′ GCC TCT AGG GCA TAA TTC CAA C.

   This primer hybridizes to LTR templates 1 nt from the left edge of the LTR, nt 310–331 in accession Z17327. This primer is complementary to the coding strand, and therefore faces, as does the direct primer, outward from the element toward the flanking DNA.
3. PCR buffer: The 10× stock contains 750 mM Tris–HCl, pH 8.8, 200 mM $(NH_4)_2SO_4$, 15 mM $MgCl_2$, 0.1% Tween-20.
4. Thermostable DNA polymerases: We have tried a range of thermostable polymerases, including *Taq* polymerase from suppliers, including, but not limited to, Promega (M1861, storage buffer “A”), Epicentre (Masteramp™ Q82100), Solis BioDyne (Tartu, Estonia, *FIRE*Pol), Finnzymes/Thermo Scientific (Espoo, Finland, DyNAzyme™), PE Applied Biosystems (Amplitaq®), and B&M Labs (Madrid, Spain, Biotools DNA polymerase from *Thermus thermophilus*) and have not found differences in the results.
5. Thermocyclers: We have used either a Mastercycler Gradient (Eppendorf-Netheler-Hinz GmbH, Germany) or a PCT-225 DNA Engine Tetrad (MJ Research, Waltham, MA, USA) but have not extensively surveyed others. When using primers in cross-species experiments, it is best to consider possible differences in ramping time for various thermocycler and tube combinations and to optimize these.

6. Agarose: High resolution over a wide range of fragment sizes is important. We have used RESolute™ Wide Range Agarose (Product 337100, BIOzymTC bv, Landgraaf, The Netherlands). Alternatively, 3:1 Nusieve® agarose (50090, FMC Bioproducts, Rockland ME, USA) may give good results.

2.4. REMAP for BARE-1 in Cereals

Materials for REMAP are the same as described above, Subheading 2.3, for IRAP with the exception of the primers.

1. *BARE*-1 reverse primer: 5′ CAT TGC CTC TAG GGC ATA ATT CCA ACA.

   This is equivalent to LTR-B, described previously (39), and is complementary to nt 309–335 of the *BARE*-1a sequence (accession Z17327), extending to the left terminus of the LTR.

2. *BARE*-1 forward primer: 5′ CTA CAT CAA CCG CGT TTA TT.

   This matches nucleotides 1,993–2,012 of *BARE*-1a, extending to 105 bp from the 3′ terminus of the LTR.

   A range of SSR primers can be used in combination with either the forward or the reverse retrotransposon primer. These are given below together with the hybridization temperature to be used in PCR:

	Hybridization temperature for PCR, °C	
SSR	***BARE*-1 reverse**	***BARE*-1 forward**
$(GA)_9C$	56	56
$(GT)_9C$	56	56
$(CA)_{10}G$	57	57
$(CT)_9G$	56	56
$(AC)_9C$	56	56
$(AC)_9G$	56	56
$(AC)_9T$	56	56
$(AG)_9C$	56	56
$(TG)_9A$	56	56
$(TG)_9C$	56	56
$(AGC)_6C$	60	60
$(AGC)_6G$	60	60
$(AGC)_6T$	60	60
$(CAC)_7A$	60	60
$(CAC)_7G$	60	60

(continued)

(continued)

SSR	Hybridization temperature for PCR, °C	
	BARE-1 reverse	*BARE*-1 forward
$(CAC)_7T$	60	60
$(ACC)_6C$	60	60
$(ACC)_6G$	60	60
$(ACC)_6T$	60	60
$(CTC)_6A$	60	60
$(CTC)_6G$	60	60
$(GAG)_6C$	60	60
$(GCT)_6A$	60	60
$(GCT)_6C$	60	60
$(GTG)_7A$	60	60
$(GTG)_7C$	60	60
$(TCG)_6G$	60	60
$(TGC)_6A$	60	60
$(TGC)_6C$	60	60

2.5. iPBS

1. Template DNA: DNA prepared by most standard methods as for IRAP above is suitable.
2. Primers: A set of primers with superior performance are given below. Primers are made in unphosphorylated, unlabeled form. For separation on sequencing systems, fluorescein or Cy5-labeled primers may be used, but the reaction conditions should be reoptimized as these dyes affect primer annealing to the template.

18-mer iPBS primers:

	Sequence	T_m (°C)[a]	CG (%)	Optimal annealing T_a (°C)
2220	ACCTGGCTCATGATGCCA	59.0	55.6	57.0
2221	ACCTAGCTCACGATGCCA	58.0	55.6	56.9
2222	ACTTGGATGCCGATACCA	55.7	50.0	53.0
2224	ATCCTGGCAATGGAACCA	56.6	50.0	55.4
2225	AGCATAGCTTTGATACCA	50.5	38.9	55.0
2228	CATTGGCTCTTGATACCA	51.9	44.4	54.0
2229	CGACCTGTTCTGATACCA	53.5	50.0	52.5
2230	TCTAGGCGTCTGATACCA	54.0	50.0	52.9

(continued)

(continued)

	Sequence	T_m (°C)[a]	CG (%)	Optimal annealing T_a (°C)
2231	ACTTGGATGCTGATACCA	52.9	44.4	52.0
2232	AGAGAGGCTCGGATACCA	56.6	55.6	55.4
2237	CCCCTACCTGGCGTGCCA	65.0	72.2	55.0
2238	ACCTAGCTCATGATGCCA	55.5	50.0	56.0
2239	ACCTAGGCTCGGATGCCA	60.4	61.1	55.0
2240	AACCTGGCTCAGATGCCA	58.9	55.6	55.0
2241	ACCTAGCTCATCATGCCA	55.5	50.0	55.0
2242	GCCCCATGGTGGGCGCCA	69.2	77.8	57.0
2243	AGTCAGGCTCTGTTACCA	54.9	50.0	53.8
2244	GGAAGGCTCTGATTACCA	53.7	50.0	49.0
2245	GAGGTGGCTCTTATACCA	53.1	50.0	50.0
2249	AACCGACCTCTGATACCA	54.7	50.0	51.0
2251	GAACAGGCGATGATACCA	54.3	50.0	53.2
2252	TCATGGCTCATGATACCA	52.7	44.4	51.6
2253	TCGAGGCTCTAGATACCA	53.4	50.0	51.0
2255	GCGTGTGCTCTCATACCA	57.1	55.6	50.0
2256	GACCTAGCTCTAATACCA	49.6	44.4	51.0
2257	CTCTCAATGAAAGCACCA	52.4	44.4	50.0
2295	AGAACGGCTCTGATACCA	55.0	50.0	60.0
2298	AGAAGAGCTCTGATACCA	51.6	44.4	60.0
2373	GAACTTGCTCCGATGCCA	57.9	55.6	51.0
2395	TCCCCAGCGGAGTCGCCA	66.0	72.2	52.8
2398	GAACCCTTGCCGATACCA	57.1	55.6	51.0
2399	AAACTGGCAACGGCGCCA	63.4	61.1	52.0
2400	CCCCTCCTTCTAGCGCCA	61.6	66.7	51.0
2401	AGTTAAGCTTTGATACCA	47.8	33.3	53.0
2415	CATCGTAGGTGGGCGCCA	62.5	66.7	61.0

3. PCR buffer and thermostable polymerases: A range of thermostable polymerases and corresponding buffers will work, for example DyNAzyme™ II (Finnzymes, Thermo Scientific) or DreamTaq™ polymerase (Fermentas, Thermo Scientific) with their proprietary buffers. Alternatively, one may use a 1× stock containing: 20 mM Tris–HCl (pH 8.8), 2 mM $MgSO_4$, 10 mM KCl, and 10 mM $(NH_4)_2SO_4$. Generally, we use a combination

of 1 unit DyNAzyme™ II and 0.04 units Pfu DNA Polymerase (both from Fermentas, Thermo Scientific), though other theromostable enzymes are suitable.

4. Thermocyclers: We have used either a PTC-100 Programmable Thermal Controller (MJ research Inc., Bio-Rad Laboratories, USA) or a Mastercycler Gradient (Eppendorf AG) with 0.2-ml tubes or 96-well plates, but have not extensively surveyed others.
5. Agarose: 1.5% (w/v) agarose gels (RESolute Wide Range, BIOzym) cast in 1× TBE electrophoresis buffer (50 mM Tris–H_3BO_3, 0.5 mM EDTA, pH 8.6).

2.6. RBIP

1. DNA: High DNA quality is not important for the success of RBIP. Miniprep plant DNA, with large amounts of contaminating RNA and polysaccharides, does not affect the success rate of the technique.
2. Reagents: PCR reagents (see Note 11): Standard proprietary PCR reagents are used. As in all PCR, success is more likely with hot-start Taq enzyme.
3. Primers: Three primers are required for a standard RBIP PCR, namely, a common primer and two allele-specific primers (Figs. 2 and 7). For gel-based detection of RBIP products, these are standard oligonucleotides roughly 18–22 bp in length. For TAM-based detection, the common primer is 5′-biotinylated and the allele-specific primers each carry a 20-bp tag linked to the 5′ end of the primer by a hydrocarbon chain spacer (27). The three tags below work well with each other.

Oligonucleotide type	Sequence (5′–3′)
Common primers	5′ biotin-labeled, locus-specific primers
Allele-specific Tag PCR primer ("a" Tag)	TCTTTGAGTTTGACCATGCA[L]N_x
Allele-specific Tag PCR primer ("c" Tag)	GCCATACAATAGTCACGTTG [L]N_x
Allele-specific Tag PCR primer ("e" Tag)	ACCGCATCCGAACATTTGTC[L]N_x
A–B Tag detector oligonucleotide	[Cy]TCTTTGAGTTTGACCATGCAACG TGAGCGACAATCAGGACGGCT ACGTGCAATACTTAGT
A′–B′ Tag detector oligonucleotide	[Cy]TCGCTCACGTTGCATGGTCA AACTCAAAGAACTAAGTATTG CACGTAGCCGTCCTGATTG
C–D Tag detector oligonucleotide	[Cy]GCCATACAATAGTCACG TTGGAGTTGGACACCTACTGAA TACACTTATACCGCTTACGAG

(continued)

(continued)

C′–D′ Tag detector oligonucleotide	[Cy]TGTCCAACTCCAACGTG ACTATTGTATGGCCTCGTAAG CGGTATAAGTGTATTCAGTAGG
E–F Tag detector oligonucleotide	[Cy]ACCGCATCCGAACATTTGT CAGTTGAGCATTCTGCCTAAG CCCACTATTCCATCAAGTCT
E′–F′ Tag detector oligonucleotide	[Cy]ATGCTCAACTGACAAATGTT CGGATGCGGTAGACTTGATGG AATAGTGGGCTTAGGCAG

(L) = C-18 hydrocarbon spacer; N_x = allele-specific primer (see Fig. 2); Cy = Cy fluorophore, typically Cy3 or Cy5. Both Cy3 and Cy5 versions of all six tag detector oligonucleotides can be used, allowing the detection of any tag with either fluorophore. An “a” tag is detected during microarray hybridization with a combination of A–B and A′–B′ Tag detector oligonucleotides, a “c” tag is detected by a combination of C–D and C′–D′ Tag detector oligonucleotides, and an “e” tag by a combination of E–F and E′–F′ Tag detector oligonucleotides.

2.7. ISPB

1. DNA: Any standard DNA extraction procedure is sufficient to perform ISBP or ISBP-SNP genotyping. High-quality DNA is not required. The DNA quantity is highly dependent on the genotyping method and is similar to that for other types of markers. For PCR amplification-based techniques, 25 ng is sufficient.
2. Reagents: Commercial reagents normally in use for other PCR methods, as well as proprietary scoring equipment, can be used for ISBP amplification or genotyping.

3. Methods

3.1. SSAP for PDR1 in Pea

Although the method described below employs a radioactive label, ^{33}P, a nonradioactive protocol using fluorescent labeling has been developed (108).

1. DNA digestion: Digest c. 0.5 μg genomic DNA in RL buffer with 5 U restriction endonuclease *Taq*I in a total volume of 40 μL. Incubate at 65°C for 2–3 h (see Note 2).
2. Adapter ligation: To the 40 μL digest from step 1, add 12.5 pmol *Taq* adapter (from 50 pmol/μL stock). Make up to 1 mM ATP, add 1 U T4 DNA ligase, and adjust the total volume to 50 μL in 1× RL. Incubate at 37°C overnight.

3. Template preparation and storage: Dilute the ligated SSAP template DNA from step 2 by addition of 100 μL of TE, pH 8, and store at –20°C. (Use 3 μL of this diluted template for a 10 μL PCR volume).

4. Labeling reaction: Kinase label the sequence-specific primer in bulk and later dispense the labeled primer among the reactions. The quantity depends on the number of reactions required; the example shown is designed for 30 reactions. The label used here, ^{33}P, is safer and more convenient than ^{32}P, but ensure that appropriate shielding, transport, and disposal procedures are followed.

Primer (100 ng/μL)	4.5 μL
[γ-^{33}P]ATP	2.0 μL (370 kBq/μL)
10× T4 polynucleotide kinase buffer	2.0 μL
Water (see Note 3)	11.0 μL sterile distilled
5 U T4 polynucleotide kinase (10 U/μL)	0.5 μL
Total volume	20.0 μL

Incubate at 37°C for at least 1 h.

Assemble the reaction components, except for the [γ-^{33}P] ATP, together in a clearly marked screw-capped 1.5-ml Eppendorf tube; dispense the [γ-^{33}P]ATP in a laboratory appropriately equipped for work with radioactivity according to local safety guidelines. Incubate the labeling reaction at 37°C in a heating block designated for radioactive work.

5. Labeled PCR: Assemble as follows for 30 reactions of 10 μL. Each reaction uses 3 μL of template, so 7 μL of the reaction mix must be added to each. Therefore, in this example, 210 μL reaction mix must be prepared for aliquoting.

Labeled primer	20 μL (from 4)	These should be equimolar
Adapter primer (7.5 ng/μL)	60 μL	
10× PCR buffer	30 μL	
1 mM dNTP	60 μL (200 μM each final concentration)	
Taq DNA polymerase	6 U	
Water (sterile) to make	210 μL	

Dispense 7 μL to each 3 μL template sample and set up the PCR according to Vos and coworkers (5): ten cycles of 94°C for 30 s, 55°C (reducing by 1°C per cycle) for 30 s, and 72°C for 60 s, followed by 20 cycles of 94°C for 30 s, 45°C for 30 s,

and 72°C for 60 s. The reaction is completed with a final extension step for 72°C for 7 min.

Check the PCR machine with the Geiger counter before and after use.

6. Stopping the reaction: Add 10 μL of stop solution to each 10 μL PCR; denature by heating to 95°C for 3 min, and cool on ice. Store the reactions at –20°C until ready to load onto a gel. Use care; formamide is a mutagen.
7. Setting up of the polyacrylamide gels: Prepare the sequencing gel apparatus and cast the gel according to standard procedures suited for your specific apparatus (see Note 4).
8. Running and processing the gel: Mount the gel/glass plate assembly on the electrophoresis unit; add TBE buffer to top and bottom trays; clean out the wells with buffer using a syringe and needle; connect up to a power pack and pre-run the gel for ca. 30 min at 1,500–1,600 V to warm up. Disconnect the electrophoresis unit, flush out the wells with buffer, and load the denatured samples into the wells (1 μL sample is generally enough). Continue running the gel for the desired time at 1,500–1,600 V (c. 2 h). Discard the buffer down a drain designated for disposal of low-grade radioactive liquid waste. When the plates have cooled down, remove one of the side spacers. Pry the plates apart using a thin spatula placed in the gap between the plates at a corner. This is a hazardous procedure as glass fragments may break off or plates may crack and shatter. The gel should remain attached to the nonsilanized plate and can be transferred onto 3-MM paper with an extra sheet for backing; trim the excess paper surround close to the gel. Place a piece of cling film over the gel to protect the gel drier cover from contamination. Dry for 1–2 h at 80°C in the vacuum gel drier. Expose the dried gels to an X-ray film or phosphoimager plates. An example SSAP gel for *Pisum* is shown in Fig. 3.

3.2. SSAP for *BARE1* in Barley

1. DNA digestion: Total genomic DNA from the plant of interest is completely digested using two restriction enzymes, one a rare cutter and the other a frequent cutter. The rationale for this is explained by Vos and coworkers (5), and is summarized below.

   The frequent cutter generates small DNA fragments, which amplify well by PCR and are in the correct size range for separation on a denaturing or sequencing gel. The number of fragments amplified can be reduced by using a combination of rare- and frequent-cutting restriction enzymes, allowing amplification of fragments with a rare cutter site at one end and a frequent cutter site at the other, to the exclusion of the other fragments. Presumably, it also decreases the chance of a fragment ligating to itself. In this example, we used *Mse*I and *Pst*I, as these had been previously used in barley (38), although any combination of rare and frequent-cutting enzymes could be tried (see Note 2).

*Mse*I cuts	T TAA AAT T
*Pst*I cuts	CTGCA G G ACGTC

Prepare a digest as follows:

Total genomic DNA	1.0 μg
*Mse*I	5 U
*Pst*I	5 U
10× RL buffer	2 μL
H_2O	To 20 μL total volume

Digest at 37°C for at least 1 h.

2. *Mse*I/*Pst*I adaptor ligation: Take digested DNA (1 μg in 20 μL) and add the following:

1.0 μL	*Mse*I adaptors (40 pmol)
0.5 μL	*Pst*I adaptor (20 pmol)
1.0 μL	10 m*M* γATP
0.4 μL	RL buffer
0.5 μL	T4 ligase

Incubate at 37°C for 3 h, and then store template DNA (at a final concentration of 40 ng/μL) at –20°C.

3. Preamplification PCR: This is useful when working with large genome sizes to reduce the restriction fragments to a manageable number (see Note 5). The PCR conditions are the preferred ones for our Techne Genius PCR machine, and should be adjusted as appropriate to others. Prepare the reaction in a total volume of 25 μL:

H_2O	to 25 μL
10× PCR buffer	2.5 μL
dNTPs	0.2 mM final concentration
*Mse*I primer	75 ng
*Pst*I primer	75 ng
Template	0.75 μL (~30 ng)
Taq DNA polymerase	1 U (0.2 μL)

We use the following PCR program: 1 min 95°C warm up; 30 cycles of 1 min 94°C denaturing, 1 min 60°C annealing step, 1 min 72°C extension step; 7 min 72°C final extension. After the reaction is complete, add 55 μL T0.1E and store at −20°C.

4. End labeling of the *BARE*-1 oligo: This oligo complements the start of the *BARE*-1 5′ LTR. The final A on this primer is a selective base, designed to anneal to and amplify only the fraction of fragments in which the first nucleotide of the flanking sequence is an A. Also, this A is one of the two nucleotides which causes mismatches to the 3′ LTR, thus reducing the chance of priming into the retrotransposon from this LTR. A total of 1 μL of labeled oligo is made per PCR. We have mainly used [γ-^{32}P]ATP, but ^{33}P label may be used (see Note 3).

   Prepare end-labeling reactions:

[γ-^{32}P]ATP	1 μCi (3,000 Ci/mMol)
BARE1 oligo (50 ng/μL)	0.13 μL
10× kinase buffer	0.1 μL
T4 polynucleotide kinase	0.25 U (0.025 μL)
H_2O	to 1 μL total volume

   Incubate at 37°C for at least 30 min. Denature kinase at 70°C for 10 min, and then place on ice immediately. Spin at 16,000×*g* for 15 s on desktop microcentrifuge. Store at −20°C.

5. Labeled SSAP PCR: Generally carried out with selective primers (see Note 6).

   Add the following per PCR:

[γ-^{32}P]ATP-labeled *BARE*-1 oligo	1 μL
Unlabeled *BARE*-1 oligo (50 ng/μL)	0.5 μL
Selective *Mse*I or *Pst*I primer (50 ng/μL)	0.6 μL[a]
10× PCR buffer	2 μL
dNTPs	0.2 mM final concentration
Preamplified DNA (from step 3)	2 μL
H_2O	To a total volume of 20 μL
Taq DNA polymerase	0.5 U (0.1 μL)

[a]Selective primers, as described in Subheading 2.2

6. The PCR program is as follows: 36 cycles in total: 94°C, 1 min; 13 cycles of 65°C for 1 min, imposing a −0.7°C decrease per cycle ("touchdown PCR"), 72°C for 1 min, and 94°C for

1 min; 22 cycles of 56°C for 1 min, 72°C for 1 min, and 94°C for 1 min; a final extension of 72°C for 7 min.

7. Running samples on a denaturing gel: Gels are set up as in step 7 in Subheading 3.1 above. See Note 4 for further considerations. Add 20 μL of sequencing stop buffer to each sample, and mix well. Denature by incubation at 90°C for 5 min, and then place on ice immediately. Load each sample onto a 6% denaturing polyacrylamide gel. Load an amount appropriate to the size of combs you are using. We use shark's-tooth combs, but larger well-forming combs can be used. Samples usually take 1 h 45 min to 2 h to run. It is also useful to run a marker alongside. Fix gel if necessary. Gels are exposed with X-ray film for 1 to 5 days. Do not use an intensifying screen for ^{32}P gels. If the procedure is working reasonably well, you should get a visible result in a day or two. An alternative is to use a phosphoimager and imaging plates rather than X-ray film.

3.3. IRAP for BARE1 in Cereals

The technique is presented as developed for barley.

1. Set up the PCR: The reaction here is designed for 20-μL tubes, but can be scaled down for use in microtiter plates (see Note 7). Each reaction contains:

10× PCR buffer	2 μL
Template DNA (10 ng/μL)	20 ng
PCR primers (one, the other, or both)	200 nM each final concentration
dNTPs	200 μM final concentration
1 U *Taq* DNA polymerase	
H_2O	To 20 μL

2. Carry out the PCR: The reactions are carried out with a amplification profile consisting of 94°C 3 min; 30 cycles of 95°C for 15 s, 60°C for 30 s, and 72°C for 2 min; a final extension at 72°C for 10 min (see Note 8).

3. Electrophoretic resolution of the PCR products: Take one-fifth of the PCR, mix with loading buffer, and analyze on a wide-resolution agarose gel. We have used 2% RESolute™ agarose, but 1.2–1.5% Seakem 3:1 NuSieve® agarose is expected to work as well. Carry out the electrophoresis in a 20-cm-long gel for 7 h at 100 V in a Pharmacia GNA-200, 20×20-cm format, in standard Tris borate (50 mM Tris–H_3BO_3, 0.5 mM EDTA, pH 8.6) buffer, and visualize by staining with ethidium bromide (see Note 9).

3.4. REMAP for BARE1 in Cereals

The example given is for barley.

1. Set up the PCR: The reaction here is designed for 20-μL tubes, but can be scaled down for use in microtiter plates.

Each reaction contains:

10× PCR buffer (as for IRAP)	2 μL
Template DNA (10 ng/μL)	20 ng
PCR primers (one, the other, or both)	200 nM each final concentration
dNTPs	200 μM final concentration
1 U *Taq* DNA polymerase	
H_2O	To 20 μL

2. Carry out the PCR: The reactions are carried out with a program consisting of 94°C for 3 min; 28–32 cycles of 95°C for 15 s, 56–60°C (according to primer pair, see Subheading 2) for 30 s, and 72°C for 2 min; a final extension at 72°C for 10 min.
3. Electrophoretic resolution of the PCR products: As for IRAP, Subheading 3.3, step 3, above. An example REMAP gel is shown in Fig. 5 (see Note 10).

3.5. iPBS

1. Set up the PCR: The reaction here is for 25-μL tubes, but can be scaled down for use in microtiter plates.

   Each reaction contains:

10× DreamTaq™ PCR buffer	2.5 μL
Template DNA (10 ng/μL)	25–50 ng
iPBS primer	1 μM final concentration
dNTPs	200 μM final concentration
1 U DreamTaq™ polymerase	
0.04 units Pfu DNA polymerase	
H_2O	To 25 μL

2. Carry out the PCR: The reactions are carried out with a amplification profile consisting of 1 cycle at 95°C for 3 min; 28–30 cycles of 95°C for 15 s, 50–60°C (see above) for 60 s, and 72°C for 60 s; a final extension step of 72°C for 10 min.
3. Electrophoretic resolution of the PCR products: Take one-fifth of the PCR, mix with loading buffer, and analyze on a wide-resolution agarose gel. We have used 1.7% (w/v) agarose gels (RESolute Wide Range, BIOzym) with 1× STBE electrophoresis buffer (50 mM Tris–H_3BO_3, 0.5 mM EDTA, pH 8.6) at 80 V for 7 h and visualized by staining with ethidium bromide. Gels were scanned on a FLA-5100 imaging system (Fuji Photo Film (Europe) GmbH.) scanner at a resolution of either 50 or 100 μM. An example is shown in Fig. 6.

3.6. RBIP for PDR1 in Pisum sativum

1. Set up the PCR: The amount of template DNA here is based on use with pea (*Pisum sativum*). Each reaction contains:

10× PCR buffer (Promega)	2 μL
Template DNA	10 ng
PCR primers	40 ng each
dNTPs	200 μM final concentration
1 U *Taq* DNA polymerase	
H_2O	To 20 μL

2. Carry out the PCR: This program was constructed for a Techne Genius machine but can be adapted to others. It consists of 95°C for 1 min; 35 cycles of 94°C for 1 min, 55°C for 1 min, and 72°C for 1 min; a final extension at 72°C for 7 min; maintenance at 4°C.
3. Analyze the RBIP products: For gel-based analysis, the products are electrophoretically separated on 1.5% agarose gels containing ethidium bromide in TBE buffer (Fig. 7b). Nylon filter-based detection of RBIP products has been replaced by fluorescence-based detection of RBIP PCRs by the Tagged Microrray Marker (TAM) method (96), which is fully described in (92). The PCRs contain one biotin-labeled common primer, which can amplify both the occupied and unoccupied alleles, together with two tagged allele-specific primers ("occupied" and "unoccupied" allele primers, respectively), only one of which can produce a PCR product in a homozygous sample (see Subheading 2.5 above). The tags are 20-bp sequence extensions, each of which is recognized by its corresponding fluorescently labeled tag detector oligonucleotide set.

The RBIP PCR products are spotted onto a streptavidin-coated microarray slide using a robotic microarrayer (53). The slide is then prehybridized in 4× SSC+Denhardt's solution+0.01% SDS for 30 min at 30°C, before hybridization under coverslip to 20 μl of fluorescently labeled detector oligonucleotide mix (100 ng/μl per oligonucleotide) at 37°C for 30 min. The slide is then washed in 0.5× SSC, 0.01% SDS at 30°C for 10 min, followed by two washes in 0.2× SSC 5 min. The slide is then dried at room temperature and scanned by any fluorescence microarray scanner (see Note 12).

3.7. High-Resolution Melting Analysis of ISBP in Wheat

Several techniques have already been described to genotype ISBP markers. The following example is a new HRM approach that has been developed for SNP genotyping on a Roche LightCycler® 480 in wheat.

1. Set up the PCR:

Template DNA	25 ng
AmpliTaq Gold® 360 Master mix (Applied Biosystems)	5 μL
360 GC enhancer (Applied Biosystems)	0.4 μL
Syto® 9 (Molecular Probes)	2 pmol
PCR primers	6 pmol

Add H_2O for a final volume of 10 μL

2. PCR parameters:

Program name	**Preincubation**						
Cycles	**1**	**Analysis mode**	**None**				
Target (°C)	**Acquisition mode**	**Hold (hh:mm:ss)**	**Ramp rate (°C/s)**	**Acquisition (per °C)**	**Sec target (°C)**	**Step size (°C)**	**Step delay (cycles)**
95	None	00:10:00	4.8	–	0	0	0

Program name	**Amplification**						
Cycles	**55**	**Analysis mode**	**Quantification**				
Target (°C)	**Acquisition mode**	**Hold (hh:mm:ss)**	**Ramp rate (°C/s)**	**Acquisition (per °C)**	**Sec target (°C)**	**Step size (°C)**	**Step delay (cycles)**
95	None	00:00:10	4.8	–	0	0	0
62	None	00:00:30	2.5	–	55	1	1
72	Single	00:00:30	4.8	–	0	0	0

Program name	**High-resolution melting**						
Cycles	**1**	**Analysis mode**	**Melting curves**				
Target (°C)	**Acquisition mode**	**Hold (hh:mm:ss)**	**Ramp rate (°C/s)**	**Acquisition (per °C)**	**Sec target (°C)**	**Step size (°C)**	**Step delay (cycles)**
95	None	00:01:00	4.8	–	0	0	0
40	None	00:01:00	2.5	–	0	0	0
65	None	00:00:01	4.8	–	0	0	0
95	Continuous	–	0.02	25	0	0	0

Program name	Cooling						
Cycles	1	Analysis mode	None				
Target (°C)	Acquisition mode	Hold (hh:mm:ss)	Ramp rate (°C/s)	Acquisition (per °C)	Sec target (°C)	Step size (°C)	Step delay (cycles)
40	None	00:00:30	2.5	–	0	0	0

3. Data analysis:
   Data are analyzed using the LightCycler® 480 Software v1.5. Genotypes are called automatically and classified into different groups (homozygous AA, BB, or heterozygous AB) according to their normalized and temperature-shifted melting curves.

4. Notes

1. Rapid ways exist for obtaining the other side of any given retrotransposon insertion. The first of these relies upon the fact that retrotransposons generate a duplication of the host insertion site sequence when they insert. For Ty1-*copia* group retrotransposons, this is a random 5-bp sequence which can be obtained from sequencing the SSAP, IRAP, or REMAP band. This same 5-bp sequence is present on each side of the insertion and these can be used as selective bases at the 3′ end of a primer which is specific for the other (unsequenced) end of the insertion. The SSAP, IRAP, or REMAP amplification with this primer on accessions containing the particular insertion and accessions lacking it (this information is available from the marker data) usually yields a very small number of candidate bands corresponding to the other side of the insertion. The correct band can be chosen by its cosegregation with the original marker in a set of samples that are polymorphic for the band. This band can then be sequenced to give the other side of the insertion and that is all that is needed for the RBIP marker.

   Alternatively, a genome-walking approach can be employed (e.g., GenomeWalker™ kit, BD Biosciences Clonetech, Palo Alto, USA) (38, 55). This is similar to SSAP in principle, but uses a specific primer derived from the host DNA flanking the insertion rather than from the retrotransposon itself, oriented for synthesis toward the insertion site. Sequence analysis of the fragments obtained from accessions lacking the insertion reveals the sequence at the other side of the insertion.

4.1. SSAP for PDR1 in Pea

2. Step 1, DNA digestion: On occasion, the digestion step does not run to completion, presumably as a consequence of some contaminant in the DNA prep. This results in a track with extra bands on the final gel so that the sample appears exceptional in element number and also distantly related to the other samples (because many bands are not shared). The presence of incomplete digestion can be checked by digesting some of the final sample to be run on the gel: bands will disappear revealing the presence of amplification products with internal *Taq*I sites. Alternatively, a specific enzyme digestion buffer can be used and changed for the ligation step. However, this is a little tedious and does not often appear to be necessary. Enzymes other than *Taq*I, or two enzymes, could be used in this step.

   This type of behavior can be exploited in studies of DNA methylation. For example, *Sau* 3A does not cut C-methylated sites, but *Mbo*I does (97), so the comparison of *Sau* 3A and *Mbo*I SSAPs is informative. Some enzymes are blocked by C-methylation; this may not occur at a symmetric sequence, and there may be no convenient isoschizomer control (e.g., *Hin*dIII). In such cases, the comparison of the SSAP products with *Hin*dIII-digested SSAPs can be a useful alternative.

3. Step 4: ^{33}P poses a hazard mainly as a consequence of contamination. The β particle emission is low energy compared to ^{32}P. Follow safety guidelines appropriate for handling of radioactive materials.

4. Step 7: Gradient gels (98) or high-salt bottom buffers can be used to compress the banding toward the bottom of the gel, maximizing the information content yield from each run.

4.2. SSAP for BARE-1 in Barley

5. Step 3: The primers in this step carry no selective bases. The adapter/primer configuration is as described in Subheading 2.

6. Step 5: The selective primers used here gave us the most polymorphism with the *BARE*-1 primer and a manageable number of strong bands with the least background on the film. The number of selective bases has to be optimized for each retrotransposon family in a given species. It should be remembered that for any given combination of restriction enzymes (in this example, *Pst*I/*Mse*I) and selective primers only a subset of the retrotransposon family is amplified. Although this is an inevitable consequence of the limits of PCR amplification and gel-based fragment resolution, additional combinations of digests, adapters, and primers allow analysis of other subsets of the potential integration sites.

4.3. IRAP for BARE-1 in Cereals

7. Step 1: If the primer is not fully complementary to the template retrotransposon (as would be the case in unconserved regions of a retrotransposon or in divergent families of

elements), the PCR buffer, in particular salt and pH, but not the polymerase, may influence the results.

8. Step 2: The number of reaction cycles, template quantity, primer concentration, and enzyme quantity may need to be optimized for specific retrotransposon families and plant species. We use up to 1.2 U enzyme and up to 35 cycles in some cases. The annealing temperature has to be adjusted to match the primers used.
9. Step 3: The IRAP reaction generates a complex mixture of fragments of wide size range. Slow electrophoresis as described improves the fragment resolution, as does longer separation distances and high-quality agarose. We routinely use a 20×20-cm Pharmacia gel box (GNA-200) and combs having 1-mm thickness. An example IRAP gel is shown in Fig. 4.

4.4. REMAP for BARE-1 in Cereals

Generally, the same comments for IRAP apply here.

10. Step 2: If there is high background in the lanes, the amount of template can be reduced to 10 ng.

4.5. RBIP

11. Several different proprietary PCR buffers (PE, Promega, Qiagen) have been tried and all have worked. Primers should follow the normal rules for good primer design. In particular, they should be carefully screened against the possibility of primer dimer artifacts and we have always been careful to keep the T_m of all primers used in a single reaction within 2°C of each other. Typically, we use primers of around 20 bases with 40–50% G/C content. Fluorescently labeled tag detector oligonucleotides are purified by 10% SDS-PAGE.
12. Failed PCRs generate low or nonexistent signals for both fluorophores. Typical failure rates are between 3 and 5% in our experience. In addition, some reactions yield signals for both fluorophores, even if only one allele is known to be present. These correspond to background artifactual amplification at a different genomic site (either occupied or unoccupied backgrounds are possible). This tends to be a property of particular RBIP markers and the best way is to discard such markers from the analysis (28).

5. Prospects

Retrotransposons are highly useful as molecular markers, in the analysis of genome structure, and as tools for the reverse-genetic characterization of gene function (98–101, 109)—as both makers and markers of genetic diversity. The protocols presented here have been built around specific retrotransposon families and particular plants.

However, retrotransposons throughout the eukaryotes share common structures and life cycles, permitting adaptation to a wide range of research materials. Key considerations for adaptation of the method to the plant of interest are the LTR length, copy number of the retrotransposon family for which the PCR primers are designed, and the genome structure of the plant. Long LTRs necessitate primers near the termini, whereas LTRs of only several hundred base pairs allow more flexibility in this regard. Retrotransposons in high copy number may produce too many bands for efficient amplification or gel resolution in all methods, except RBIP and ISBP. This problem can be overcome by increasing the number of selective bases in SSAP or by designing the retrotransposon primer in IRAP or REMAP to bridge the joint between the LTR and the flanking region and to carry selective bases at its 3′ end. Genome organization, regarding the nesting of retrotransposons' insertion sites and the proximity of microsatellites to retrotransposons, affects the relative efficacy of IRAP, REMAP, and SSAP.

A valuable aspect of retrotransposon marker systems is that the phylogenetic resolution is dependent on the activity of any particular retrotransposon family. The more active the family, the better the resolution in closely related germplasm. The many examples of explosions in retrotransposon copy number in particular clades of plants (13, 19, 26–29, 110) show that certain retrotransposon families can be phylogenetically diagnostic as well. To take advantage of this feature, if sequence data from a reference genome are not available, one must employ a general method for the isolation of new retrotransposon families.

The internal domains of retrotransposons contain conserved motifs necessary for carrying out the replicative life cycle, which form the basis of methods to isolate retrotransposons. In particular, the RNase domain for superfamily *Copia* and the integrase domain for *Gypsy* are sufficiently close to the 3′ LTR to permit an SSAP or genome-walking method to be used, employing a PCR primer anchored in either of these regions to isolate the 5′ termini of LTRs of almost any retrotransposon from most eukaryotes (55). The recent iPBS method (38) is particularly suited to the isolation of new retrotransposon families, particularly TRIMs and LARDs, because it is independent of the presence of an intact open reading frame. In this way, novel elements can be applied to IRAP, REMAP, and SSAP and then in turn the integration sites developed for RBIP and ISBP.

High-throughput sequencing data, if in hand, can be mined for retrotransposons to be developed into primers for IRAP, REMAP, or SSAP or for locus-specific primers for ISBP and RBIP. However, shotgun sequencing for the sake of finding retrotransposons per se is less efficient than the targeted methods described above. For example, a single read on the GS FLX Titanium platform of Roche 454 Life Sciences (102) will yield approximately 4×10^8 nt, of which 3.2×10^8 (80%) is likely from retrotransposons. The PBS and LTR termini (TIR and conserved flanking region) comprise 40 nt, or only 0.4% of each retrotransposon and 0.32% of the total sequence output, which is further divided into many individual retrotransposon families. From the shotgun sequences of small portions of genomes (10% or less), moderately abundant or rare retrotransposons (500 copies per genome or less) are difficult to recognize reliably (102).

In recent years, SNP marker methods have become subject to enormous technological development (9), in parallel with the rapid growth in the efficiency and decrease in cost of high-throughput sequencing (103, 104). The advent of high-throughput sequencing together with in silico SNP mining, however, has not undermined the place of retrotransposon marker systems among the tools of genetic analysis. First, SNPs are most often mined from a set of cultivars; when these SNPs are applied to landraces or wild relatives, the ascertainment bias—the cultivars' SNPs are absent but replaced by others—can interfere with comparative analyses of genetic diversity (105). Retrotransposon marker systems, such as SSAP, IRAP, REMAP, and iPBS, are not subject to this problem. Second, nongenic retrotransposon insertional polymorphisms are far more likely to be selectively neutral than SNPs from genes; neutral markers complement markers under selection. Third, retrotransposons access a different genomic compartment than do genic markers. When markers are used for genome sequencing projects in which genetic and physical maps need to be linked, having markers in regions of low gene density can be highly important for extending and improving the assembly.

Moreover, even as an increasing number of plant species have been subject to EST sequencing from multiple accessions, permitting SNP mining, a vast diversity of "orphan" crops and wild species remain, studied by labs who need access to inexpensive analytical approaches useful for low and medium throughput but scalable to high throughput. For example, the RBIP-TAM method has been successfully adapted for efficient, high-throughput analyses of both retrotransposon insertions and SNP detection. This has allowed phylogenetic analysis of a complete large germplasm collection of pea (106). For these reasons, we expect that retrotransposon marker systems will find continuing use in phylogenetic studies, fingerprinting, genetic mapping, and germplasm characterization for the foreseeable future.

Acknowledgments

Development of the methods described in this chapter was funded by contracts BIO-4-CT-960508, QLK5-CT-2000-01502, and FOOD-CT-2005-513959 to the Commission of the European Communities as well as by the Academy of Finland, Grant 120810, by Project Exbardiv of the ERA-NET Plant Genomics program and by the Ministry of Education of the Czech Republic project MSM2678424601. We are grateful to Ruslan Kalendar, Maggie Knox, and Steven Pearce for material contributions to the methods presented here.

References

1. Botstein D, et al (1980) Construction of a genetic linkage map using restriction fragment length polymorphisms. Am J Hum Genet 32:314–331
2. Quraishi UM, et al (2009) Genomics in cereals: from genome-wide conserved orthologous set (COS) sequences to candidate genes for trait dissection. Funct Integr Genomics 9:473–484
3. Williams JG, et al (1990) DNA polymorphisms amplified by arbitrary primers are useful as genetic markers. Nucl Acids Res 18:6531–6535
4. Zhao X, Kochert G (1993) Phylogenetic distribution and genetic mapping of a $(GCG)_n$ microsatellite from rice (*Oryza sativa*). Plant Mol Biol 21:607–614
5. Zietkiewicz E, Rafalski A, Labuda D (1989) Genome fingerprinting by simple sequence repeat (SSR)-anchored polymerase chain reaction amplification. Genomics 20:176–183
6. Yamamoto K, Sasaki T (1997) Large-scale EST sequencing in rice. Plant Mol Biol 35:135–144
7. van Orsouw NJ, et al (2007) Complexity reduction of polymorphic sequences (CRoPS): a novel approach for large-scale polymorphism discovery in complex genomes. PLoS ONE 2:e1172
8. Wenzl P, et al (2004) Diversity Arrays Technology (DArT) for whole-genome profiling of barley. Proc Natl Acad Sci USA 101:9915–9920
9. Ding C, Jin S (2009) High-throughput methods for SNP genotyping. Methods Mol Biol 578:245–254
10. Knox MR, Ellis THN (2001) Stability and Inheritance of Methylation States at PstI Sites in *Pisum*. Mol Gen Genet 265:497–507
11. Eickbush TH, Jamburuthugoda VK (2008) The diversity of retrotransposons and the properties of their reverse transcriptases. Virus Res 34:221–234
12. Wicker T, et al (2007) A unified classification system for eukaryotic transposable elements. Nat Rev Genet 8:973–982
13. Schulman AH, Kalendar R (2005) A movable feast: Diverse retrotransposons and their contribution to barley genome dynamics. Cytogenetics and Genome Research 110:598–605
14. Kumar A, Bennetzen J (1999) Plant retrotransposons. Annu Rev Genet 33:479–532
15. Frankel AD, Young JA (1998) HIV-1: Fifteen proteins and an RNA. Annu Rev Biochem 67:1–25
16. Vicient CM, Kalendar R, Schulman AH (2001a) Envelope-containing retrovirus-like elements are widespread, transcribed and spliced, and insertionally polymorphic in plants. Genome Res 11:2041–2049
17. Sandmeyer SB, Menees TM (1996) Morphogenesis at the retrotransposon-retrovirus interface: gypsy and copia families in yeast and *Drosophila*. Curr Top Microbiol Immunol 2124:261–296
18. Mills RE, et al (2007) Which transposable elements are active in the human genome? Trends Genet 23:183–191
19. Vitte C, Panaud O (2005) LTR retrotransposons and flowering plant genome size: emergence of the increase/decrease model. Cytogenetics and Genome Research 110:91–107
20. Flavell AJ, et al (1992) *Ty1-copia* group retrotransposons are ubiquitous and heterogeneous in higher plants. Nucl Acids Res 20:3639–3644
21. Suoniemi A, Tanskanen J, Schulman AH (1998) Gypsy-like retrotransposons are widespread in the plant kingdom. Plant J 13:699–705
22. Voytas DF, et al (1992) *Copia*-like retrotransposons are ubiquitous among plants. Proc Natl Acad Sci USA 89:7124–7128
23. Heslop-Harrison JS, et al (1997) The chromosomal distributions of *Ty1-copia* group retrotransposable elements in higher plants and their implications for genome evolution. Genetica 100:197–204
24. Kumar A, et al (1997) The Ty1-copia group of retrotransposons in plants: genomic organisation, evolution, and use as molecular markers. Genetica 100:205–217

25. Suoniemi A, et al (1996) Retrotransposon *BARE*-1 is a major, dispersed component of the barley (*Hordeum vulgare* L.) genome. Plant Mol Biol 30:1321–1329
26. Hawkins JS, et al (2008) Phylogenetic determination of the pace of transposable element proliferation in plants: *Copia* and LINE-like elements in *Gossypium*. Genome 51:11–18
27. Park M, et al (2011) Comparative analysis of pepper and tomato reveals euchromatin expansion of pepper genome caused by differential accumulation of *Ty3/Gypsy*-like elements. BMC Genomics 12:85
28. Vicient CM, et al (1999) Retrotransposon *BARE*-1 and its role in genome evolution in the genus *Hordeum*. Plant Cell 11: 1769–1784
29. Zedek F, et al (2010) Correlated evolution of LTR retrotransposons and genome size in the genus *Eleocharis*. BMC Plant Biol 10:265
30. Hawkins JS, et al (2009) Rapid DNA loss as a counterbalance to genome expansion through retrotransposon proliferation in plants. Proc Natl Acad Sci USA 106:17811–17816
31. International Brachypodium Initiative (2010) Genome sequencing and analysis of the model grass Brachypodium distachyon. Nature 463:763–768
32. Shirasu K, et al (2000) A contiguous 66 kb barley DNA sequence provides evidence for reversible genome expansion. Genome Res 10:908–915
33. Wicker T, Keller B (2007) Genome-wide comparative analysis of *copia* retrotransposons in Triticeae, rice, and *Arabidopsis* reveals conserved ancient evolutionary lineages and distinct dynamics of individual *copia* families. Genome Res 17:1072–1081
34. Rowold DJ, Herrara RJ (2000) *Alu* elements and the human genome. Genetica 108:57–72
35. Shimamura M, et al (1997) Molecular evidence from retroposons that whales form a clade within even-toed ungulates. Nature 388:666–670
36. Soleimani VD, Baum BR, Johnson DA (2005) Genetic diversity among barley cultivars assessed by sequence-specific amplification polymorphism. Theor Appl Genet 110:1290–1300
37. Tam SM, et al (2009) LTR-retrotransposons Tnt1 and T135 markers reveal genetic diversity and evolutionary relationships of domesticated peppers. Theor Appl Genet 119:973–989
38. Kalendar R, et al (2010) iPBS: A universal method for DNA fingerprinting and retrotransposon isolation. Theor Appl Genet 121:1419–1430
39. Kalendar R, et al (2010) Analysis of plant diversity with retrotransposon-based molecular markers. Heredity 106:520–530
40. Kalendar R, Schulman A (2006) IRAP and REMAP for retrotransposon-based genotyping and fingerprinting. Nature Protoc 1:2478–2484
41. Paux E, et al (2010) Insertion site-based polymorphism markers open new perspectives for genome saturation and marker-assisted selection in wheat. Plant Biotechnol J 8:196–210
42. Schulman AH, Flavell AJ, Ellis THN (2004) The application of LTR retrotransposons as molecular markers in plants. Methods Mol Biol 260:145–173
43. Syed NH, Flavell AJ (2006) Sequence-specific amplification polymorphisms (SSAPs): a multi-locus approach for analyzing transposon insertions. Nature Protoc 1:2746–2752
44. Waugh R, et al (1997) Genetic distribution of *BARE*-1-like retrotransposable elements in the barley genome revealed by sequence-specific amplification polymorphisms (S-SAP). Mol Gen Genet 253:687–694
45. Ellis THN, et al (1998) Polymorphism of insertion sites of Ty1-copia class retrotransposons and its use for linkage and diversity analysis in pea. Mol Gen Genet 260:9–19
46. Itoh T, et al (2007) Curated genome annotation of Oryza sativa ssp. japonica and comparative genome analysis with Arabidopsis thaliana. Genome Res 17:175–183
47. Manninen OM, et al (2006) Mapping of major spot-type and net-type net-blotch resistance genes in the Ethiopian barley line CI 9819. Genome 49:1564–1571
48. Kalendar R, et al (2000) Genome evolution of wild barley (*Hordeum spontaneum*) by *BARE*-1 retrotransposon dynamics in response to sharp microclimatic divergence. Proc Natl Acad Sci USA 97:6603–6607
49. Yu G-X, Wise RP (2000) An anchored AFLP- and retrotransposon-based map of diploid *Avena*. Genome 43:736–749
50. Korswagen HC, et al (1995) Transposon Tc1-derived, sequence-tagged sites in *Caenorhabditis elegans* as markers for gene mapping. Proc Natl Acad Sci USA 93:14680–14685
51. Van den Broeck D, et al (1998) Transposon Display identifies individual transposable elements in high copy number lines. Plant J 13:121–129
52. Vogel JM, Morgante M (1992) A microsatellite-based, multiplexed genome assay.

In: Plant Genome III Conference. San Diego, CA USA

53. Vos P, et al (1995) AFLP: A new technique for DNA fingerprinting. Nucl Acids Res 21:4407–4414

54. Lee D, et al (1990) A *copia*-like element in *Pisum* demonstrates the uses of dispersed repeated sequences in genetic analysis. Plant Mol Biol 15:707–722

55. Pearce SR, et al (1999) Rapid isolation of plant Ty1-copia group retrotransposon LTR sequences for molecular marker studies. Plant J 19:711–717

56. Vershinin AV, Ellis TH (1999) Heterogeneity of the internal structure of PDR1, a family of Ty1/copia-like retrotransposons in pea. Mol Gen Genet 262:703–713

57. Kalendar R, et al (1999) IRAP and REMAP: Two new retrotransposon-based DNA fingerprinting techniques. Theor Appl Genet 98:704–711

58. Bento M, et al (2010) Genome merger: from sequence rearrangements in triticale to their elimination in wheat-rye addition lines. Theor Appl Genet 121:489–497

59. Boyko E, et al (2002) Combined mapping of *Aegilops tauschii* by retrotransposon, microsatellite, and gene markers. Plant Mol Biol 48:767–790

60. Baumel A, et al (2002) Inter-retrotransposon amplified polymorphism (IRAP), and retotransposon-microsatellite amplified polymorphism (REMAP) in populations of the young allopolyploid species *Spartina angelica* Hubbard (Poaceae). Mol Biol Evol 19:1218–1227

61. Bernet GP, Asins MJ (2004) Identification and genomic distribution of gypsy like retrotransposons in *Citrus* and *Poncirus*. Theor Appl Genet 108:121–130

62. Branco CJ, et al (2007) IRAP and REMAP assessments of genetic similarity in rice. J Appl Genet 48:107–113

63. Breto MP, et al (2001) The diversification of *Citrus clementina* Hort. ex Tan., a vegetatively propagated crop species. Mol Phylogenet Evol 21:285–293

64. Pereira HS, et al (2005) Genomic analysis of Grapevine Retrotransposon 1 (Gret 1) in *Vitis vinifera*. Theor Appl Genet 111:871–87895.

65. Smykal P (2006) Development of an efficient retrotransposon-based fingerprinting method for rapid pea variety identification. J Appl Genet 47:221–230

66. Smýkal P, et al (2011) Genetic diversity of cultivated flax (Linum usitatissimum L.) germplasm assessed by retrotransposon-based markers. Theor Appl Genet 122: 1385–1397

67. Teo CH, et al (2005) Genome constitution and classification using retrotransposon-based markers in the orphan crop banana. J Plant Biol 48:96–105

68. Vicient CM, et al (2001) Active retrotransposons are a common feature of grass genomes. Plant Physiol 125:1283–1292

69. Vukich M, et al (2009) Genetic variability in sunflower and in the Helianthus genus as assessed by retrotransposon-based molecular markers. Theor Appl Genet 19:1027–1038

70. Kankaanpää J, Mannonen L, Schulman AH (1996) The genome sizes of *Hordeum* species show considerable variation. Genome 39: 730–735

71. Panstruga R, et al (1998) A contiguous 60 kb genomic stretch from barley reveals molecular evidence for gene islands in a monocot genome. Nucl Acids Res 26:1056–1062

72. Devos K (2010) Grass genome organization and evolution. Curr Opin Plant Biol 13: 139–145

73. SanMiguel P, et al (1996) Nested retrotransposons in the intergenic regions of the maize genome. Science 274:765–768

74. McCouch SR, et al (1997) Microsatellite marker development, mapping and applications in rice genetics and breeding. Plant Mol Biol 35:89–99

75. Saghai Maroof MA, et al (1994) Extraordinarily polymorphic microsatellite DNA in barley: Species diversity, chromosomal locations, and population dynamics. Proc Natl Acad Sci USA 91:5466–5470

76. Varshney RK, Graner A, Sorrells ME (2005) Genic microsatellite markers in plants: features and applications. Trends Biotechnol 23:48–55

77. Ramsay L, et al (1999) Intimate association of microsatellite repeats with retrotransposons and other dispersed repetitive elements in barley. Plant J 17:415–425

78. Provan J, et al (1999) *Copia*-SSR: A simple marker technique which can be used on total genomic DNA. Genome 42:363–366

79. Chadha S, Gopalakrishna T (2005) Retrotransposon-microsatellite amplified polymorphism (REMAP) markers for genetic diversity assessment of the rice blast pathogen (*Magnaporthe grisea*). Genome 48: 943–945

80. Jääskeläinen M, et al (1999) Retrotransposon *BARE*-1: Expression of encoded proteins and formation of virus-like particles in barley cells. Plant J 20:413–422

81. Manninen O, et al (2000) Application of *BARE*-1 retrotransposon markers to map a

major resistance gene for net blotch in barley. Mol Gen Genet 264:325–334

82. Tanhuanpää P, et al (2007) A major gene for grain cadmium accumulation in oat (*Avena sativa* L.). Genome 2007:588–594
83. Tenhola-Roininen T, et al (2011) A doubled haploid rye linkage map with a QTL affecting α-amylase activity. J Appl Genet doi: 10.1007/s13353-011-0029-1
84. Kelly NJ, Palmer MT, Morrow CD (2003) Selection of retroviral reverse transcription primer is coordinated with tRNA biogenesis. J Virol 77:8695–8701
85. LeGrice SFJ (2003) "In the beginning": Initiation of minus strand DNA synthesis in retroviruses and LTR-containing retrotransposons. Biochemistry 42:14349–14355
86. Mak J, Kleiman L (1997) Primer tRNAs for reverse transcription. J Virol 71:8087–8095
87. Marquet R, et al (1995) tRNAs as Primer of Reverse Transcriptases. Biochemie 77: 13–124
88. Hizi A (2008) The reverse transcriptase of the Tf1 retrotransposon has a specific novel activity for generating the RNA self-primer that is functional in cDNA synthesis. J Virol 82:10906–10910
89. Kalendar R, et al (2008) *Cassandra* retrotransposons carry independently transcribed 5S RNA. Proc Natl Acad Sci USA 105: 5833–5838
90. Leigh F, et al (2003) Comparison of the utility of barley retrotransposon families for genetic analysis by molecular marker techniques. Mol Genet Genomics 269:464–474
91. Flavell AJ, et al (1998) Retrotransposon-based insertion polymorphisms (RBIP) for high throughput marker analysis. Plant J 16:643–650
92. Jing R, Bolshakov V, Flavell AJ (2007) The tagged microarray marker (TAM) method for high-throughput detection of single nucleotide and indel polymorphisms. Nature Protoc 2:168–177
93. Paux E, et al (2006) Characterizing the composition and evolution of homoeologous genomes in hexaploid wheat through BAC-end sequencing on chromosome 3B. Plant J 48:463–474
94. Bartoš J, et al (2008) A first survey of the rye (*Secale cereale*) genome composition through BAC end sequencing of the short arm of chromosome 1R. BMC Plant Biol 8:95
95. Paux E, et al (2008) A physical map of the 1Gb bread wheat chromosome 3B. Science 322:101–104
96. McClelland M, Nelson M, Raschke E (1994) Effect of site-specific modification on restriction endonucleases and DNA modification methyltransferases. Nucl Acids Res 22: 3640–3659
97. Flavell AJ, et al (2003) A microarray-based high throughput molecular marker genotyping method: the tagged microarray marker (TAM) approach. Nucl Acids Res 31:e115
98. Biggin MD, Gibson TJ, Hong GF (1983) Buffer gradient gels and 35S label as an aid to rapid DNA sequence determination. Proc Natl Acad Sci USA 80:3963–3965
99. Cheng X, et al (2011) Reverse genetics in medicago truncatula using *Tnt1* insertion mutants. Methods Mol Biol 678:179–190
100. Kumar A, Hirochika H (2001) Applications of retrotransposons as genetic tools in plant biology. Trends Plant Sci 6:127–134
101. Okamoto H, Hirochika H (2000) Efficient insertion mutagenesis of Arabidopsis by tissue culture-induced activation of the tobacco retrotransposon *Tto1*. Plant J 23:291–304
102. Macas J, Neumann P, Navrátilová A (2007) Repetitive DNA in the pea (*Pisum sativum* L.) genome: comprehensive characterization using 454 sequencing and comparison to soybean and *Medicago truncutata*. BMC Genomics 8:427
103. Chan EY (2009) Next-generation sequencing methods: impact of sequencing accuracy on SNP discovery. Methods Mol Biol 578:95–111
104. Kircher M, Kelso J (2010) High-throughput DNA sequencing – concepts and limitations. BioEssays 32:524–536
105. Moragues M, et al (2010) Effects of ascertainment bias and marker number on estimations of barley diversity from high-throughput SNP genotype data. Theor Appl Genet 120:1525–153487
106. Jing R, et al (2010) The genetic diversity and evolution of field pea (Pisum) studied by high throughput retrotransposon based insertion polymorphism (RBIP) marker analysis. BMC Evol Biol 10:44
107. Turpeinen T, Kulmula J, Nevo E (1999) Genome size variation in *Hordeum spontaneum* populations. Genome 42:1094–1099
108. Knox M, et al (2009) High-throughput retrotransposon-based fluorescent markers: improved information content and allele discrimination. Plant Methods 5:10
109. Piffanelli P, et al (2010) Large-scale characterization of Tos17 insertion sites in a rice T-DNA mutant library. Plant Mol Biol 65:587–560
110. SanMiguel P, Bennetzen JL (1998) Evidence that a recent increase in maize genome size was caused by the massive amplification of intergene retrotransposons. Ann Bot 82:37–44

Chapter 8

Individual Analysis of Transposon Polymorphisms by AFLP

Susanta K. Behura

Abstract

The DNA polymorphisms caused by insertion and excision of transposable elements (TEs) are applicable in studying genome dynamics, genetic diversity, and molecular evolution, generating genome-wide molecular maps and investigating functional attributes of transposons in epigenetics and diseases. Identification of individual mutations caused by TEs using the principles of amplified fragment length polymorphism assay is a reliable and cost-effective approach. The method relies upon selective polymerase chain reaction (PCR) of flanking regions of TE insertion sites in the genome. A detailed procedure is described in this chapter that outlines each step starting from the preparation of PCR template to identification and isolation of the polymorphic bands. The approach outlined in this protocol can be adopted to identify individual polymorphisms caused by any transposon in any organism.

Key words: Insertion/excision polymorphism, Transposon display, Sequence-specific amplification polymorphism, Amplified length polymorphism, Molecular markers, Genome dynamics

1. Introduction

Transposons are ubiquitous mobile genetic elements and constitute considerable fraction of all genomes (1–7). The DNA polymorphisms caused by insertion and excision of transposable elements are important aspects in many molecular biology researches. Identification of TE-induced genetic variation is important as increasing evidences now suggest that they may have significant roles in genome dynamics (8), population diversity and speciation (9–14), disease traits (15, 16), and other functional attributes, such as epigenetic regulation (17) and gene expression (18). TE-associated polymorphisms are also used as reliable molecular markers for genome mapping and genetic diversity studies in insects (19–22), plants (23–26), and microbes (27).

Previously, hybridization-based methods, such as restriction fragment length polymorphism (RFLP) assays, were employed to identify insertion and excision polymorphisms of TEs (28–30).

Yves Bigot (ed.), *Mobile Genetic Elements: Protocols and Genomic Applications*, Methods in Molecular Biology, vol. 859,
DOI 10.1007/978-1-61779-603-6_8, © Springer Science+Business Media, LLC 2012

There are several disadvantages of this traditional method. For example, isolation of a polymorphic DNA is not feasible from this approach. Now, direct sequencing of DNA is an obvious alternative. Although cost of DNA sequencing has substantially decreased, sequencing TEs in genome-wide manner of hundreds of individuals of a population is, however, not a feasible approach at this moment. Thus, alternative experimental tools, mostly based on polymerase chain reaction (PCR) assay, have been popular choices to identify sequence polymorphisms associated with transposable elements (31). Amplified Fragment Length Polymorphism (AFLP), a PCR-based genetic fingerprinting assay developed some 15 years ago (32), has been a powerful and sensitive method for detecting DNA polymorphisms. Over the past decade or so, several PCR formats have been developed based on the core principles of AFLP to detect polymorphisms associated with specific loci or genes of interest (19, 22, 25, 26, 31, 33). The AFLP-PCR method targets multiple loci in the genome (between 50 and 100 fragments) and has high reproducibility, resolution, and sensitivity (34). In addition, no prior sequence information is needed for amplification (35). As a result, AFLP has become a useful and popular molecular technique in the study of taxa, including bacteria, fungi, insects, and plants, where much is still unknown about the genomic makeup of various organisms.

The basic principle of applying AFLP to detect TE polymorphisms is the replacement of one AFLP adapter-specific primers with a TE-specific primer (Fig. 1). The genomic DNA is digested with a restriction enzyme (usually, a 6 cutter) whose recognition site is absent in the internal sequence of the element. A reference complete sequence of the element is generally used to assess the presence or absence of this site. To minimize self-ligation of the restricted DNA by single enzyme, a second restriction enzyme (usually, a 4 cutter) is also used. The recognition site of the second restriction enzyme may not be present within TE. However, the presence of this restriction site in the 5′ end of the TE-specific primer does not interfere with the PCR amplification (Fig. 1). The TE-anchored primer is generally designed to have sequence complementary either to one of the highly conserved domains of the putative transposase or the terminal repeats of the element (22, 25, 33). The second primer that is employed in these PCR assays is specific to the adapter attached to the flanking restriction site end. This modified approach of AFLP-PCR is also often referred to as "Sequence-Specific Amplified Polymorphism" (SSAP) (33) or as "Transposon Display" (TD) (36).

Here, we describe a detailed protocol of AFLP-PCR for identification of individual polymorphisms associated with transposable elements. The AFLP procedure described here utilizes *Eco*RI and *Mse*I restriction site to selectively amplify polymorphisms of *mariner*-like elements (MLEs) that are widespread in insects (37).

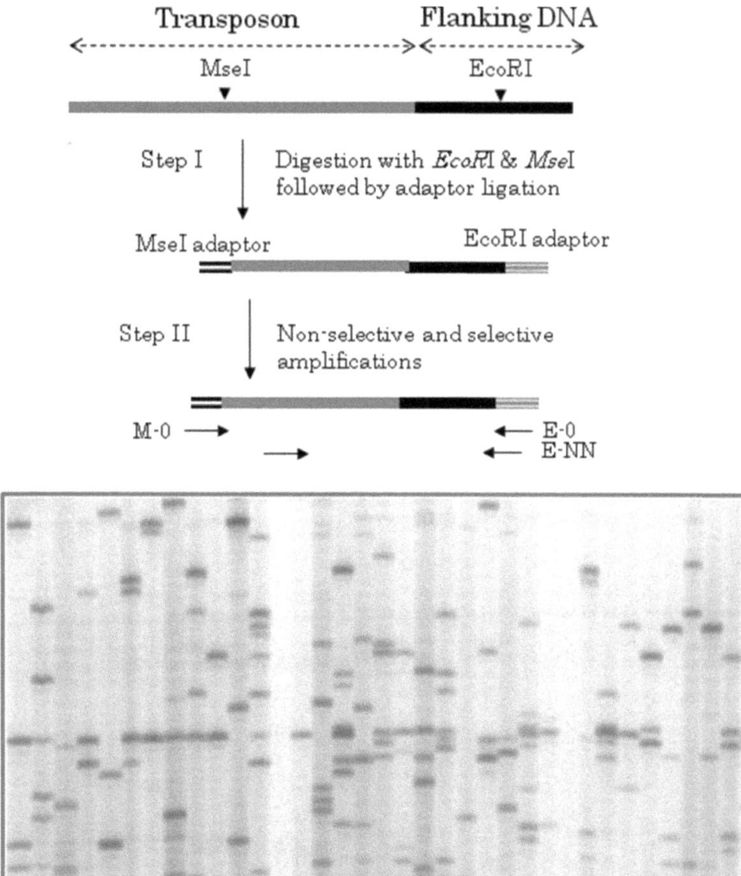

Fig. 1. Principles of AFLP-PCR to identify TE-induced polymorphisms. The *black line* represents flanking DNA of transposon. The transposon is shown in *gray line*. The *Eco*RI and *Mse*I sites are shown. After restriction digestion of genomic DNA with *Eco*RI and *Mse*I enzymes and ligation of digested DNA with linkers, a preamplification step is performed with nonselective primers E-0 and M-0 to generate template for selective PCRs. Selective PCRs are then performed with a primer that is specific to the transposase and another primer (E-NN) that is specific to the flanking *Eco*RI linker but contains additional selective nucleotides (NN) at the 3′-end. An example of AFLP-PCR amplification is shown with a portion of autoradiography of the polyacrylamide gel.

The *Eco*RI restriction site is not present within the particular MLE sequence. The *Mse*I site, although present within MLE, is localized outside the targeted amplification region as shown in Fig. 1. This method has been successfully tested in identifying polymorphisms associated with MLEs in the genome of rice gall midge (*Orseolia oryzae*) and Hessian fly (*Mayetiola destructor*), which are agriculturally important pests of rice and wheat, respectively (19, 22, 38). The AFLP-PCR method is a versatile technique as the selectivity of target amplification can be enhanced or relaxed by the user. For example, prior to selective PCR, a second

PCR step is performed with the preamplified DNA as template using a nested primer specific to the TE and a nonselective primer specific to the linker (Fig. 2). This highly selective AFLP-PCR amplifies only a few targeted amplicons that, in some cases, can be visualized by 2% agarose gel instead of requiring a polyacrylamide gel (Fig. 2). The AFLP-PCR methods can be useful for identifying

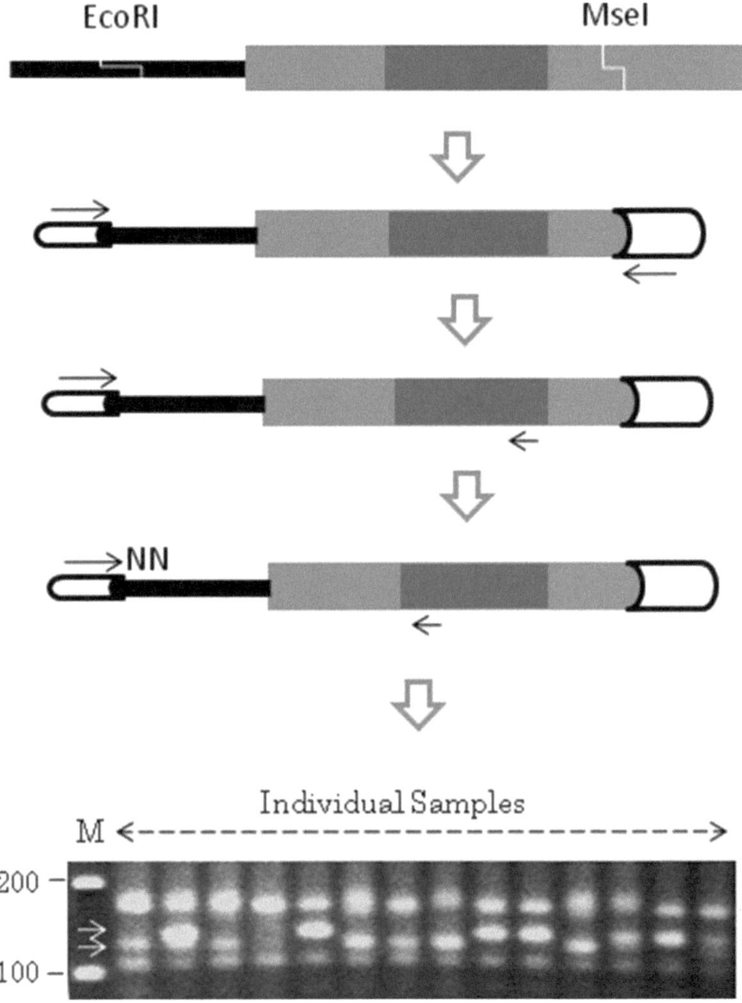

Fig. 2. Selective AFLP-PCR using transposon-specific nested primers. The *gray box* represents a mariner-like element (MLE) and the *black line* represents the flanking DNA. The *dark gray box* within the element shows a highly conserved domain of the putative transposase. The restriction sites are indicated as *Eco*RI and *Mse*I. After restriction digestion and linker ligation, three rounds of PCR are performed. The first PCR is done for preamplification of the ligated DNA using nonselective primers that are specific to the linkers. The second PCR step is performed with a TE-specific primer and another primer that is specific to the flanking linker. In the third PCR step, a nested primer specific to the domain sequence is used along with selective primers anchoring to flanking linker. This PCR method is highly selective and produces a few numbers of amplicons that may be resolved in a 2% agarose gel.

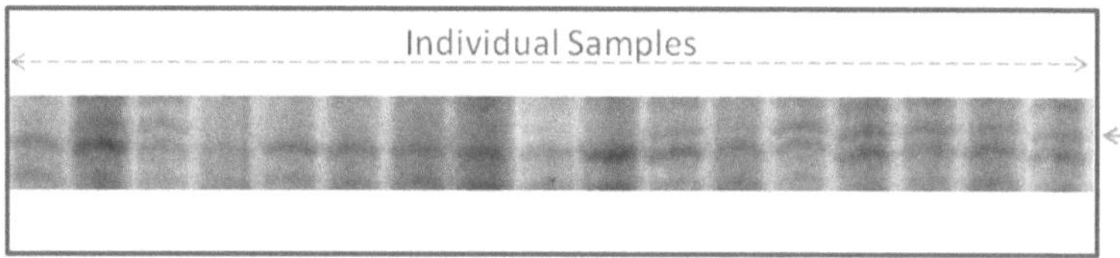

Fig. 3. Identification of TE-induced polymorphisms in a segregating mapping population of Hessian fly. The *arrow on the right* shows the polymorphic band in a denaturing polyacrylamide gel of AFLP-PCR products. *Each lane* represents an individual progeny of the mapping population.

as well as reliably scoring TE-associated polymorphisms among individual samples for genetic mapping of TE integration sites in the genome (Fig. 3). The method may be adopted to profile polymorphisms associated with any transposable elements in any genome, of course, with appropriate design of TE-specific primers.

2. Materials

2.1. Primers

Purchase primers required for preparation of *Eco*RI and *Mse*I linkers. The sequences of these primers are *Eco*RI-F: 5′-CTC GTA GAC TGC GTA CC-3′; *Eco*RI-R: 5′-AAT TGG TAC GCA GTC TAC-3′ and *Mse*I-F: 5′-GAC GAT GAG TCC TGA G-3′; *Mse*I-R: 5′-TAC TCA GGA CTC AT-3′. Also, purchase the linker-specific primers (nonselective) which are *Eco*RI-0: 5′-GACTGCGTACCA ATTC-3′ and *Mse*I-0: 5′-GATGAGTCCTGAGTAA-3′. The linker-specific primers consist of a core sequence along with a specific sequence that represents the recognition site of the restriction enzyme (32). Additionally, design and purchase TE-specific primers and *Eco*RI-selective primers that are required for selective amplification steps. The *Eco*RI-selective primers contain one to three additional random bases after the 3′ end so that only a subset of fragments that are complementary to these selective bases are amplified from the genomic DNA. The TE-specific oligonucleotide is complementary to either the inverted termini sequences or a highly conserved domain of the transposon. The design of the TE-specific primer varies depending upon the type of transposon being investigated.

2.2. Enzymes

We purchased enzymes from New England Biolabs.

1. *Eco*RI restriction enzyme.
2. *Mse*I restriction enzyme.
3. T4 DNA Ligase.
4. T4 Polynucleotide Kinase.
5. Taq DNA polymerase.

2.3. Molecular Biology Products and Chemicals

1. Agarose LE.
2. Deoxyribonucleotide triphosphates (dNTPs).
3. Promega 100 bp molecular weight ladder.
4. SequaMark™ size marker (for sequencing gel) (Invitrogen).
5. TEMED.
6. Urea: Molecular biology quality.
7. X-ray films.

2.4. Solutions and Kits

All solutions are prepared with autoclaved nanopure water.

1. 5× RL buffer (restriction ligation buffer): 10 mM Tris.HAc, pH 7.5, 10 mM MgAc, 50 mM KAc, 5 mM DTT, and 50 ng/μl BSA.
2. 10× TBE buffer: 108 g Tris base, 55 g boric acid, and 9.3 g Na4EDTA (make final volume to 1 L with water).
3. Formamide stop buffer: 98% formamide, 10 mM EDTA, pH 8.0, and 0.1% Bromo Phenol Blue and 0.1% Xylene cyanol as tracking dyes.
4. 10:1 TE buffer: 10 mM Tris–HCl, pH 7.0, and 1 mM EDTA.
5. 0.1% Ethidium bromide.
6. 6% Acrylamide:bis acrylamide solution (19:1).
7. 10% Ammonium persulfate.
8. 10 M NaOH.
9. 3 M sodium acetate, pH 5.2.
10. Glycogen, 10 mg/ml.

For DNA purifications, we used the following:

1. The QIAGEN DNeasy Kit for genomic DNA purification.
2. The QIAquick PCR Purification Kit to purify or elute amplification products.

3. Methods

3.1. Restriction Digestion of Genomic DNA

1. Take 0.5 μg of purified genomic DNA from each individual and digest them separately in 30-μl reaction volume that contains 6 μ of 5× RL buffer, 5 U each of *Mse*I and *Eco*RI enzyme, and sterile water to make the final volume (see Note 1).
2. Incubate the reactions at 37°C for 3 h.
3. Run an aliquot (5 μl) of each digest on an 0.8% agarose gel stained with ethidium bromide to check that each samples have been digested properly (you should observe smears in each lane) (see Note 2).

3.2. Preparation of Likers and Ligation with Digested DNA

1. Take 1.7 µg *Eco*RI-F oligo and 1.5 µg *Eco*RI-R oligo in a final volume of 60 µl (with sterile water) to prepare 5 pMol/µl of *Eco*RI linker. Similarly, add 16 µg *Mse*I F oligo with 14 µg *Mse*I R oliog and adjust H_2O to 60 µl to prepare 50 pMol/µl of *Mse*I linker.
2. Incubate the mixtures at 65°C for 10 min. Then, incubate the tubes at 37°C for 10 min followed by incubation at room temperature for 20 min.

   The *Eco*RI double-stranded linker structure is

   5'-CTCGTAGACTGCGTACC-3'

   3'-CATCTGACGCATGGTTAA-5'

   The *Mse*I double-stranded linker structure is

   5'-GACGATGAGTCCTGAG -3'

   3'-TACTCAGGACTCAT-5'

   Annealed linkers can be stored at −20°C.
3. Ligate the *Eco*RI and *Mse*I linkers with the digested DNA. Add the remaining 25 µl of the digested DNA of each simple with 1.0 µl each of the *Mse*I and *Eco*RI linker, 7.0 µl 5× RL buffer, 1.0 µl of 10 mM ATP, and 0.5 µl of T4 DNA Ligase (400 U per µl) and make the final volume to 35 µl with sterile water.
4. Incubate the reaction at 16°C overnight followed by heat killing the enzyme at 65°C for 10 min.

3.3. Preamplification of Ligated DNA

A preamplification of the ligated material is performed. It generates template for selective amplification. The preamplification is also helpful in generating as much template as required for many selective PCRs (see Notes 3 and 4). The preamplifications are generally done by using the nonselective linker-specific primers (*Eco*RI + 0 and *Mse*I + 0). In some cases, especially with complex genomes, an additional selective nucleotide is added to these primers in the preamplification step to reduce the number of bands. The preamplification steps are as follows.

1. Take 2 µl (~30 ng DNA) of the ligated DNA; add 2.5 µl of PCR buffer (10×), 30 ng of each primer (*Eco*RI + 0 and *Mse*I + 0), 3 µl of 1.5 mM dNTP mix, 1 unit *Taq* DNA polymerase, and add sterile water to make final reaction volume 25 µl.
2. Perform PCR with 1 cycle of 95°C for 5 min followed by 25 cycles of 95°C for 30 s, 60°C for 45 s, and 72°C for 1 min. A final extension for 7 min at 72°C is added at the end.
3. After PCRs are over, take 5 µl of each product and check them on a 2% agarose gel stained with ethidium bromide. You should observe smears in each sample in the range of ~200–1,000 bp.

4. Dilute the preamplified products tenfold with sterile water and store them in small aliquots at 4°C for immediate use and the stock at –20°C for future PCRs.

3.4. Primer Labeling

1. The TE-specific primer is end labeled with γ^{33}P-ATP. The labeling reaction contains 5 ng of TE-specific primer, 0.2 μl of 10 μCi/μl γ^{33}P-ATP, 0.1 μl of 10× T4 kinase buffer, 0.02 μl T4 polynucleotide kinase (10 units/μl), and remaining volume by sterile water to make the reaction volume 1 μl. This 1 μl labeled primer is required for one single selective PCR. Therefore, the reaction should be scaled up depending upon the number of samples (see Notes 5 and 6).
2. Incubate at 37°C for 1 h.
3. Heat kill the enzyme at 70°C for 10 min.

3.5. Selective Amplification

1. Set up selective amplification for each sample by mixing 2 μl of tenfold diluted preamplified DNA, 2 μl of 10× PCR buffer, 2.0 μl of 1.5 mM dNTP mix, 30 ng selective *Eco*RI linker-specific primer, 5 ng of labeled TE-specific primer (from above step), 1 unit of *Taq* DNA Polymerase, and make the final volume to 20 μl with sterile water.
2. Perform PCR using "touch-down" thermocycling conditions with 95°C for 5 min (1 cycle), then 1 min at 94°C, 1 min annealing at temperature 0.7°C decreasing from 65 and 72°C for 1 min (20 cycles), followed by another 20 cycles with conditions 94°C for 1 min, 56°C for 1 min, and 72°C for 1 min. A final extension at 72°C for 5 min is also recommended.
3. After PCR is completed, add equal volume of formamide stop buffer to each sample.
4. Denature the amplified products by incubation at 90°C for 5 min, followed by putting the tube on ice immediately. Samples are now ready for loading into a denaturing polyacrylamide gel.

3.6. Run Denaturing Polyacrylamide Gel

Prior to completion of selective PCR, a 5% denaturing polyacrylamide gel should be made ready. The gel casting procedure varies among gel systems. A stepwise procedure is described here using a BRL S2 gel system.

1. Clean glass plates with 10 M NaOH and then with soap under tap water. Rinse the plates with distilled water. After the plates are dried, siliconize the short glass plate with Rain-X.
2. Put the spacers (thickness 0.4 mm) between the clean glass plates and attach the clamps.
3. Seal both sides with Scotch electrical tape (yellow tape). Make sure that you press the tape firmly at the bottom corners of the plates.

4. To prepare 5% denaturing polyacrylamide gel mix, add 100 ml 10× TBE, 125 ml of 40% acrylamide:bisacrylamide solution (19:1), 450 g of urea, and make it to 1 L with water. Filter it with Millipore filtration system and store at 4°C.
5. Before pouring the gel mix, add 400 μl of 10% fresh APS (w/v) and 80 μl of TEMED to 100 ml of 5% denaturing polyacrylamide gel solution (the 100 ml gel solution should be at room temperature).
6. Pour the gel mix immediately into the sandwiched plates and place the comb (shark tooth) with the teeth upward. Upon gel polymerization, it creates a straight running front at the top of the gel.
7. After the gel is polymerized (may take 30 min to 1 h depending on temperature), remove the comb and mount the gel on the electrophoresis apparatus.
8. Pre-run the gel with 1× TBE as running buffer for 45 min.
9. After the pre-run, rinse the well with TBE buffer using a syringe and place the comb with the teeth downward. Load 2–5 μl of the PCR samples.
10. Run the electrophoresis at 80 W. The temperature of the plate should be approximately 55°C.
11. Stop the electrophoresis when the bromophenol blue is just running off and the xylene cyanol is at about two-third down from the top.
12. Disassemble the gel. Pour some ice on the plates (keep for about 10 min). Remove the tape from both sides and gently take the smaller plate off while the gel is still attached to the large plate. Place a 3MM Whatman paper and slowly lift the paper. The gel should now be attached to the paper. Put a Saran Wrap on the gel and dry the gel on a standard slab dryer at 80°C for about 2 h.
13. After the gel is dried, expose it to standard X-ray film. Put several florescent rulers (Stratagene) on the sides of the dried gel if you want to isolate polymorphic bands from the gel.

3.7. Excision of Individual Polymorphic Bands from the Gel

1. Identify the polymorphic bands from the expose X-ray film that you want to isolate (see Notes 7–9).
2. Align the exposed X-ray film on the florescent rulers (Stratagene) pasted on the sides of the dried gels. Try to be as precise as possible.
3. Use a sterile razor blade to excise the required bands and collect them in autoclaved 1.5-ml tubes. The gel is again exposed to X-ray film to make sure that the required bands had been excised.
4. Each of the excised fragments is added with 100 μl of sterile water. Incubate at room temperate for at least 10 min and then

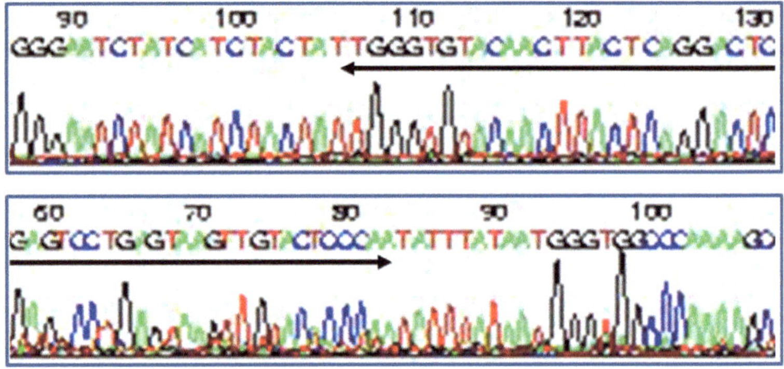

Fig. 4. Validation of AFLP-PCR by sequencing. The sequences are shown for the insertion sites of a mariner-like element in Hessian fly genome. The sequences on the *top* and *bottom panels* represent the inverted terminal repeats of the mariner element and the sequences after that represent the flanking DNA. The "TA" duplications at the junction sites are characteristic features of mariner transposition events.

heat the tubes in a boiling water bath for 2 min followed by centrifugation at 14,000 rpm (15,339 × g) for 8 min.

5. The supernatant is added with 3 M sodium acetate, (pH 5.2) (to a final concentration of 0.3 M), 5 ml of glycogen (10 mg/ml), and 2 volumes of cold ethanol.
6. The mixture is kept overnight at –70°C. Centrifuge the contents at 14,000 rpm for 30 min.
7. Wash the pellet with 70% ethanol, dry it, and dissolve the DNA in 20 μl of sterile water.
8. An aliquot of the DNA (5 μl) is then used as template to reamplify the band by the same primer pairs that were used to amplify them in the AFLP-PCR assay.
9. Check 5 μl of PCR product on a 2% agarose gel to confirm amplification of a single amplicon. Purify the PCR product using QIAGEN'S QIAquick PCR Purification Kit.
10. Sequence the fragment either directly (using the TE-specific primer) or after cloning it into a plasmid vector (a clone can be stored as glycerol stock for future usage). The sequence information is used to confirm authenticity of the PCR amplification (Fig. 4) (see Note 10).

4. Notes

1. Restriction digestion is poor if the quality of the DNA is not good. This generally happens when phenol:chloroform step is not performed properly. In that case, repurify the DNA. Use of a commercial kit is also recommended, if the problem persists. Also check the ratio of optical density of each sample at 260

and 280 nm by a spectrophotometer. Make sure that initial amounts of DNA of each sample used for restriction digestion are nearly equal.

2. When DNA quantity that can be obtained from individual is critical, as little as 200 ng of DNA can be used for restriction digestion. It is possible that the reproducibility of bands may be compromised as we tend to use low amount of starting material.
3. If the selective PCR produces too many bands, additional selective nucleotides (up to 6 selective bases) may be added to the AFLP primers.
4. Selective bases can be added both to linker-specific primer as well as to the transposon-specific primer.
5. Use of ^{32}P for labeling is also a possibility. But it is recommended to use ^{33}P as it gives tighter bands in autoradiography and also it is relatively safer to handle than ^{32}P.
6. The choice of restriction enzymes should be based on the reference sequence information of the transposable element under study. That is necessary for the design of TE-specific primer. Also, make sure that the sequence portion between the TE-specific primer and TE end (in the 3′ direction of the TE-specific primer) does not contain a recognition site of the restriction enzymes you use.
7. The AFLP-PCR is a flexible technique. You can use many enzyme pairs and large number of primer combinations to explore polymorphisms.
8. When copy number of TE is low in the genome, highly selective AFLP-PCR amplifies only few amplicons that can be visualized by 2% agarose gel instead of running a polyacrylamide gel (Fig. 2).
9. While running a denaturing polyacrylamide gel, take some additional precautions that can avoid a bad banding pattern even for a good PCR product. For example, make sure that the running front of the gel is regular; otherwise, distorted banding pattern may be obtained. Also, add 0.5 M sodium acetate to the 1× TBE buffer in the lower buffer compartment as it helps improve resolution of the bands. It creates a salt gradient in the gel and works similar to a gradient gel electrophoresis.
10. Sequencing of AFLP band is recommended to validate the selective PCR. When the TE-specific primer is designed based on the long terminal repeats (LTRs) of LTR elements, the sequence of the amplicon should contain a portion of the terminal repeat sequences (sequence after the 3′ end of the primer-binding region). For non-LTR elements, such as *mariner*-like elements (MLEs), the presence of "TA" next to

the inverted terminal repeat (ITR) sequence, should indicate the authenticity of the TE-anchored amplification of the bands (Fig. 4). When the AFLP primer is specific to the internal region (transposase), a portion of the TE sequences should be manifested in the amplicon sequences.

Acknowledgments

I am thankful to Dr. David W. Severson at the University of Notre Dame for encouragement, support, and help. I also gratefully acknowledge Dr. Jeffery J. Staurt at Purdue University for invitation and encouragement to write this chapter.

References

1. Berg DE, Howe MM (1989) Mobile DNA. American Society of Microbiology, Washington, DC
2. McDonald JF (1993) Evolution and consequences of transposable elements. Curr Opin Genet Dev 3:855–864
3. Kidwell MG, Lisch D (1997) Transposable elements as sources of variation in animals and plants. Proc Natl Acad Sci USA 94: 7704–7711
4. Miller WJ, McDonald JF, Pinsker W (1997) Molecular domestication of mobile elements. Genetica 100:261–270
5. Finnegan DJ (1990) Transposable elements and DNA transposition in eukaryotes. Curr Biol 2:471–477
6. Kempken F, Kück U (1998) Transposons in filamentous fungi – facts and perspectives. Bioessays 20:652–659
7. Pritham EJ (2009) Transposable elements and factors influencing their success in eukaryotes. J Hered 100:648–655
8. Feschotte C, Pritham EJ (2007) DNA transposons and the evolution of eukaryotic genomes. Annu Rev Genet 41:331–368
9. Langley CH, Brookfield JFY, Kaplan N (1983) Transposable elements in Mendelian populations. 1. A theory. Genetics 104:457–471
10. Biémont C (1992) Population genetics of transposable DNA elements. A Drosophila point of view. Genetica 86:67–84
11. Santolamazza F, et al. (2008) Insertion polymorphisms of SINE200 retrotransposons within speciation islands of Anopheles gambiae molecular forms. Malar J 7:163
12. Lee YC, Langley CH (2010) Transposable elements in natural populations of Drosophila melanogaster. Philos Trans R Soc Lond B Biol Sci 365:1219–1228.
13. Capy P, et al. (1994) Horizontal transmission versus ancient origin: mariner in the witness box. Genetica 93:161–170
14. Rebollo R, et al. (2010) Jumping genes and epigenetics: Towards new species. Gene 454: 1–7
15. Belancio VP, Deininger PL, Roy-Engel AM (2009) LINE dancing in the human genome: transposable elements and disease. Genome Med 1:97
16. O'Donnell KA, Burns KH (2010) Mobilizing diversity: transposable element insertions in genetic variation and disease. Mob DNA 1:21.
17. Huda A, Jordan IK (2009) Epigenetic regulation of Mammalian genomes by transposable elements. Ann N Y Acad Sci 1178:276–284.
18. Wessler SR (1998) Transposable elements and the evolution of gene expression. Symp Soc Exp Biol 51:115–122
19. Behura SK, Nair S, Mohan M (2001) Polymorphisms flanking the mariner integration sites in the rice gall midge (*Orseolia oryzae* Wood-Mason) genome are biotype-specific. Genome 44:947–954
20. Marcus JM (2005) Jumping genes and AFLP maps: transforming lepidopteran color pattern genetics. Evol Dev 7:108–114
21. Behura SK (2006) Molecular marker systems in insects: current trends and future avenues. Mol Ecol 15:3087–3113
22. Behura SK, Shukle RH, Stuart JJ (2010) Assessment of structural variation and molecular mapping of insertion sites of Desmar-like elements in the Hessian fly genome. Insect Mol Biol 19:707–715

23. Pearce SR, et al. (1999) Rapid isolation of plant Ty1-copia group retrotransposon LTR sequences for molecular marker studies. Plant J 19:711–717
24. Kumar A, Hirochika H (2001) Applications of retrotransposons as genetic tools in plant biology. Trends Plant Sci 6:127–134
25. Park KC, et al. (2003) A new MITE family, Pangrangja, in Gramineae species. Mol Cells 15:373–380
26. Casa A, et al. (2000) The MITE family Heartbreaker (Hbr): Molecular markers in maize. Proc Natl Acad Sci USA 97: 10083–10089
27. Hayes F (2003) Transposon-based strategies for microbial functional genomics and proteomics. Annu Rev Genet 37:3–29
28. Nakayashiki H, et al. (1999) Transposition of the retrotransposon MAGGY in heterologous species of filamentous fungi. Genetics 153:693–703
29. Eto Y, et al. (2001) Comparative analyses of the distribution of various transposable elements in Pyricularia and their activity during and after the sexual cycle. Mol Gen Genet 264:565–577
30. Arcà B, Savakis C (2000) Distribution of the transposable element Minos in the genus Drosophila. Genetica 108:263–267
31. Syed NH, Flavell AJ (2006) Sequence-specific amplification polymorphisms (SSAPs): a multilocus approach for analyzing transposon insertions. Nat Protoc 1:2746–2752
32. Vos P, et al. (1995) AFLP: a new technique for DNA fingerprinting. Nucleic Acids Res 23:4407–4414
33. Waugh R, et al. (1997) Genetic distribution of Bare-1-like retrotransposable elements in the barley genome revealed by sequence-specific amplification polymorphisms (S-SAP). Mol Gen Genet 253:687–694
34. Mueller UG, Wolfenbarger LL (1999) AFLP genotyping and fingerprinting. Trends Ecol Evol (Amst) 14:389–394
35. Meudt HM, Clarke AC (2007) Almost forgotten or latest practice? AFLP applications, analyses and advances. Trends Plant Sci 12: 106–117
36. Van den Broeck D, et al. (1998) Transposon display identifies individual transposable elements in high copy number lines. Plant J 13:121–129
37. Robertson HM (1993) The mariner transposable element is widespread in insects. Nature 362:241–245
38. Harris MO, et al. (2003). Grasses and gall midges: Plant defense and insect adaptation. Annu Rev Entomol 48: 549–577

Chapter 9

Construction of a Library of Random Mutants in the Spirochete *Leptospira Biflexa* Using a *Mariner* Transposon

Leyla Slamti and Mathieu Picardeau

Abstract

In comparison to other bacterial species, genetics of leptospires are in their infancy. Recently, we developed a system for random transposon mutagenesis in the saprophyte *Leptospira biflexa* and then applied this approach to the pathogen *L. interrogans*. Thousands of random mutants can be readily obtained in *L. biflexa* by random insertion of *Himar*1 in the genome, thereby generating extensive libraries of mutants that could be screened for phenotypes affecting diverse aspects of the biology of the bacterium. This system should be particularly useful for the identification of new genes of unknown function in *Leptospira* spp. This chapter describes a procedure for transposition in *L. biflexa* via conjugation of a plasmid delivering *Himar*1, isolation of mutants, and mapping of the insertion sites on the chromosome.

Key words: *Himar*1, *Mariner*, Mutagenesis, Spirochetes, Library, Random mutants

1. Introduction

The genus *Leptospira* belongs to the order *Spirochaetales* and includes both saprophytic and pathogenic members. The saprophyte *Leptospira biflexa* appears to be one of the fastest growing aerobic spirochetes, with 1 week required to obtain colonies on solid medium in comparison to at least 3 weeks for pathogenic *Leptospira* spp. Manipulation of nonpathogenic species does not require biosafety containment or stringent logistic constraints. Many tools have been developed for the genetic manipulation of *L. biflexa* (1), which is much more easily transformable than the pathogen *L. interrogans*. The genome sequence of *L. biflexa*, which contains 3,500 genes, shares approximately 2,000 genes with the genome sequence of pathogens (2). Certain aspects of leptospiral biology may, therefore, be more easily studied in the fast-growing nonpathogenic species *L. biflexa*. This model bacterium can be

Yves Bigot (ed.), *Mobile Genetic Elements: Protocols and Genomic Applications*, Methods in Molecular Biology, vol. 859, DOI 10.1007/978-1-61779-603-6_9, © Springer Science+Business Media, LLC 2012

used to identify the functions of core leptospiral genes, which include many genes of unknown function. Metabolic and biosynthetic pathways conserved in both pathogenic and saprophytic species could be studied in *L. biflexa* to gain insight into the biology of *Leptospira* spp. in general.

Recently, we showed that we could deliver the *Himar1 mariner* transposon in *L. biflexa* by electroporation (3) or conjugation (4). Consistent with mariner-based mutagenesis systems tested in other bacterial species, analysis of transposon integration sites showed that all *Himar1* insertions in *L. biflexa* occurred at a TA dinucleotide without any additional target site preference. Thousands of random mutants can be easily obtained in *L. biflexa*, thereby generating extensive libraries of mutants that could be screened for phenotypes affecting diverse aspects of metabolism and physiology, such as amino-acid biosynthesis and iron acquisition systems (3, 5).

This procedure describes the protocols for (1) preparing liquid and solid media, (2) delivering *Himar1* in *L. biflexa* by conjugation, (3) isolating random mutants, and (4) mapping the insertion sites on the chromosome by semirandom PCR and Blast analysis.

2. Materials

1. pCjTKS2 (Fig. 1).
2. *Escherichia coli* ß2163 (Demarre, 2005 #405) harboring pCjTKS2 (see Note 1).
3. *Leptospira biflexa* serovar Patoc strain Patoc 1 (see Note 2).
4. Leptospira Medium Base EMJH (Difco).
5. EMJH supplement (available for purchase: Leptospira enrichment EMJH, Difco).
6. Agar Noble (Difco).
7. Incubators for plates and liquid cultures (30 and 37°C).
8. Spectrophotometer.
9. Kanamycin, 50 mg/mL.
10. Diaminopimelic acid (DAP), 300 mM.
11. Glass filtration units (Millipore).
12. Aspirator or vacuum pump.
13. Cellulose-acetate filters (pore size 0.1 μm; diameter 25 mm; Millipore).
14. DNA extraction kit (Maxwell Cell Purification Kit, Promega) (optional).
15. Oligonucleotide primers.

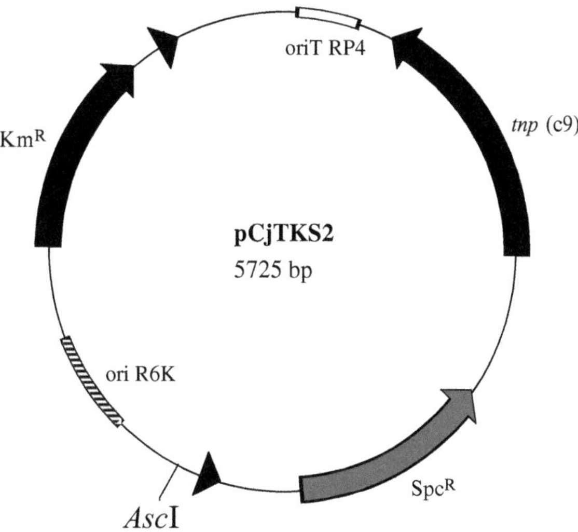

Fig. 1. Schematic representation of plasmid pCJTKS2 used for random transposon mutagenesis. The suicide conjugative plasmid contains (1) the *Himar1* transposon carrying a kanamycin-resistant cassette and oriR6K, (2) the C9 hyperactive transposase (11), (3) the *ori*T of RP4, and (4) a spectinomycin-resistant cassette. The *Asc*I restriction site can be used to clone DNA fragments (*gfp*, etc.) in the transposon.

Name	Target	Sequence (5′→3′)
TnK1	KanR Transposon	CTTGTCATCGTCATCCTTG
Tnk2	KanR Transposon	GTGGCTTTATTGATCTTGGG
Deg1	Degenerate primer with 5′ tag	GGCCACGCGTCGACTAGTAC NNNNNNNNNNGATAT
Deg2	Degenerate primer with 5′ tag	GGCCACGCGTCGACTAGTAC NNNNNNNNNNTCTT
TnkN1	Nested primer for KanR Transposon	CGTCATGGTCTTTGTAGTC TATGG
TnKN2	Nested primer for KanR Transposon	TGGGGATCAAGCCTGATTGGG
Tag	Binds to tag on Deg1/Deg2	GGCCACGCGTCGACTAGTAC

16. Thermal cycler.
17. Taq polymerase and buffer provided by the manufacturer (GE Healthcare).

3. Methods

3.1. Preparation of EMJH Liquid Medium (6, 7)

1. Stock solutions, 1.5% $MgCl_2$, 0.4% $ZnSO_4$, 0.5% $FeSO_4$, 1.5% $CaCl_2$, 0.02% vitamin B12, and 10% pyruvic acid sodium salt, are prepared in distilled water, autoclaved, and stored at 4°C.
2. EMJH supplement can be purchased (Leptospira enrichment EMJH, Difco) or prepared according to the following protocol. To prepare 100 mL of supplement mix, dissolve 10 g bovine serum albumin (BSA) into distilled water, avoiding foaming when stirring. Add the following: 1 mL of each stock solutions, 0.4 g glycerol, 1.25 g Tween 80. Bring the final volume to 100 mL with water and then store indefinitely at –20°C.
3. For 1 L of EMJH liquid medium, dissolve 2.3 g EMJH Base into 800 mL distilled water, and sterilize by autoclaving. Add 100 mL of Leptospira enrichment EMJH or add 100 mL albumin supplement and 0.05% $FeSO_4$. Adjust pH to 7.4 with NaOH or HCl as necessary. Bring the final volume to 1 L with water. Sterilize by filtration through a 0.22-μm filter, and then aliquot as needed for experiments. Store at 4°C.

3.2. Preparation of EMJH Solid Medium

1. For 1 L of medium, dissolve 12 g agar in 340 mL water and sterilize by autoclaving.
2. Cool the autoclaved solution to ~55°C.
3. Dissolve 2.3 g EMJH Base into 500 mL distilled water, and then sterilize by autoclaving.
4. Add 100 mL of Leptospira enrichment EMJH or add 100 mL albumin supplement and 0.05% $FeSO_4$.
5. Adjust pH to 7.4 with NaOH or HCl if necessary.
6. Bring the final volume to 660 mL with water.
7. Sterilize by filtration through a 0.22-μm filter.
8. Warm the solution to ~55°C before adding 340 mL agar solution.
9. When necessary, add antibiotics.
10. Pour individual Petri dishes (~30 mL) preferentially under a hood to minimize contamination. Once plates have solidified, store them inverted in plastic containers at 4°C until needed.

3.3. Generation of a Random Transposon Library in L. biflexa

One conjugation experiment requires 0.5 mL of *E. coli* donor cells and 5 mL of *Leptospira* recipient cells.

1. Grow Leptospira cells at 30°C until midexponential phase (OD_{420} ~ 0.3) in liquid EMJH.
2. Grow *E. coli* cells carrying the conjugative plasmid harboring the transposon overnight in LB supplemented with the appropriate antibiotics. In the morning, inoculate EMJH, supplemented

with 0.3 mM DAP only, with the *E. coli* cells (1/50 of the final culture volume). Grow cells at 37°C until an OD_{420} ~0.3.

3. Prepare the filtration unit by depositing an acetate-cellulose filter onto the filter holder. Secure the funnel on top. Pipette 0.5 mL *E. coli* onto the filter. Add 5 mL *Leptospira*. Open the vacuum and let the liquid filter through. Once the filter is dry, turn off the vacuum and gently remove the funnel from the holder. Place the filter (bacteria-side up) on top of an EMJH plate supplemented with 0.3 mM DAP. Incubate overnight at 30°C (agar-side down).
4. Remove the filter from the plate and place it in a 15-mL tube containing 1 mL of EMJH. Vortex the tube until the bacteria from the filter are resuspended in the medium. Spread the cells on EMJH plates (see Note 5) supplemented with kanamycin (final concentration 50 μg/mL) and incubate at 30°C until colonies are visible (1 week for the saprophytic strains and approximately 4 weeks for the pathogenic strains). Cells of the donor strain *E. coli* ß2163, which is auxotroph for DAP, should be eliminated after plating. Alternatively, the transposon can be delivered to Leptospira cells by electroporation (see Note 4).
5. Screen the library for the appropriate phenotype.

3.4. Identification of the Transposon Insertion Site in Each Mutant

Once the library has been screened and the mutants of interest have been isolated, the transposon insertion site is identified using the semirandom PCR technique.

1. The PCR can be performed directly on colonies by picking the colony from the plate with a pipette and mixing it with 25 μL of water in an Eppendorf tube. The mixture is then heated at 95°C for 15 min. His crude extract is then used for the PCR. Otherwise, genomic DNA is extracted and used as a template for subsequent PCR (Fig. 2).

   Set up the following reaction for the first PCR round:

Reagent	1× reaction
dNTPs, 25 mM	0.5
10× PCR	2.5
$MgCl_2$, 25 mM	1.5
Primer 1, 10 mM (TnKl or TnK2)	1.25
Primer 2, 10 mM (Degl or Deg2)	1.25
Taq	0.2
Water	16.5
Template (~40 ng DNA total)	1.3
Total	25

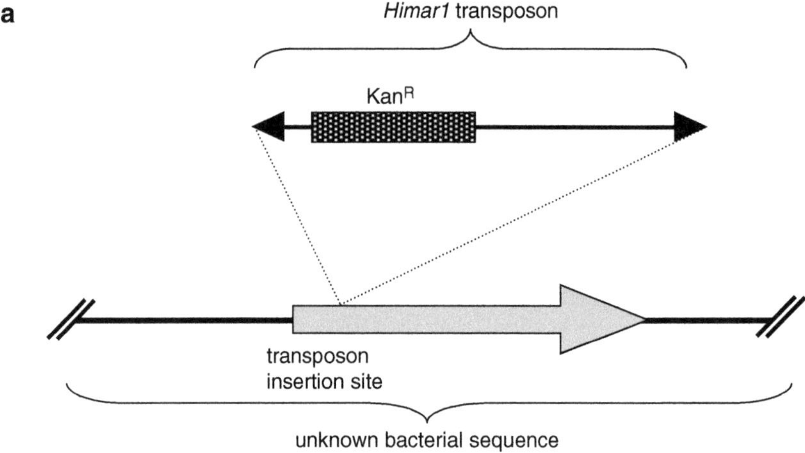

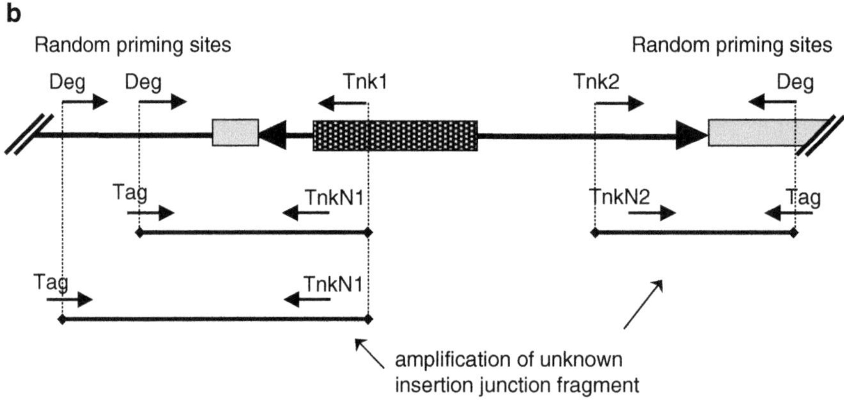

Fig. 2. Identification of the transposition insertion sites by semirandom PCR.

2. Run the following program on the thermal cycler: denaturation: 95°C 5 min; amplification (40 cycles) 95°C 15 s/40°C 1 min/72°C 2 min; final extension: 72°C 10 min. This round of PCR rarely produces an amplification product visible on a gel.
3. Set up the following reaction for the second PCR round:

Reagent	1× reaction
dNTPs, 25 mM	0.5
10× PCR	2.5
MgCl2, 25 mM	1.5
Oligo 3, 100 mM (TnkN1 or TnkN2)	0.25
Oligo 4, 100 mM (Tag)	0.25
Taq	0.2
Water	19
Template-PCR round 1	0.8
Total	25

4. Then, run the following program on the thermal cycler: denaturation: 95°C 5 min; amplification (35 cycles): 95°C 15 s/55°C 30 s/72°C 2 min; final extension: 72°C 10 min.
5. Represent the various amplification products obtained with degenerate primers that annealed in the vicinity of the transposon (Fig. 2).
6. The PCR are then sent to sequencing, according to the guidelines of the service provider, using the nested primer that anneals in the transposon (TnKN1 or TnKN2 in the case of pCjTKS2).
7. The transposon insertion site is then identified by comparing the resulting sequence with the *Leptospira* genome of interest by BLAST analysis (http://blast.ncbi.nlm.nih.gov/) or using the SpiroScope (http://www.genoscope.cns.fr/agc/mage) database (8). Localization of the insertion is confirmed by PCR with primers annealing to the region flanking the expected transposon insertion site on the chromosome of the mutant strain as well as the wild-type strain. The size of the amplification product for the mutant strain should be in agreement with the size of the amplicon for the wild-type strain plus the size of the transposon.

4. Notes

1. Spontaneous Spc^R clones appear more frequently than Kan^R clones (9). It is, therefore, recommended to use the transposon carrying the kanamycin-resistance gene.
2. Strains other than *L. biflexa*, including pathogenic strains, can be used as recipient cells (4).
3. Spreading 50 μL/plate usually yields a good number of isolated colonies/plate. However, this depends on the efficiency of the conjugation.
4. Optimum transformation efficiency was obtained with field strengths of 10–15 kV/cm (10).

Acknowledgments

We thank Gerald Murray (Monash University, Australia) for sharing the protocol of semirandom PCR.

References

1. Louvel, H., and Picardeau, M. (2007) Genetic Manipulation of *Leptospira biflexa*, J. Wiley and Sons, Hoboken, NJ
2. Picardeau, M., et al. (2008) Genome sequence of the saprophyte *Leptospira biflexa* provides insights into the evolution of *Leptospira* and the pathogenesis of leptospirosis, PLoS ONE 3, e1607
3. Louvel, H., Saint Girons, I., and Picardeau, M. (2005) Isolation and characterization of FecA- and FeoB-mediated iron acquisition systems of the spirochete *Leptospira biflexa* by random insertional mutagenesis, J. Bacteriol. 187, 3249–3254
4. Picardeau, M. (2008) Conjugative transfer between *Escherichia coli* and *Leptospira* spp. as a new genetic tool, Appl. Environ. Microbiol. 74, 319–322
5. Louvel, H., et al. (2006) Comparative and functional genomic analyses of iron transport and regulation in *Leptospira* spp., J. Bacteriol. 188, 7893–7904
6. Ellinghausen, H. C., and McCullough, W. G. (1965) Nutrition of *Leptospira* pomona and growth of 13 other serotypes: fractionation of oleic albumin complex and a medium of bovine albumin and polysorbate 80, Am. J. Vet. Res. 26, 45–51
7. Johnson, R. C., and Harris, V. G. (1967) Differentiation of pathogenic and saprophytic leptospires, J. Bacteriol. 94, 27–31
8. Vallenet, D., et al. (2009) MicroScope: a platform for microbial genome annotation and comparative genomics, Database (Oxford) bap021
9. Poggi, D., et al. (2010) Antibiotic resistance markers for genetic manipulations of *Leptospira* spp., Appl. Environ. Microbiol. 76, 4882–4885
10. Murray, G. L., et al. (2009) Genome-wide transposon mutagenesis in pathogenic *Leptospira* spp., Infect. Immun. 77, 810–816
11. Lampe, D. J., et al. (1999) Hyperactive transposase mutants of the *Himar1* mariner transposon, Proc Natl Acad Sci U S A. 96, 11428–11433

Chapter 10

Ac–Ds Solutions for Rice Insertion Mutagenesis

Emmanuel Guiderdoni and Pascal Gantet

Abstract

Rice is the model plant for monocotyledons. Since the completion of the high-quality sequence of its genome, the international community is deploying efforts to identify the function of the 30–40,000 non-transposable element genes of rice. These efforts comprise the creation of large collections of rice mutants accessible to the international scientific community. In addition to loss of function mutants, insertion mutagenesis using *Agrobacterium*-mediated transformation and engineered mobile elements allows the identification of genes through enhancer or gene trapping or activation tagging. The maize transposable element *Ac–Ds* is known to be active in rice since the early 1990s and it does not interfere with endogenous rice transposons. This is the guaranty that induced mutation obtained with the mobility of the *Ds* element will be stable when the source of *Ac* transposase is removed from the mutated genome. In this paper, we describe single- or double-component T-DNA constructs that have been used to introduce a functional *Ac–Ds* system in rice and allowed the generation and selection of different type of *Ds* insertion mutations in the rice genome.

Key words: Rice, Functional genomics, *Ac–Ds*, Insertion mutagenesis, Enhancer trapping, Gene trapping, Activation tagging

1. Introduction

Since its adoption in 1985 by the international community as the monocot model plant, the rice genome has been sequenced and annotated (1, 2). The challenge is now to assign a function to the 30–40,000 nontransposable element rice genes. Mutagenesis, followed by phenotypic screens, is the more straightforward approach to reach that objective. Chemical, physical, and insertion mutagenesis have been used to generate rice mutants. Chemical agents, such as ethyl methanesulfonate (EMS), or physical agents, such as γ-ray or fast neutrons, have been used to generate mutant populations (3). Aside their use in forward genetics, these populations have become

Yves Bigot (ed.), *Mobile Genetic Elements: Protocols and Genomic Applications*, Methods in Molecular Biology, vol. 859, DOI 10.1007/978-1-61779-603-6_10, © Springer Science+Business Media, LLC 2012

in the last decade hopeful tools for reverse genetics as well since identification of point mutations in a given sequence in DNA pools or of deletions in a determined region is now possible through PCR-based methods, such as TILLING and respectively hybridization on DNA arrays, respectively (4, 5). An alternative way is to use insertion mutagenesis that allows to tag the mutation with a known DNA sequence and to sequence insertion points in the genome. Since the late 1990s, different insertion mutagens have been used in rice. The *Ty1-copia* retrotransposon *Tos17* was first extensively used. *Tos17* exists in low copy number in rice (between 1 and 10 copies by genome). Under natural conditions, *Tos17* is inactive, but the element can be activated by in vitro callus culture (6). Due to these favorable features, *Tos17* has been used for the genome-wide mutagenesis of the standard *japonica* cultivar Nipponbare and an average of 10 stably integrated copies scattered over the genome are observed in plants regenerated from 5-month-old callus cultures. A collection of 51,625 regenerated rice lines carrying 510,000 insertions has been produced. To date, most of the current successful examples of forward and reverse analysis of rice insertion mutants involved the use of the *Tos17* library (7). Thanks to the progresses in *Agrobacterium*-mediated transformation of rice, T-DNA libraries are being generated worldwide, notably in Korea, China, Taiwan, and France. The Chinese rice functional genomics program has produced more than 230,000 insertion lines and the current goal is to produce 100,000 flanking sequenced tags (FSTs) (8). *Agrobacterium*-mediated transformation was also used to mobilize in the rice genome engineered two-component maize transposable element systems, such as *En/Spm-dSpm* (9, 10) and *Ac–Ds* (11–13). Further studies used preferentially the *Ac–Ds* transposable element which has no interference with endogenous rice transposable element, which may be the case with *En/Spm-dSpm* (14). Aside its disruption potential, insertion mutagenesis with T-DNA and engineered transposable elements allows further gene identification through enhancer/gene trapping or activation tagging. An enhancer trap comprises a reporter gene fused to a minimal promoter, containing a TATA box and a transcriptional start site not transcriptionally active but whose transcription can be triggered by neighboring chromosomal enhancer elements. The minimal promoter is controlling a reporter gene and allows detection of the activity of gene promoters during development or in response to signals. These enhancer trap insertions tend to conduct to a high frequency of gene detection but do not always correspond to a disruption of the detected gene. This is an advantage to characterize expression profile of essential genes for which disruptive mutation and K.O. can be lethal. A gene trap contains a promoterless reporter gene, the expression of which occurs only when insertion is within a transcriptional unit and in correct orientation. The presence of one or more splice acceptor sites aligned in all reading frames, preceding the reporter gene, allows expression if insertion occurs in an intron. Frequency of expression is generally lower compared to enhancer trap, but corresponds to insertion within genes and more likely to knockouts. When insertion occurs in an intron or an exon, translational fusions are generated between the reporter gene and upstream exons of the interrupted gene. This may create translational fusions which provide information about protein localization (12, 15). An Activation Tag typically contains multimerized transcriptional enhancers (e.g., −343 to −90 fragment from the CaMV 35S promoter) conducting to ectopic activation of expression of the neighboring gene as gain of function mutation (16).

The maize *Ac* transposable element was shown active in rice since the early 1990s when its functionality was demonstrated in rice protoplasts (17). Since this period, sophisticated double-component systems using an *Ac* transposase source dissociated from a nonautonomous *Ds* element have been engineered. Several studies reported the use of a single-T-DNA to carry both AcTpase source and *Ds* element into the rice genome (11–13) or using two parent lines, one transformed with a T-DNA carrying the AcTpase source and the other one transformed with a T-DNA carrying an engineered *Ds* element (13, 18). In this latter case, the two components of the transposable element are joined in the same genome by crossing. After the mobilization of the *Ds* element, it can be segregated out from the transposase source in subsequent generation to stabilize the insertion at a given genome site.

In this paper, we illustrate and comment representative constructs and screening methods that have been used to generate different type of insertion mutants using the *Ac–Ds* transposable element for rice gene function discovery.

2. Methods

2.1. Constructions Used for Ac–Ds-Mediated Enhancer Trapping: A Case Study

A single-T-DNA/double-component system was created at Plant Research International, Wageningen, The Netherlands, and used in the frame of a European consortium for rice insertion mutagenesis EU-OSTID (11, 19). From the T-DNA left border (LB) to the T-DNA right border (RB), the construction contained the following (Fig. 1):

- The hygromycin phosphotransferase (*Hpt*) gene placed under the control of the *p35S* promoter. This gene is used for selection of transformed cells and is indicative of the presence of an expressed T-DNA on the basis of hygromycin resistance.
- The *su1* gene from *Streptomyces grisoleus* (20) encoding a cytochrome P450 enzyme that can convert the pro-herbicide R7402 (DuPont, Wilmington, Del.) into a cytotoxic form (21). This gene is driven by the *PSsu* rubisco small subunit

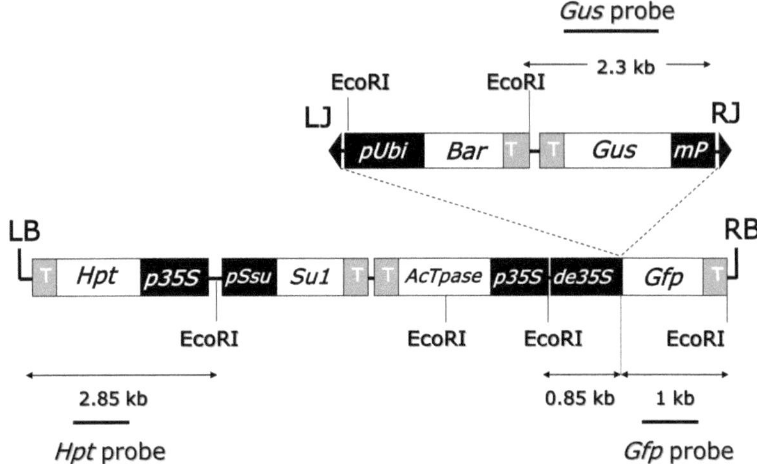

Fig. 1. Single-T-DNA construct used for enhancer trap Ac–Ds insertion mutagenesis (11). The position of *Eco*R1 sites is indicated as well as the probes used for Southern blot analysis and the expected size of the hybridized fragment. *AcTpase* gene encoding the *Ac* transposase, *Bar* gene encoding the phosphinothricin acetyltransferase, *de35S* double-enhancer CaMV 35S promoter, *Gfp* gene encoding the green fluorescent protein, *Gus* gene encoding the β-glucuronidase, *Hpt* hygromycin phosphotransferase gene, *LB* left T-DNA border, *LJ* left junction of the artificial *Ds* element, *mP* minimal promoter, *p35S* CaMV 35S promoter, *pSsu* rubisco small subunit promoter, *pUbi* promoter of the maize ubiquitin gene, *RB* right T-DNA border, *RJ* right junction of the artificial *Ds* element, *Su1* cytochrome P450 gene used as negative marker of selection, *T* terminator.

promoter. This gene is used as a negative selection marker to eliminate among progeny plants those still carrying the AcTpase source, and select for stabilized *Ds* inserts.

- The *AcTpase* gene under the control of the constitutive promoter *p35S*, as a source of AcTpase.
- The green fluorescent protein (*Gfp*) gene placed under the control of the *CaMV 35S* promoter with a doubled enhancer sequence (*de35S*). In the complete T-DNA, the engineered *Ds* element is inserted between the *Gfp* coding sequence and the *de35S* promoter. When the *Ds* element is excised by the AcTpase, the *Gfp* gene falls under the control of the *de35S* and is expressed. As GFP expression is a visual marker of *Ds* excision, it can be used to manually select transformed hygromycin-resistant cell lines, where the *Ds* element has transposed. These GFP-positive cell lines can regenerate plants with reinserted *Ds* element(s).
- The *Ds* element contains between its right (*Rj*) and left (*Lj*) junctions the phosphinothricin acetyltransferase gene (*Bar*) under the control of the constitutive promoter of the maize ubiquitin gene (*PUbi*), and the *Gus* gene encoding the β-glucuronidase under the control of a minimal promoter (*mP*). *Bar* gene expression allows to select on the basis of the

Basta herbicide resistance plants bearing a *Ds* element still in its T-DNA launching pad or reinserted anywhere in the genome. The *Gus-mP* can be transcriptionally activated by enhancer elements of nearby genes.

This T-DNA construct allows to select primary regenerants following *Agrobacterium* cocultivation on the basis of cell resistance to hygromycin for the integration of T-DNA and on the basis of GFP fluorescence to select transgenic calli, where the *Ds* element was excised. Following molecular characterization of active lines in T0 generation, T1 progeny plants still containing the T-DNA carrying the AcTpase source can be eliminated on the basis of their susceptibility to the pro-herbicide R7402, and those containing a stabilized unlinked Ds insertion can be selected on the basis of their resistance to the herbicide Basta.

*Eco*R1 restriction of plant genomic DNA and DNA blot hybridization with a probe specific to the *Gfp* gene is expected to reveal a diagnostic 1.85 or 3.3 kbp hybridizing fragments when *Ds* has excised from the T-DNA launching pad or remained in its T-DNA launching pad, respectively. Using a probe specific to the *Gus* coding sequence, detection of a diagnostic 3.3 kb fragment and of fragments of sizes larger than 2.3 kb is anticipated in case of an inactivity of the *Ds* element and *Ds* insertion at various sites in the genome, respectively. The *Hpt* probe reveals fragments bigger than 2.85 kb. Their number is function of the number of T-DNA insertion sites in the genome.

Using this construction, a collection of transposon *Ac–Ds* enhancer trap lines have been developed in rice. Transposition events occurred in 90% of the T1 plants and corresponded mostly (90%) to single *Ds* insertion events. Seventy percent of the events was stable in further generation. Eighty percent of the tested insertion occurred in genes, but few lines among 536 T1 plants tested revealed GUS activity (see Note 1). This collection is now available for the scientific community, with 10,000 independent insertion lines, including 5,000 lines for which *Ds* FSTs have been determined (7).

2.2. Constructions Used for Ac–Ds-Mediated Gene Trapping: A Case Study

A two-T-DNA, double-*Ac–Ds*-component system was developed by V. Sundaresan group in Temasek, Singapore, to induce Gene Trap *Ds* insertion mutant rice collection (13). In this case, the AcTpase source is introduced in a parent line while the inactive *Ds* element is introduced in a second parent line. *Ds* mobilization is induced by crossing the two parent lines, and can be stopped in the progeny by segregating out the T-DNA carrying the AcTpase source in progeny plants.

The T-DNA carrying the AcTpase contains, from the T-DNA left border to the T-DNA right border, the following elements (Fig. 2a):

- The *Hpt* gene placed under the control of the *p35S* promoter to select transformed cells on the basis of hygromycin resistance.

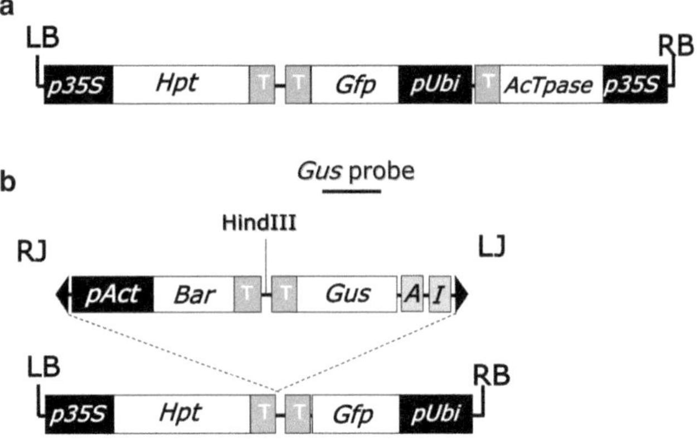

Fig. 2. Two-T-DNA, double-Ac–Ds-component system used for gene trap Ac–Ds insertion mutagenesis (13). (**a**) AcTpase carrier T-DNA, (**b**) *Ds* element carrier T-DNA. The position of *Hin*dIII site is indicated as well as the probe used for Southern blot analysis. *A* triple splice acceptor, *AcTpase* gene encoding the *Ac* transposase, *Bar* gene encoding the phosphinothricin acetyltransferase, *Gfp* gene encoding the green fluorescent protein, *Gus* gene encoding the β-glucuronidase, *Hpt* Hygromycin phosphotransferase gene, *I* intron from the *Arabidopsis* G protein gene, *LB* left T-DNA border, *LJ* left junction of the artificial *Ds* element, *p35S* CaMV 35S promoter, *pAct* promoter of the rice actin gene, *pUbi* promoter of the maize ubiquitin gene, *RB* right T-DNA border, *RJ* right junction of the artificial *Ds* element, *T* terminator.

- The *Gfp* gene placed under the control of *pUbi* promoter: This gene is used for a negative selection of the plants carrying the AcTpase source.
- The *AcTpase* gene under the control of the *p35S* promoter.

The T-DNA carrying the *Ds* element contains, from the T-DNA left border to the T-DNA right border, the following elements (Fig. 2b):

- The *Hpt* gene placed under the control of the *p35S* promoter to select transformed cells on the basis of hygromycin resistance.
- The *Gfp* gene placed under the control of *pUbi* promoter: This gene is used for a negative selection of the plants carrying the empty T-DNA after *Ds* transposition.

The *Ds* element, flanked by its right and left junctions (*RJ* and *LJ*, respectively), is inserted between the *Hpt* and the *Gfp* gene. It comprises the following:

- The *Bar* gene under the control of the constitutive promoter of rice actin gene (*pAct*): In this context, the *Bar* gene is used as a positive marker to select plants carrying a *Ds* element.
- The *Gus* gene preceded by an intron from the *Arabidopsis* G protein gene (*I*) and a triple splice acceptor (*A*): These two last

components promote the fusion of the *Gus* sequence into an open reading frame and the translation of a functional GUS protein. Following GUS staining, this allows to visualize the expression pattern of the gene in which the *Ds* element is inserted.

The *Hin*dIII restriction site located between the *Bar* and *Gus* genes is used to reveal *Ds* element insertion after cutting the genomic DNA by *Hin*dIII and DNA blotting using a probe specific to the *Gus* coding sequence. Different sizes of hybridized fragments reveal different *Ds* insertion and the number of hybridized fragments corresponds to the number of new *Ds* inserts.

The two T-DNAs are stably integrated in independent parent lines. Transformed cells are selected on the basis of resistance against hygromycin. The presence of the T-DNA in the plant can be checked also on the basis of GFP expression. After crossing a plant carrying the AcTpase source with a plant carrying the *Ds* element, F2 seedlings derived from seeds harvested on F1 plants carrying a mobilized *Ds* element can be selected first by a negative selection based on GFP expression to eliminate plants still containing the T-DNA bearing the source of AcTpase, and second on the base of their resistance to Basta to select plants containing a *Ds* element. Southern blot analysis confirmed the presence of at least two new *Ds* inserts in 80% of the T2 selected families. Analysis of 2,057 *Ds* flanking sequences in individual plants showed that 88% contained a unique *Ds* insertion. FST analysis revealed that the *Ds* insertions were distributed randomly throughout the genome and a preference of *Ds* insertion into gene-coding regions. Nevertheless, chromosomes 4 and 7 presented two fold as many insertions as that expected and a hot spot of insertion within a 40-kpb region of chromosome 7 was identified (see Note 2). Twenty thousand of these *Ds* lines were further analyzed, and mutations conferring an increased yield or resistance to abiotic stresses, such as drought, salinity, and cold, were identified (22). Among 2,852 lines analyzed, 8% presented GUS staining pattern in various organs and at different stages of development (see Note 1). Interestingly, assays showed that for some plants presenting an interesting phenotype due to a new *Ds* insert remobilization of the element can be achieved by crossing with plants expressing AcTpase, still allowing to recover plants presenting a favorable phenotype in the progeny devoid of any foreign DNA sequences. In this case, mutation results from the footprint left upon excision of the *Ds* element. Mutants of this collection are available for the international community (7).

A variant of this two-T-DNAs, double-*Ac–Ds*-component system, using a basta-resistance excision marker system, was also used in a large-scale field screening procedure to generate successfully a gene trap *Ds* insert mutant library (23).

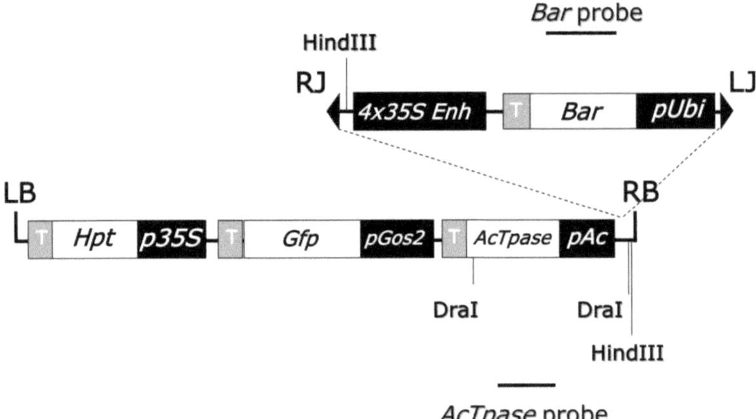

Fig. 3. Single-T-DNA double-component construct used for Ac-Ds activation tagging insertion mutagenesis (Van Arkel and Pereira, unpublished data). *4x35SEnh* tetramerized enhancers of the CaMV 35S promoter, *AcTpase* gene encoding the *Ac* transposase, *Bar* gene encoding the phosphinothricin acetyltransferase, *Gfp* gene encoding the green fluorescent protein, *Hpt* hygromycin phosphotransferase gene, *LB* left T-DNA border, *LJ* left junction of the artificial *Ds* element, *p35S* CaMV 35S promoter, *pAc* native promoter of the *AcTpase* gene, *pGos2* promoter of rice *Gos2* gene, *pUbi* promoter of the maize ubiquitin gene, *RB* right T-DNA border, *RJ* right junction of the artificial *Ds* element, *T* terminator.

2.3. Constructions Used for Ac–Ds Activation Tagging: A Case Study

This single-T-DNA/double-component system devoted to Activation Tagging *Ds* insertion mutagenesis was prepared at Plant Research International, Wageningen, The Netherlands (G. Van Arkel, A. Pereira, pers. com), and introduced into rice. From the T-DNA left border (LB) to the T-DNA right border, the construct contained the following (Fig. 3):

- The *Hpt* gene placed under the control of the *p35S* promoter to select transformed cells on the basis of hygromycin resistance.
- The *Gfp* gene placed under the control of the constitutive *pGos2* promoter. This gene cassette allows first a positive selection of transgenic callus expressing the GFP and containing the T-DNA, and seconds a negative selection of plants still carrying the T-DNA and the AcTpase source after *Ds* transposition.
- The *AcTpase* gene under the control of its native maize *pAc* promoter.

These elements are followed by the *Ds* mobile element which comprises, from the right junction to the left junction, the following:

- A tetramerized enhancer element of the CaMV 35S promoter which acts as a transcriptional activator of expression for genes in which or in the vicinity of which the *Ds* element inserts.

- The *Bar* gene under the control of the constitutive promoter *pUbi*. The *Bar* gene is a positive marker to select plants which carry a *Ds* element either at its T-DNA position or inserted in the plant genome.

The *Hin*dIII restriction sites located just after the RJ inside the *Ds* element and just after the LJ in the T-DNA allow generation of diagnostic fragments for excision/insertion of the *Ds* element. Genomic DNA restricted with *Hin*dIII and hybridized by Southern blot to a *Bar*-specific probe will reveal a hybridized fragment of 3.6 kbp whether the *Ds* element is still present in the T-DNA or fragments of various sizes if the *Ds* element has transposed to other sites in the genome. In this last case, the number of fragments of different size corresponds to the number of *Ds* insertions present in the genome at different sites. Blot hybridization of a *Dra*I DNA digest to an *AcTpase* probe reveals a diagnostic fragment of 3.8 or 7.6 kbp if the *Ds* element has or has not excised from the T-DNA, respectively.

Transformed cells are selected on the basis of resistance to hygromycin and of GFP expression. T1 plants exhibiting GFP expression are eliminated and remaining plants are selected on the base of Basta resistance to select for stable *Ds* insertions. Plants harboring stable new *Ds* insertion without the T-DNA AcTpase source can be further analyzed at the phenotypic and genotypic levels. A collection using this construction has been created and is under evaluation (Meynard et al., unpublished results).

3. Notes

1. In both enhancer trapping or gene trapping insertion mutagenesis, only a small part of the plants was shown to exhibit GUS activity in the tested organs, with a frequency ranging from 1 to 8% (11, 13). Similar results were previously reported in *Ac–Ds* mutagenesis in rice using other constructs (12, 24). This rather low frequency compared to T-DNA enhancer/gene trapping (25, 26) is apparently a limiting step for the use of engineered *Ds* elements for gene detection. On the contrary, high frequency of GUS expression in T-DNA enhancer trap lines might be artifact and due to integration of rearranged copies of T-DNA placing enhancer elements of the T-DNA itself as activator of the minimal promoter of the enhancer trap. Also, the *Ds* inserted element might particularly be prone to silencing mechanisms which are commonly involved in the defense of genomes against invasive transposable elements. In rice, it seems that the presence of several *Ds* elements or maintenance for more than one generation of an active AcTpase in the presence of *Ds* element increases the silencing of the genes carried by the *Ds* element. For this reason, it is likely preferable

to remove as soon as possible the source of AcTpase and to select lines carrying only one or a few number of *Ds* elements inserted in the genome. The incidence of *Ds* silencing on the activation tagging mutants has not been yet evaluated.

2. Examination of distribution of *Ds* inserts in the rice genome revealed that this element preferentially inserts into genes. This is a significant advantage of the *Ac–Ds* system to generate mutation in rice genes since T-DNA mutagenesis creates insertions both in genic and intergenic regions with a similar frequency. Nevertheless, in the study conducted by Kolesnik et al. (13), it is reported that the *Ds* insertion was not homogeneous throughout the genome. For example, chromosome 9, 11, and 12 exhibit a low frequency of *Ds* inserts that could be related to the heterochromatic structure of these chromosomes (27, 28). The chromosome 1 was enriched in *Ds* insertions. The presence of two *Ds* T-DNA launching pads in this chromosome in this study could explain this by a high rate of intrachromosomal insertions, frequent for the *Ac–Ds* system that often transposes in the vicinity of its launching pad (29). Nevertheless, interchromosomal transposition events of the *Ds* element occurred more frequently in chromosome 7 and 4. In chromosome 7, a hot spot of *Ds* insertion was identified in a region of 40 kpb while interchromosomal clusters were also observed in other chromosomes. It was hypothesized that these clusters could result from a physical proximity of donor and acceptor sites during the replication process (30). These biases of insertion among the rice genome stress another limiting feature of *Ac–Ds* mutagenesis in rice. It also stresses the necessity to use different types of mutagen exhibiting complementary insertion preferences to insure a complete coverage of the genome and to get a chance to generate a mutation in any rice gene.

References

1. Itoh T, et al. (2007) Curated genome annotation of Oryza sativa ssp. japonica and comparative genome analysis with Arabidopsis thaliana. Genome Res 17:175–183
2. Yu J, et al. (2002) A draft sequence of the rice genome (Oryza sativa L. ssp. indica). Science 296:79–92.
3. Hirochika H, et al. (2004) Rice mutant resources for gene discovery. Plant Mol Biol 54:325–334.
4. Wu JL, et al. (2005) Chemical- and irradiation-induced mutants of indica rice IR64 for forward and reverse genetics. Plant Mol Biol 59:85–97.
5. Till BJ, et al. (2007) Discovery of chemically induced mutations in rice by TILLING. BMC Plant Biol 7:19.
6. Hirochika H (2001) Contribution of the Tos17 retrotransposon to rice functional genomics. Cur Opin Plant Biol 4:118–122.
7. Krishnan A, et al. (2009) Mutant resources in rice for functional genomics of the grasses. Plant Physiol 149:165–170.
8. Xue,Y, Li J, Xu Z (2003) Recent highlights of the China Rice Functional Genomics Program. Trends Genet 19:390–394.
9. Greco R, et al. (2004) Transcription and somatic transposition of the maize En/Spm transposon system in rice. Mol Gen Genomics 270:514–523.
10. Kumar CS, Wing RA, Sundaresan V (2005) Efficient insertional mutagenesis in rice using the maize En/Spm elements. Plant J 44:879–892.

11. Greco R, et al. (2003) Transpositional behaviour of an Ac/Ds system for reverse genetics in rice. Theor Appl Genet 108:10–24.
12. Chin HG, et al. (1999) Molecular analysis of rice plants harboring an Ac/Ds transposable element-mediated gene trapping system. Plant J 19:615–623.
13. Kolesnik T, et al. (2004) Establishing an efficient Ac/Ds tagging system in rice: large-scale analysis of Ds flanking sequences. Plant J 37:301–314.
14. Greco R, et al. (2004) Transcription and somatic transposition of the maize En/Spm transposon system in rice. Mol Genet Genomics 270:514–523.
15. Jeon JS et al. (2000) T-DNA insertional mutagenesis for functional genomics in rice. The Plant J 22:561–570.
16. Jeong DH, et al. (2002) T-DNA insertional mutagenesis for activation tagging in rice. Plant Physiol 130:1636–1644.
17. Enoki H, et al. (1999) Ac as a tool for the functional genomics of rice. Plant J 19:605–613.
18. Nakagawa Y, et al. (2000) Frequency and pattern of transposition of the maize transposable element Ds in transgenic rice plants. Plant Cell Physiol 41:733–742.
19. van Enckevort LJ, et al. (2005) EU-OSTID: a collection of transposon insertional mutants for functional genomics in rice. Plant Mol Biol 59:99–110.
20. O'Keefe DP et al. (1994) Plant Expression of a Bacterial Cytochrome P450 That Catalyzes Activation of a Sulfonylurea Pro-Herbicide. Plant Physiol 105:473–482.
21. Tissier AF, et al. (1999) Multiple independent defective suppressor-mutator transposon insertions in Arabidopsis: a tool for functional genomics. Plant Cell 11:1841–1852.
22. Jiang SY, et al. (2007) Ds insertion mutagenesis as an efficient tool to produce diverse variations for rice breeding. Plant Mol Biol 65:385–402.
23. Luan WJ, et al. (2008) An efficient field screening procedure for identifying transposants for constructing an Ac/Ds-based insertional-mutant library of rice. Genome 51:41–49.
24. Izawa T, et al. (1997) Transposon tagging in rice. Plant Mol Biol 35219–229.
25. Johnson AA, et al. (2005) Spatial control of transgene expression in rice (Oryza sativa L.) using the GAL4 enhancer trapping system. Plant J 41:779–789.
26. Sallaud C, et al. (2004) High throughput T-DNA insertion mutagenesis in rice: a first step towards in silico reverse genetics. Plant J 39:450–464.
27. Chen M, et al. (2002) An integrated physical and genetic map of the rice genome. The Plant Cell 14:537–545.
28. Zhao Q, et al (2002) A fine physical map of the rice chromosome 4. Genome Research 12:817–823.
29. Healy J, et al. (1993) Linked and unlinked transposition of a genetically marked Dissociation element in transgenic tomato. Genetics 134:571–584.
30. Dooner HK, et al. (1994) Distribution of unlinked receptor sites for transposed Ac elements from the bz-m2(Ac) allele in maize. Genetics 136:261–279.

Chapter 11

Engineering the *Caenorhabditis elegans* Genome by *Mos1*-Induced Transgene-Instructed Gene Conversion

Valérie J.P. Robert

Abstract

Mos1-induced *t*ransgene-*i*nstructed gene *c*onversion (*Mos*TIC) is a technique of choice to engineer the genome of the nematode *Caenorhabditis elegans*. *Mos*TIC is initiated by the excision of *Mos1*, a DNA transposon of the Tc1/Mariner super family that can be mobilized in the germ line of *C. elegans*. *Mos1* excision creates a DNA double-strand break that is repaired by several cellular mechanisms, including transgene-instructed gene conversion. For *Mos*TIC, the transgenic repair template used by the gene conversion machinery is made of sequences that share DNA homologies with the genomic region to engineer and carries the modifications to be introduced in the genome. In this chapter, we present two *Mos*TIC protocols routinely used.

Key words: Genome engineering, Knock-in allele, Knockout allele, *Caenorhabditis elegans*, Tc1/Mariner family, Protein tagging

1. Introduction

Knock-in and knockout alleles are very useful tools to study gene function. Their engineering is based on homologous recombination between a transgene carrying modifications (tags, deletions, or point mutations) to introduce into the genome and the endogenous locus to modify. In the model organism *Caenorhabditis elegans*, such techniques turned out to be very inefficient. To explain this, it was hypothesized that standard transgenes that, in *C. elegans*, are made by germ line microinjection and are highly repeated concatemerized molecules with specific chromatin structure might be unsuitable substrates for recombination. Based on this hypothesis, two independent strategies were employed to try to improve transgene–genome recombination efficiency. The first one relies on the use of low-copy transgenes generated by biolistic transformation

Yves Bigot (ed.), *Mobile Genetic Elements: Protocols and Genomic Applications*, Methods in Molecular Biology, vol. 859, DOI 10.1007/978-1-61779-603-6_11, © Springer Science+Business Media, LLC 2012

and that are expected to have a chromatin structure more suitable for recombination (1). The second one is based on the observation that, in the genome of the *C. elegans* germ line, a targeted DNA double-strand break (DSB) induced by the mobilization of preexisting insertions of homologous or heterologous transposable elements of the Tc1/Mariner superfamily can be repaired by transgene-instructed gene conversion (2–4). Using this strategy, *Mos1*-induced *t*ransgene-*i*nstructed gene *c*onversion (*Mos*TIC) (Fig. 1), an efficient and easy-to-implement technique, was recently developed to engineer the *C. elegans* genome and added to the *C. elegans* genetic and genomic toolbox (5, 6). In this chapter, we describe *Mos*TIC.

To initiate *Mos*TIC, the formation of a DSB is experimentally induced in the germ line DNA by the mobilization of a preexisting insertion of the heterologous DNA transposon *Mos1* (7) present in the region to engineer (Fig. 1). DSB formation occurs in a genetic

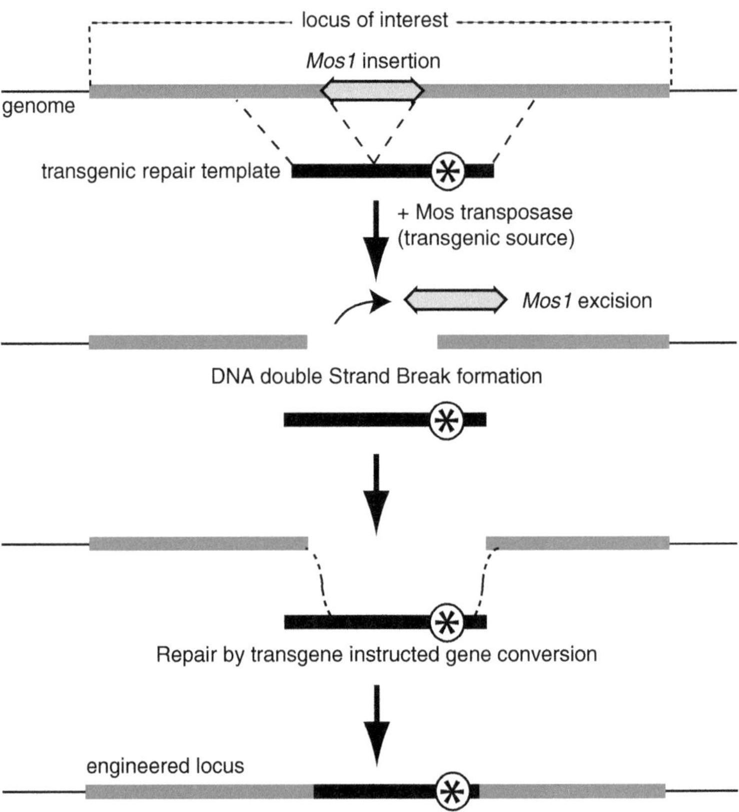

Fig. 1. Principle of *Mos*TIC. *Mos*TIC is induced by the excision of a preexisting insertion of *Mos1*, a Drosophila member of the Tc1/Mariner family, experimentally introduced in *C. elegans*. *Mos1* insertion makes a DNA double-strand break that can be repaired by gene conversion using a transgenic repair template carrying the modifications to be introduced in the genome.

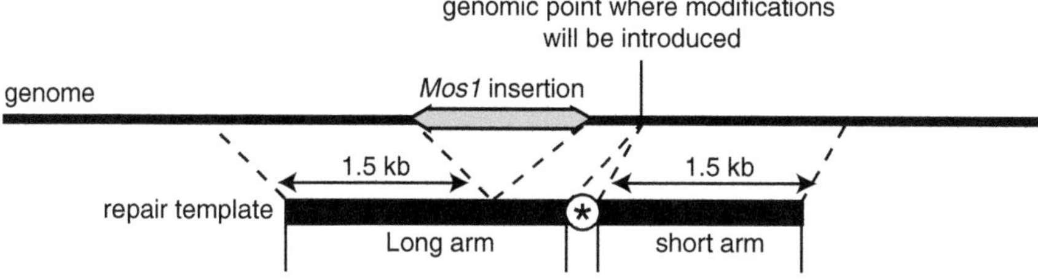

Fig. 2. The repair template contains (1) the modifications to introduce in the genome (represented with an *asterisk*); (2) genomic region localized between the genomic point, where modifications have to be introduced and the *Mos1* insertion point; and (3) two arms that are homologous to the DNA region to engineer and are at least 1.5 kb long.

background containing a transgene repair template (Fig. 2), which shares DNA homologies with the broken genomic region and carries modifications (deletions, tags, point mutations…) that have to be introduced in the genome. During DSB repair in the germ line, this transgene can be used as a repair template and animals carrying an engineered genome are recovered in the progeny of the worms in which DSB was induced.

2. Materials

2.1. Solutions and Buffers

1. Normal growth medium (NGM). See ref. 8.

   Equipment and reagents

   - NaCl.
   - Agar.
   - Peptone.
   - 5 mg/ml cholesterol in ethanol (do not autoclave).
   - 1 M KPO_4 buffer, pH 6.0 (108.3 g KH_2PO_4, 35.6 g K_2HPO_4, H_2O to 1 L).
   - 1 M $MgSO_4$.
   - Petri plates.

   Methods

   Mix 3 g NaCl, 17 g agar, and 2.5 g peptone in a 2-L Erlenmeyer flask. Add 975 ml H_2O. Cover the mouth of flask with aluminum foil. Autoclave for 50 min. Cool flask in 55°C water bath for 15 min. Add 1 ml 1 M $CaCl_2$, 1 ml 5 mg/ml cholesterol in ethanol, 1 ml 1 M $MgSO_4$, and 25 ml 1 M KPO_4 buffer. Swirl to mix well.

   Dispense the NGM solution into petri plates. Fill plates two-third full of agar.

   Leave plates at room temperature for 2–3 days before use to allow for detection of contaminants and to allow excess moisture to evaporate. Plates stored in an airtight container at room temperature is usable for several weeks.

2. M9 buffer: 22 mM KH_2PO_4, 22 mM Na_2HPO_4, 85 mM NaCl, 1 mM $MgSO_4$. Sterilize by autoclaving. Store at room temperature. Alternatively, you can make a 10× stock that you will keep sterile and dilute when necessary.
3. Lysis buffer: 50 mM KCl, 10 mM Tris-HCl, pH 8.2, 25 mM $MgCl_2$, 0.45% NP-40, 0.45% Tween-20, 0.01% gelatin. Filter sterilize, aliquot in 1.5-ml tubes, and store at −20°C.
4. Proteinase K at 20 mg/ml (Eurobio GEXPRK00-6R).

2.2. Identification of a C. elegans Strain with a Mos1 Insertion in the Region of Interest

*Mos*TIC is initiated by the excision of a preexisting insertion of the Drosophila mauritiana transposon *Mos1* that was experimentally introduced in *C. elegans* (9, 10). Before starting *Mos*TIC, you must isolate a *C. elegans* strain carrying a *Mos1* insertion located in a 1-kb-long region centered on the genomic position, where modifications will be introduced (see Note 1). Such strains can be identified in two ways.

1. A *C. elegans* collection containing about 80,000 *Mos1* insertions has been generated by the NemaGENETAG consortium ((11–13) and http://elegans.imbb.forth.gr/nemagenetag/). About 13,000 of these insertions were mapped and the insertion position of each of them is annotated in Wormbase (http://www.wormbase.org/) (see Note 2). *C. elegans* stocks carrying these insertions are distributed upon request (http://www.cgmc.univ-lyon1.fr/cgmc_info_celeganstp.php). These insertions were also organized in a PCR-screenable library that is being characterized (Robert and Bessereau, unpublished data) and will be made publicly available for screening.
2. A *Mos1* mutagenesis (12, 14) can be performed using the detailed protocol provided in refs. 15, 16 (see Note 3).

They were not reported so far, but we expect that some genes or genomic regions might be difficult to target with *Mos1*. If no *Mos1* insertion is recovered in the region of interest after screening the NemaGENETAG collections and mutagenesis, a technique also based on *Mos1* excision and that might be useful to express tagged version of the protein of interest was developed to insert single-copy transgenes in the *C. elegans* genome ((17) and Note 4).

2.3. Design of the Repair Template

A standard repair template carries modifications to introduce into the genome flanked by two "arms" homologous to the genomic region broken by *Mos1* excision (Fig. 2). Each of them contains at least 1.5 kb of homology (see Note 5). In addition, one arm (the long one) contains the genomic DNA sequence comprised between the *Mos1* insertion point and the genomic point, where modifications will be introduced. The repair template is constructed using classical molecular biology techniques (such as PCR fusion (18)) and cloned into a standard plasmid.

2.4. Transposase Expression Vectors

Two Mos transposase plasmids are available. They both contain the Mos transposase coding sequences modified for optimal expression in *C. elegans* (9). In *pJL44*, Mos transposase expression is driven by the heat-shock inducible promoter *Phsp-16.48*. In *pJL43*, Mos transposase expression is driven by the germ line constitutive promoter *Pglh-2*. Both plasmids are available at addgene (http://www.addgene.org).

2.5. pPD118. 33 (Pmyo-2::GFP)

pPD118. 33 (*Pmyo-2*::GFP) is available at addgene.

3. Methods

In the germ line of *C. elegans*, the mobilization of *Mos1* is achieved by using a modified version of its transposase cloned under the control of either Phsp-16.48, a heat-shock inducible promoter, or Pglh-2, a constitutively germ line-expressed promoter. Depending on which strategy is used to promote Mos transposase expression, two protocols have been developed (Fig. 3). See Note 6 for advice to choose the protocol.

3.1. Protocol 1

In this protocol, transgenic lines are constructed by germ line microinjection of a DNA mix containing both the repair template and the *pJL44* plasmid carrying the Mos transposase sequences expressed under the control of the heat-shock promoter. Mos transposase expression is induced by heat-shock treatment of transgenic animals and engineered animals are scored in the progeny of heat-shocked animals, where they are recovered with a frequency varying from 10^{-3} to 5×10^{-5} animals per progeny.

1. Outcross the *C. elegans* strain carrying the insertion of interest (see Note 7).
2. Use a standard germ line microinjection setup (19, 20) to inject, in the outcrossed line homozygous for the selected *Mos1* insertion, a DNA mix containing (a) the repair template at 50 ng/μl, (b) *pJL44* (*Phsp-16.48::MosTase*) at 50 ng/μl, and (c) *pPD118. 33* (*Pmyo-2*:: GFP) at 5 ng/μl as a transformation marker (see Note 8). Inject about 20 animals and keep them at 20°C prior to transgenic F1 screening. Isolate F1 transgenic animals with a mild-to-strong pharyngeal GFP expression and screen their progeny to identify transgenic lines. Select one to five independent lines with an intermediate-to-high transgene transmission rate. Amplify the transgenic populations at 25°C to minimize potential transgene silencing (see Note 9). If the *Mos1*-containing strain and/or transgenic lines are unhealthy at 25°C, rather use the second protocol (see Note 6).

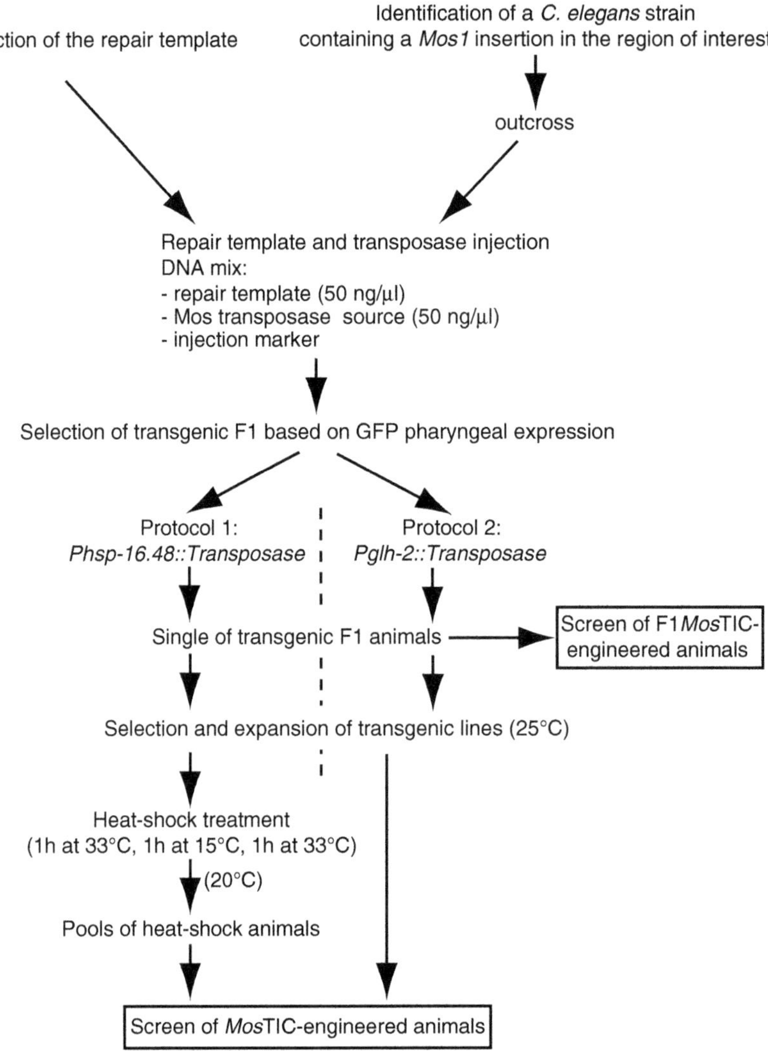

Fig. 3. Two protocols have been developed to engineer the *C. elegans* genome by *Mos*TIC. They both start with the identification of a *C. elegans* strain containing a *Mos1* insertion in the region to engineer and the construction of the repair template. They differ on the strategy used to induce *Mos1* excision. In Protocol 1, excision is induced by heat-shock treatment in established transgenic lines, whereas in Protocol 2 excision occurs directly in the germ line of injected animals. Both screening by phenotypic reversion and PCR can be used with each protocol.

3. To induce Mos transposase expression and *Mos1* excision, immerse parafilm-sealed NGM plates containing about 200 young transgenic adults in a water bath setup at 33°C for 1 h, let the worms recover for 1 h at 15°C, and repeat immersion at 33°C for 1 h. At this step, we usually use transgenic animals derived from one to five independent transgenic lines. After one night at 20°C, transfer heat-shocked animals (P0) to fresh plates (see Note 10). Depending on the fertility of the heat-shocked transgenic animals and on the screening strategy you will use to identify *Mos*TIC alleles (see below), put one to five animals on the same plate.

4. In some cases, locus engineering causes phenotypic changes that could be used to identify *Mos*TIC events. For example, *Mos*TIC might be designed to cause a loss of function of the gene to engineer. In this situation, screening has to be performed in F2 animals and we recommend to put one heat-shocked P0 per plate. Alternatively, *Mos*TIC could revert the mutant phenotype caused by the *Mos1* insertion. In that case, screening can be performed in F1 and we recommend to put three to five animals per plate. Phenotypic screening strategies proved to be very efficient in some of our experiments. However, be aware that, in addition to gene conversion, other mechanisms, including end joining, are at work in the *C. elegans* germ line and can sometimes regenerate functional alleles after *Mos1* excision or introduce deleterious footprints at the excision site (5). Therefore, animals selected based on phenotypic changes of the starting strain must be analyzed at the molecular level to identify bona fide *Mos*TIC-engineered strains. We usually performed such analysis by PCR.
5. In most cases, screening relies on PCR identification of the molecular changes introduced in the engineered locus. In this case, set up the transfer of heat-shocked P0 to fresh plates in such a way that pools contain about 200 F1 genomes (100 animals). Choose one primer (P1) in the repair template and the second one (P2) in the genome but outside of the repair template (Fig. 4a). To increase the sensitivity of the PCR screening, it is possible to perform a second step of nested PCR. Using this strategy, a PCR product should be generated only if homologous recombination happens. However, a PCR product having the same size as the specific one and generated by a process known as “jumping PCR” can sometimes be amplified from transgenic animals containing a nonengineered locus. In this case, optimization of PCR conditions is required to minimize “jumping PCR” (see Note 11 for advice on PCR conditions’ optimization). For PCR identification of *Mos*TIC-engineered alleles, wash plates containing starved populations derived from heat-shocked animals using M9 buffer and transfer the animals to a 1.5-ml tube. Let the worms sediment on ice, discard supernatant, and transfer the worms to PCR tubes or plates. Perform lysis at 65°C for 2–3 h in 50 ml of lysis buffer freshly complemented with proteinase K (final concentration: 1 mg/ml). After lysis, inactivate proteinase K by a 20-min incubation at 95°C. Perform PCR on 1 ml of lysate according to the optimized conditions (see Note 11). If additional analysis or optimization is required, lysates can be stored at −80°C for a few weeks. If a nested PCR is required, dilute PCR#1 100 times to set up PCR#2. Once a positive PCR signal is identified, chunk the corresponding plate to a fresh plate (Fig. 4b).

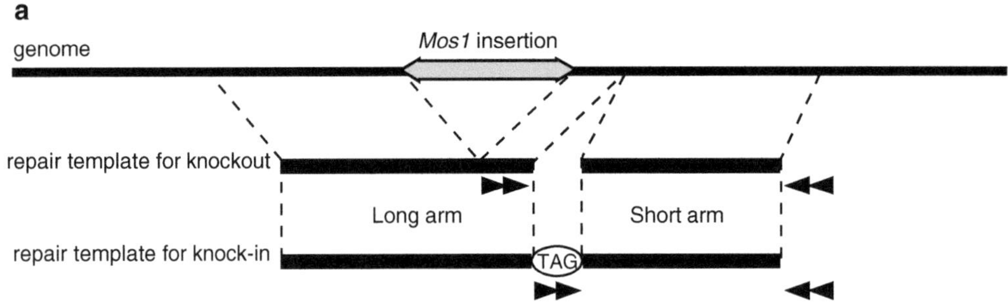

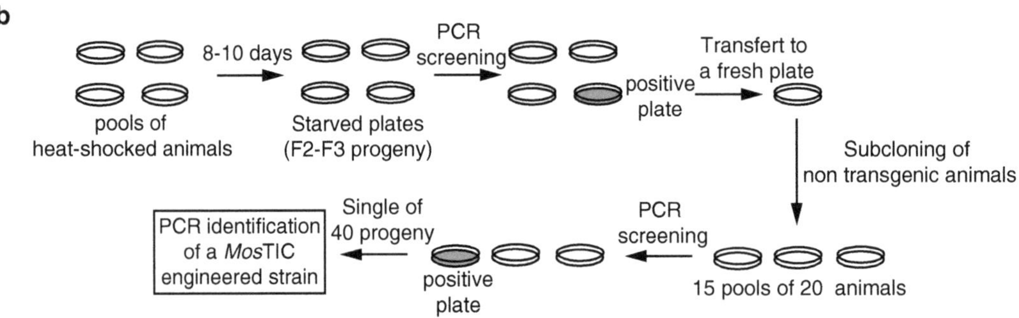

Fig. 4. PCR screening of *Mos*TIC-engineered alleles. (**a**) Primers' design. For each PCR, one primer is picked in the repair template and the other one outside. For knockout alleles, primers are chosen in such a way that they flank the deleted region. In this case, two PCR fragments can be amplified, the shorter one corresponding to the engineered allele. For knock-in alleles, the primer that recognizes the repair template can be chosen inside the tag sequence. In this situation, a PCR product should be amplified only if *Mos*TIC occurs. (**b**) Screening strategy. A first round of PCR is performed on mixed transgenic/nontransgenic populations derived of transgenic and nontransgenic worms derived from heat-shocked animals. At this step, jumping PCR (see Note 11) can occur and PCR conditions should be optimized. Once a positive population is identified, only nontransgenic worms derived from this population are further analyzed until a single *Mos*TIC-engineered worm is isolated.

From the developing population, make 15 pools of 20 nontransgenic animals (see Note 12) and analyze them as described above. From one positive subpool, clone 40 individuals to single plates to identify the *Mos*TIC-engineered strain.

3.2. Protocol 2

*Mos*TIC-engineered animals can be directly recovered at a significant rate in the F1 progeny of animals injected with a mix of the repair template and the *Pglh-2::MosTase* (*pJL43*) expression vector. Both phenotypic and PCR screening strategies described above can be used to identify alleles engineered with this protocol. Efficiency of this strategy is highly dependent on the quality of the microinjection. With our setup, *Mos*TIC-engineered allele recovery frequency varies between 1 *Mos*TIC-engineered allele for 7 successfully injected animals to 1 *Mos*TIC-engineered allele for 15 successfully injected animals (see Note 13).

1. Outcross the *C. elegans* strain carrying the insertion of interest (see Note 7).

2. Inject in the germ line of outcrossed animals homozygous for the selected *Mos1* insertion a DNA mix containing (a) the repair template at 50 ng/μl, (b) *pJL43* (*Pglh-2::MosTase*) at 50 ng/μl, and (c) *pPD118. 33* (*Pmyo-2*::GFP) at 5 ng/μl used as a transformation marker (see Note 8). Clone injected animals on NGM plates and keep them at 20°C prior to F1 screening.
3. To identify knock-in alleles by phenotypic reversion of mutant animals, select P0 derived plates containing transgenic F1 animals expressing GFP in their pharynx. Perform screening procedure for *Mos*TIC-engineered animals directly on those plates. Screening 200 F1 transgenic progeny should suffice to identify *Mos*TIC-engineered animals (see Note 14).
4. For phenotypic identification of *Mos*TIC-engineered loss-of-function alleles, clone 200 transgenic F1 progeny on NGM plates. Score for animals carrying a loss-of-function allele in F2 progeny (see Notes 14 and 15).
5. For PCR identification of *Mos*TIC-engineered animals, clone 200 transgenic F1 progeny on NGM plates. Perform PCR screening as described for protocol 1. Since *Mos*TIC is thought to occur in the germ line syncytium of the injected P0 animals, sibs selection should not be required (see Notes 14 and 15).

4. Notes

1. A study of the *Mos*TIC conversion track (5) demonstrated that *Mos*TIC efficiency is related to the distance between the *Mos1* insertion point and the genomic point, where the modifications are introduced. If the *Mos*TIC allele has to be screened using a PCR strategy, we consider that the *Mos1* insertion has to be located 500 bp away or less from the genomic point, where modifications will be introduced. If the *Mos*TIC allele can be screened using phenotypic reversion, the distance between the *Mos1* insertion site and the genomic point, where modifications will be introduced, can be increased depending on the feasibility of phenotypic screening.
2. *Mos1* insertions generated by the NemaGENETAG consortium were isolated by PCR screening and mapped by inverse PCR. Many are not mutagenic and knock-in alleles engineered from the excision of such insertions have to be screened using a PCR strategy, which can sometimes request a time-consuming step of PCR conditions' optimization. In this situation, it might be better to perform a *Mos1* mutagenesis to identify mutagenic *Mos1* insertion(s) in the region to engineer.
3. In *Mos1* mutagenesis, *Mos1* insertions of interest are identified based on phenotypic analysis. It means that it is necessary to

know which phenotypes are associated with the inactivation of your favorite gene(s). In addition, these phenotypes have to be specific enough to minimize the size of the mutagenesis. This strategy should be preferentially considered when several genes, whose inactivation leads to the same specific phenotype, have to be engineered by *Mos*TIC.

4. With *Mos*SCI, single-copy transgenes are integrated by gene conversion at intergenic genomic sites that contain a *Mos1* insertion. Four independent genomic sites were tried so far (17, 21). They were chosen because they are far enough of any coding sequences to assume that integrating additional DNA at these sites would not interfere with any biological function. *Mos*SCI is a strategy of choice to integrate single-copy transgenes to express tagged proteins under endogenous or tissue-specific promoters.
5. We increased the length of the homologous regions and showed that it did not improve *Mos*TIC efficiency (5). On the same study, decreasing the size of one of the homologous arm to 700 bp seems to significantly decrease *Mos*TIC efficiency. When screening *Mos*TIC-engineered alleles by PCR, take care to keep one of the two homologous arms short enough to be able to design one primer outside of the repair template (Fig. 4a).
6. Protocol 1 requires to generate and carefully maintain transgenic *C. elegans* lines that can be used to perform many *Mos*TIC experiments. Lines' establishment takes about 1 week and the transgene they carry can be submitted to silencing (see Note 9). In Protocol 2, *Mos*TIC occurs directly in successfully injected animals. This protocol goes faster but requires more microinjection work than Protocol 1, which is time consuming. The choice of the protocol also depends on the *C. elegans* strain that has to be injected. For instance, if this line is sensitive to heat-shock, Protocol 2 will be preferred. On the contrary, if the line to be injected has germ line morphology defects that make the injections difficult, Protocol 1 will be chosen.
7. *Mos1* mobilization in the *C. elegans* germ line generates more than one *Mos1* insertion per haploid genome (14, 22). These insertions are usually not linked. Since the remobilization of *Mos1* is not limited to a specific insertion, we recommend outcrossing the line carrying the insertion of interest in order to reduce the number of background insertions as much as possible. Background insertions can be identified by inverse PCR (for protocol, see refs. 15, 16) and their elimination after outcross can be checked by direct PCR using primers flanking the insertion points.

8. *pPD118. 33* (*Pmyo-2*::GFP) contains the GFP coding sequences cloned under the control of the *myo-2* promoter. It is widely used as a transformation marker since it allows a robust pharyngeal GFP expression and an easy identification of transgenic worms. However, the presence of this marker can be toxic for the worms and it is important to keep its concentration at 5 ng/μl or less in the DNA injection mix. Alternatively, it is also possible to use other transformation markers.
9. A strong transgene silencing process, which might be associated with heterochromatin formation, is observed in the *C. elegans* germ line (23). It might reduce *Mos*TIC efficiency by inhibiting the Mos transpsosase expression or preventing the homologous recombination machinery from accessing the repair template. To minimize this effect, we carefully maintain the transgenic lines containing the repair template and the transposase source by regularly transferring transgenic worms to fresh plates and growing them at 25°C. In addition, we freeze transgenic lines quickly after their identification and thaw new stocks every 6 weeks until a *Mos*TIC event is identified.
10. In *Mos1* mutagenesis, optimal germ line transposition is observed 24–30 h after induction of the transposase expression (14), suggesting that transposition mostly occurs in nuclei in pachytene arrest or late stages of mitosis. For this reason, we recommend to heat-shock worms when they are still young adults and to collect them 20 h later when they start to lay eggs derived from nuclei that were still in the distal region when transposition was induced.
11. While screening by PCR for a *Mos*TIC-engineered allele, false-positive *Mos*TIC signals can arise from annealing between single-stranded DNA generated from the transgenic array on the one hand and the genome on the other hand. This process, known as "jumping PCR" (24), can be minimized by optimizing PCR conditions on worm lysates prepared from non-heat-shocked transgenic lines. In our experiments, jumping PCR could be minimized by reducing the annealing time, increasing the annealing temperature, and/or diluting 100 times the worm lysates before starting PCR. "True"-positive PCR signals are the one that are still detected when PCR conditions become more stringent. For PCR optimization, it can be useful to construct a positive PCR control carrying an insert with the same sequence as the one expected at the *Mos*TIC-engineered locus. This plasmid can be serially diluted into worm lysates to verify whether the PCR conditions that minimize jumping PCR remain sensitive enough to detect rare recombination events.
12. At this step, picking nontransgenic worms allows to get rid of the jumping PCR artifacts.

13. To control the quality of injection, we use the *pPD118.33* (*Pmyo-2*::GFP) plasmid as a transformation marker in the mix containing the repair template and the transposase plasmid. We call "successfully injected animals" those which have progeny with pharyngeal GFP expression.
14. If no *Mos*TIC allele is recovered, select transgenic lines. Maintain them at 25°C for a few generations and redo selection.
15. A positive system of selection of *Mos*TIC-engineered knockout alleles was recently developed (25). In this system, *Mos*TIC is performed in a mutant background for *unc-119* and the deleted region is replaced by a gene expressing a functional version of UNC-119. Engineered genomes are screened by selecting animals rescued for UNC-119 and that do not express the fluorescent transformation markers anymore. This strategy was used in Protocol 2 to construct deletions that were up to 50 kb. It should be also possible to use it in Protocol 1.

References

1. Berezikov E, Bargmann CI, Plasterk RH (2004) Homologous gene targeting in Caenorhabditis elegans by biolistic transformation. Nucleic Acids Res 32:e40
2. Plasterk RH, Groenen JT (1992) Targeted alterations of the Caenorhabditis elegans genome by transgene instructed DNA double strand break repair following Tc1 excision. Embo J 11:287–290
3. Robert VJ, et al. (2008) Gene conversion and end-joining-repair double-strand breaks in the Caenorhabditis elegans germline. Genetics 180:673–679
4. Barrett PL, Fleming JT, Gobel V (2004) Targeted gene alteration in Caenorhabditis elegans by gene conversion. Nat Genet 36:1231–1237
5. Robert V, Bessereau JL (2007) Targeted engineering of the Caenorhabditis elegans genome following Mos1-triggered chromosomal breaks. EMBO J 26:170–183
6. Robert VJ, Katic I, Bessereau JL (2009) Mos1 transposition as a tool to engineer the Caenorhabditis elegans genome by homologous recombination. Methods 49:263–269
7. Jacobson JW, Medhora MM, Hartl DL (1986) Molecular structure of a somatically unstable transposable element in Drosophila. Proc Natl Acad Sci USA 83:8684–8688
8. Stiernagle T (2006) Maintenance of C. elegans. In: e. WormBook (ed) Wormbook.
9. Bessereau JL, et al. (2001) Mobilization of a Drosophila transposon in the Caenorhabditis elegans germ line. Nature 413:70–74
10. Robert VJ, Bessereau JL (2009) Manipulating the Caenorhabditis elegans genome using mariner transposons. Genetica 138:54154–9.
11. Duverger Y, et al. (2007) A semi-automated high-throughput approach to the generation of transposon insertion mutants in the nematode Caenorhabditis elegans. Nucleic Acids Res 35:e11
12. Granger L, Martin E, Segalat L (2004) Mos as a tool for genome-wide insertional mutagenesis in Caenorhabditis elegans: results of a pilot study. Nucleic Acids Res 32:e117
13. Bazopoulou D, Tavernarakis N (2009) The NemaGENETAG initiative: large scale transposon insertion gene-tagging in Caenorhabditis elegans. Genetica 137:39–46
14. Williams DC, et al. (2005) Characterization of Mos1-Mediated Mutagenesis in Caenorhabditis elegans: A Method for the Rapid Identification of Mutated Genes. Genetics 169:1779–1785
15. Boulin T, Bessereau JL (2007) Mos1-mediated insertional mutagenesis in Caenorhabditis elegans. Nat Protoc 2:1276–1287
16. Bessereau JL (2006) Insertional mutagenesis in C. elegans using the Drosophila transposon Mos1: a method for the rapid identification of mutated genes. Methods Mol Biol 351: 59–73
17. Frokjaer-Jensen C, et al. (2008) Single-copy insertion of transgenes in Caenorhabditis elegans. Nat Genet 40:1375–1383
18. Hobert O (2002) PCR fusion-based approach to create reporter gene constructs for expression

analysis in transgenic C. elegans. Biotechniques 32:728–730

19. StinchcombDT,etal.(1985)Extrachromosomal DNA transformation of Caenorhabditis elegans. Mol Cell Biol 5:3484–3496
20. Evans TC (2006) Transformation and microinjection. In: e. WormBook (ed) Wormbook.
21. Giordano-Santini R, Dupuy D (2011) Selectable genetic markers for nematode transgenesis. Cell Mol Life Sci 68:1917-1927 [Epub ahead of print].
22. Martin E, et al. (2002) Identification of 1088 new transposon insertions of Caenorhabditis elegans: a pilot study toward large-scale screens. Genetics 162:521–524
23. Kelly WG, et al. (1997) Distinct requirements for somatic and germline expression of a generally expressed Caernorhabditis elegans gene. Genetics 146:227–238
24. Paabo S, Irwin DM, Wilson AC (1990) DNA damage promotes jumping between templates during enzymatic amplification. J Biol Chem 265:4718–4721
25. Frokjaer-Jensen C, et al. (2010) Targeted gene deletions in C. elegans using transposon excision. Nat Methods 7:451–453

Chapter 12

Genome-Wide Manipulations of *Drosophila melanogaster* with Transposons, Flp Recombinase, and ΦC31 Integrase

Koen J.T. Venken and Hugo J. Bellen

Abstract

Transposable elements, the Flp recombinase, and the ΦC31 integrase are used in *Drosophila melanogaster* for numerous genome-wide manipulations. Often, their use is combined in a synergistic fashion to alter and engineer the fruit fly genome. Transposons are the foundation for all transgenic technologies in flies and hence almost all innovations in the fruit fly. They have been instrumental in the generation of genome-wide collections of insertions for gene disruption and manipulation. Many important transgenic strains of these collections are available from public repositories. The Flp protein is the most widely used recombinase to induce mitotic clones to study individual gene function. However, Flp has also been used to generate chromosome- and genome-wide collections of precise deletions, inversions, and duplications. Similarly, transposons that contain *attP* attachment sites for the ΦC31 integrase can be used for numerous applications. This integrase was incorporated into a transgenesis system that allows the integration of small to very large DNA fragments that can be easily manipulated through recombineering. This system allowed the creation of genomic DNA libraries for genome-wide gene manipulations and X chromosome duplications. Moreover, the *attP* sites are being used to create libraries of tens of thousands of RNAi constructs and tissue-specific GAL4 lines. This chapter focuses on genome-wide applications of transposons, Flp recombinase, and ΦC31 integrase that greatly facilitate experimental manipulation of *Drosophila*.

Key words: *P* element transposon, *piggyBac* transposon, *Minos* transposon, Flp recombinase, ΦC31 integrase, Deletion, Duplication, Recombineering, GAL4/UAS, RNAi

1. Introduction

In the past few decades, *Drosophila melanogaster* has maintained a prominent position in a sparse collection of model organisms that are typically used to answer questions related to numerous biological questions, including developmental pathways, behavior, neurobiology, and human disease. The sequence of its entire genome sequence (1) has opened a plethora of technological opportunities that were previously impossible (2, 3). Almost all

Yves Bigot (ed.), *Mobile Genetic Elements: Protocols and Genomic Applications*, Methods in Molecular Biology, vol. 859,
DOI 10.1007/978-1-61779-603-6_12, © Springer Science+Business Media, LLC 2012

these technologies are based on three basic tools: transposons, the Flp recombinase, and the ΦC31 integrase system.

Transposons, dominated by the *P* element transposon, have been seminal to the advancement of experimental fly biology (4). Aided by several transgenesis markers, genome-wide manipulations are now possible. For example, a collection of more than 15,000 *P* element insertions that are publicly available allows the creation of mutations in individual genes (5). To broaden the number of genes that can be manipulated, other transposable elements with different insertional specificities, such as *piggyBac* (6) and *Minos* (7) transposons, have been introduced (95). The strains are maintained at public repositories, including the Bloomington Drosophila Stock Center (BDSC) (8) and the Drosophila Genetic Resource Center (DGRC) (9).

Recombinases, predominantly the Flp recombinase, are mostly used to generate homozygous mutant clones in an otherwise heterozygous animal. The Flp-out technique is also used for conditional inactivation and Flp-in methods for conditional activation (10). The system is also extensively used in homologous recombination through ends in or ends out gene targeting (11, 12). In addition, genome-wide collections of transposons containing *FRT* sites have been generated that have been used to produce large collections of precise deletions encompassing most of the fly genome (8, 13–15).

About 7 years ago, the ΦC31 integrase was introduced into the fruit fly. The addition of this highly efficient integrase (16–17) to the arsenal of DNA engineering tools has allowed a major leap forward. Transposons containing the *attP* integration attachment recognition sequence, also called docking or landing sites, can be used to site-specifically integrate almost any DNA that carries the *attB* attachment site. Various collections of *attP* sites have been generated (16–19) and optimized for specific purposes, such as endeavors to create large collections of transgenic strains containing UAS-RNAi allowing the knockdown of almost all fly genes (20) or regulatory elements driving the binary GAL4 transactivator to interrogate the expression domain driven by those elements (21). In addition, highly receptive docking sites were chosen to introduce DNA clones from genomic libraries that allow rescue of mutations of almost all genes (18, 22, 23).

The goal of this essay is to survey the techniques that are based on transposons, the Flp recombinase, and the ΦC31 integrase and more specifically their impact on applications that permit genome-wide manipulations, i.e., sampling many genes or chromosomal areas. This article targets to introduce these genome-wide technologies to novel and less experienced members of the *Drosophila* community. We also discuss how transposons, the Flp recombinase, and the ΦC31 integrase can be integrated in various methodological approaches that are currently used.

1.1. Commonly Used Transposons and Transgenic Markers in the Drosophila Field

The most commonly used transposable elements in the fly field are the *P* element, *piggyBac* and *Minos* (24) (Fig. 1). *P* elements contain two terminal repeats that include inverted repeats of 31 base pairs as well as other sequences essential for productive transposition (25). A landmark study in the fly field described the physical separation of the *P* transposon backbone, the terminal repeats and other necessary sequences for transposition, from its embedded *P* transposase. This system is known as the binary vector/helper transposon transformation system (4). It allows controlled transposition of an engineered transposon from a plasmid backbone

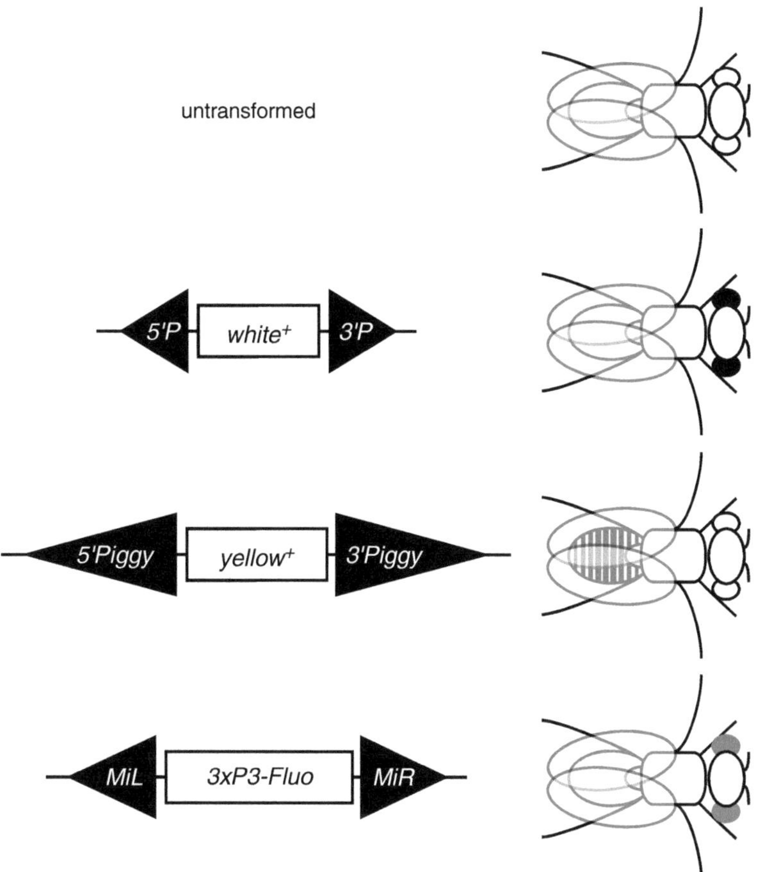

Fig. 1. Transposons and transgenic markers. Commonly used transposable elements in the *Drosophila* field are the *P* element, *piggyBac* and *Minos*. The *P* element contains *5′P* and *3′P* inverted transposon repeats of 31 base pairs while the *piggyBac* contains *5′Piggy* and *3′Piggy* inverted transposon repeats of 13 base pairs, with both transposons having other essential features near the repeats required for optimal transposition. *Minos* contains left (*MiL*) and right (*MiR*)-inverted transposon repeats of 255 base pairs. Transgenesis markers commonly used are the dominant eye marker *white*+ (*black eyes in figure*), the dominant body color marker *yellow*+ (*darkened body in figure*), and the dominant fluorescent eye marker consisting of the artificial *3xP3* promoter driving any fluorescent protein (*gray eyes in figure*).

into the host genome using the so-called helper plasmid that encodes the transposase. Later, important innovations included the generation of a genomic, stable source of transposase, allowing controlled transpositions of *P* elements through simple crosses (26, 27). This genomic stably encoded transposase opened up the road for genome-wide insertional mutagenesis experiments that have played an important role in all aspects of fly biology (28–30).

The availability of the genome sequence of *D. melanogaster* (1) opened many new avenues, including the molecular mapping of transposons at precise locations in the genome using inverse PCR or similar methods (31, 32). Although the vast majority of the *P* element insertions are homozygous viable and display no obvious phenotype, this feature is a real advantage as it ensures that mutations that are induced by imprecise excision of *P* elements (33) are likely to be caused by the imprecise excision and are not spurious in nature. Moreover, *P* elements have a unique property that has not been reported for other transposon elements. They permit transposon conversion or replacement of the DNA between the terminal inverted repeats in vivo. This is achieved by providing the donor DNA as another independent *P* element insertion, and promotes the precise swapping of both *P* elements between the respective locations catalyzed by the *P* element transposase (34, 35).

Although *P* elements have been instrumental in fly biology, this transposon also has drawbacks. First, the integration sites of *P* elements are strongly biased toward the 5′ end of genes (5, 95) and origin of replication binding sites (96). Hence, most insertions are partial loss of function mutations or do not affect the gene near which it is inserted (36). Moreover, as few insertions are in introns, few *P* element insertions can be used to create gene traps (37, 38). Second, some genes attract new *P* element integration events at a high frequency when compared to many other loci, significantly reducing the effectiveness of mutagenesis screens. Third, *P* elements do not mobilize outside of the Drosophilidae due to a host-specific factor that is required for transposition (39), hampering the use of many established technologies in other species. Hence, the use of other transposable elements has been explored.

Two of the most widely used alternative transposable elements in flies are *piggyBac* (6) and *Minos* (7). *piggyBac* is a transposon preferentially targeting TTAA sequences with terminal repeats containing 13 base pair inverted repeats among other essential sequences required for transposition. *Minos*, is a *Tc1/Mariner*-like transposon that has large terminal repeats consisting of 255 base pair inverted repeats. Both *piggyBac* and *Minos* have been incorporated into the arsenal to generate a genome-wide collection of transposable elements aiming to tag every gene with at least one transposon insertion (5, 95).

The identification of novel integration events upon hopping or integration of a transposable element is dependent on the incorporation of dominant genetic markers that can be identified. The most popular markers are the ones that make it easiest to isolate novel transgenic or mobilization events (Fig. 1). They include mini-*white*⁺, a dominant eye color marker that turns the white eye of a *white* mutant into light yellow to bright red depending on the genomic location, where the transposon is integrated (40). An alternative is mini-*yellow*⁺, a dominant body color marker that turns the yellowish body pigmentation of a *yellow* mutant body into a darker one (41). Another commonly used set of markers is based on fluorescent proteins fused to an artificial eye-specific promoter, better known as *3xP3* (42, 43). These markers are relatively easy to score but require a dissection microscope equipped with a UV light source and appropriate fluorescent filters.

1.2. Genome-Wide Transposon Collections

The Gene Disruption Project (GDP) (http://flypush.imgen.bcm.tmc.edu/pscreen/) makes all the generated transposon insertions publicly available from stock centers, such as the BDSC (http://flystocks.bio.indiana.edu/Browse/in/GDPtop.htm). The aim is to generate at least one transposon insertion in every fly gene (5, 95). The project was initially focused on using *P* elements engineered by others as well as gathering *P* element and *piggyBac* collections generated by others (30). The GDP later focused on using a *P* element, named EY, that contains a UAS site near one end of the *P* element construct to permit ectopic expression of the gene near which the transposon is inserted (5), similar to the EP *P* element previously developed (44). However, by 2004, the diminishing returns of generating additional *P* element insertions to tag new genes forced a shift. Indeed, for every 100 new insertions generated and sequenced, only 2 *P* elements in genes that were not already tagged could be identified (95). Hence, for all practical purposes, saturation had been reached with *P* elements. Currently, about 52% of all annotated *Drosophila* genes contain a *P* element insertion and it is likely that this figure will not change much in the future as no project aims at generating additional *P* element insertions.

As an alternative to *P* elements, the GDP explored other transposable elements and settled on a *Minos* transposable element, MiET (45). Based on a small pilot screen, MiET appeared to insert at random in the genome (45). A much larger screen based on ~12,000 insertions revealed that *Minos* inserts much more randomly than *P* elements, but some genes and complexes like the Bithorax and Antennapedia complex, seem to be refractory to *Minos* insertions, similar to *P* elements and *piggyBacs* (95). Based on the MiET data, a new version of the *Minos* transposon, called Minos-mediated integration cassette (MiMIC) was designed. MiMIC incorporates two inverted *attP* sites for recombinase-mediated cassette exchange (RMCE) (see below), allowing a broad array of versatile applications (97).

Other important collections that have been generated during the past decade that were based on *P* elements and *piggyBac* (46, 47) carried *FRT* sites (13–15, 48). These vectors have been used for genome-wide applications, including the development of molecularly mapped deletion and duplication kits (see next sections).

1.3. The Flp Recombinase and ΦC31 Integrase

Recombinases and integrases are site-specific DNA modification enzymes. They are especially useful for engineering if they function independently of other proteins or cofactors. Their only requirement is a pair of recombination or integration sites that are used by the respective enzymes to catalyze an exchange reaction between both sites, resulting in a recombination event or integration.

The most commonly used recombinase in the fly field is the Flp recombinase (49). Indeed, the Cre recombinase, the most widely used recombinase in the mouse community, is barely used in the fly community because of toxicity issues that are not well characterized (50). Flp recombinase recognizes a minimal *FRT* (Flp recombinase target) site of 34 base pairs, consisting of two 13 base pair inverted repeats flanking an 8 base pair asymmetric spacer. This spacer confers the directionality of the recombination reaction (51) (Fig. 2a).

The most commonly used integrase in the fly field is the ΦC31 integrase (16, 17). This integrase system has permeated every single aspect of technology development relevant to genetics and genomics in fruit flies. Initially, the system relied on coinjections consisting of a plasmid containing an *attB* integration site and mRNA encoding the integrase (16). However, the system was entirely reengineered and optimized using a germ line genomic source of ΦC31 integrase driven by regulatory elements of *vasa* or *nanos* (17) (http://www.frontiers-in-genetics.org/flyc31/). This allows the direct injection of plasmids, avoiding ultraclean DNA preparations and generation of mRNA encoding the ΦC31 for co-injection. The ΦC31 integrase recognizes a minimal *attP* (attachment site in the phage genome) site of 39 base pairs and a minimal *attB* (attachment site in the bacterial genome) site of 34 base pairs (52) (Fig. 2a). Both attachment sites contain imperfect inverted repeats flanking a 3 base pair recombination core that provides directionality. The integration reaction results in two hybrid sites, *attL* (Left), and *attR* (Right) that are no longer substrates for the integrase, making the integration irreversible.

To produce docking sites for the Flp recombinase and the ΦC31 integrase, the respective recombination sites, *FRT* and *attP*, are incorporated into transposons which are in turn integrated into the fly genome. Docking sites for regular site-specific transgenesis contain one recombination site (Fig. 3). Although thousands of docking sites containing *FRT* sites have been generated (13–15), regular site-specific transgenesis using microinjections with the Flp recombinase is not used. The engineered transposons have been used for limited in vivo remobilizations from one location to a

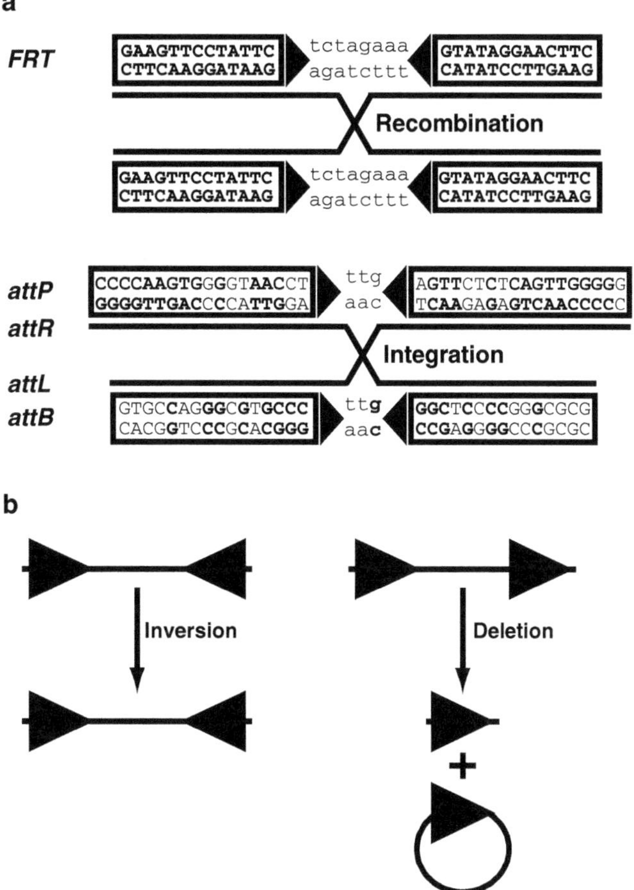

Fig. 2. Recombination site sequences for Flp recombinase and ΦC31 integrase, and recombination outcomes. (**a**) A minimal *FRT* site consists of 13 base pair inverted repeats (*indicated in bold*) flanking an 8 base pair spacer sequence determining the orientation of recombination. Recombination between two *FRT* sites results in reconstitution of the original sites, making the recombination reaction reversible. Minimal *att* sequences for ΦC31 consist of 39 base pairs (*attP*) or 34 base pairs (*attB*) that have imperfect inverted repeat structure (*indicated in bold*) flanking a 3 base pair spacer sequence determining the orientation of recombination. An integration reaction occurring between an *attP* and *attB* site, results in an *attR* and *attL* site that are no longer a substrate for the ΦC31 integrase, making the integration reaction irreversible and therefore unidirectional. (**b**) Recombination between two recombination recognition sequences can result in an inversion with recombination sites in an inverted orientation, or in a deletion with recombination sites in a direct orientation.

novel one (53). However, these transposons have been instrumental in the generation of hundreds of deletions (see below) as well as other chromosomal rearrangements, such as inversions and duplications (Fig. 2b) (54). Regular site-specific transgenesis is currently almost exclusively performed with the ΦC31 integrase. Three major sets of *attP* docking sites have been generated so far covering all chromosomes: one set is based on an *attP* site contained within

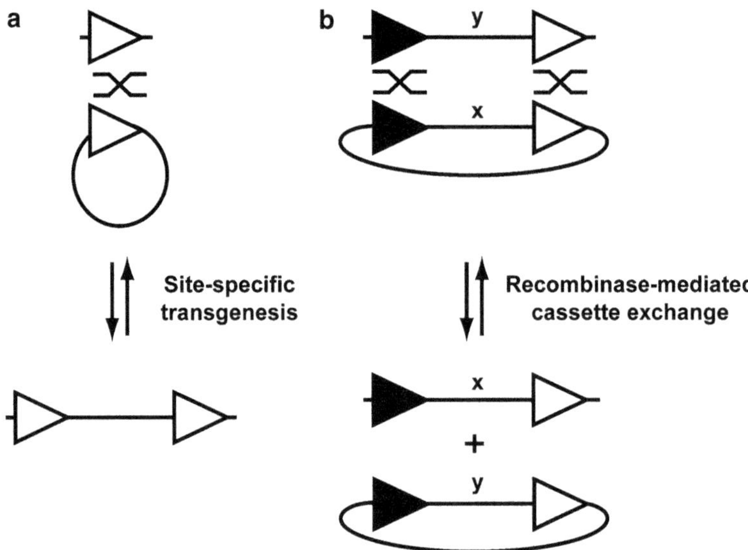

Fig. 3. Basic representations of transgene integration using one or two recombination sites. (**a**) Docking sites with a single recombination site allow the integration of entire plasmids that contain another single recombination site, including the plasmid backbone. When recombination sites are the same (*FRT* and *loxP*), immediate excision can occur after integration jeopardizing the isolation of a transgenic event. When recombination sites are different (*attP*/*attB* or *FRT*/*loxP* sites with inverted repeat variants), the integration event is irreversible and cannot be lost. (**b**) Docking sites with two recombination sites allow the swapping of material through a reaction called Recombinase Mediated Cassette Exchange (RMCE), avoiding the integration of presumably unwanted plasmid backbone. For Flp and Cre mediated RMCE, heterotypic recombination site variants that do not recombine with each other are desired to avoid elimination of the cassette between both sites (*indicated by black and white triangles*). Alternatively, an inverted orientation of recombination sites can be used (not shown). For ΦC31 mediated RMCE, this is not necessary since the recombination is unidirectional, although an inverted orientation of recombination sites is desired to ensure proper integration of the cassette only.

a *P* element containing the *mini-yellow*$^+$ marker (16, 19); the second set is based on the *piggyBac* transposon (18); the third set is based on the *Mariner* transposon and a 3xP3-driven red fluorescent eye marker (17). The latter set of docking sites has the advantage that the 3xP3-RFP marker can be removed with Cre recombinase allowing this marker to be used again in subsequent transgenic manipulations. The most useful docking sites are available from stock centers, such as the BDSC: http://flystocks.bio.indiana.edu/Browse/phiC31.htm.

To integrate the desired DNA without adjacent plasmid sequences, a second recombination site needs to be added. This allows RMCE (Fig. 3). RMCE consists of two parallel recombination reactions allowing integration of a DNA cassette flanked by a pair of recombination sites to integrate in a docking site flanked by another pair of recombination sites. Since the use of two identical sites, such as *FRT* sites, favors deletion (Fig. 2b) over RMCE, other strategies were pursued. One strategy is the use of heterotypic sites, such as *FRT3* and *FRT5*, that react preferentially with each other but not with other variants (55). Another strategy is the use of inverted recombination sites. However, this can cause cassette inversion if identical sites, such as *FRT*, are used (Fig. 2b). This problem has been circumvented with the use of two inverted *attP* sites (56). This strategy is currently the most often used since germ line genomic sources of ΦC31 are available (17), avoiding coinjections of plasmid containing the DNA to be integrated and a plasmid containing the recombinase or integrase (17). Note, however, that RMCE has been demonstrated to work for Flp with direct *FRT* and *FRT3* sites (57) as well as with Cre with several combinations of directly oriented heterotypic *loxP* sites (58).

2. Genome-wide applications of transposons, Flp recombinase and ΦC31 integrase for *Drosophila melanogaster*

Several genome-wide technologies are based on the integration of transposons, the Flp recombinase and the ΦC31 integrase. In addition, genome-wide DNA collections that encompass genomic DNA libraries, RNAi libraries, and libraries containing enhancers that drive the binary GAL4 transcriptional activator allow elegant manipulations in vivo. Here, we discuss several of the most convenient and widely used applications and tools. These include: mapping of lethal mutations with *P* elements; deletions generated with Flp and *FRT* containing transposons; screening for chemically induced mutations based on mitotic clones; genome-wide genomic DNA libraries that can be altered via recombineering; sets of X chromosome duplications to rescue and map genes; genome-wide RNAi libraries; and genome-wide libraries containing regulatory elements fused to the GAL4 transcription factor.

2.1. Genome-Wide Mapping of Lethal Mutations with P *Elements*

Forward genetic screens using chemical mutagens are still popular and may even gain in popularity as they allow the unbiased isolation of mutants with a desired phenotype (59). Genetic mapping of the resulting loci and identification of the molecular nature associated with these chemically induced mutations are often labor-intensive. One could argue that meiotic mapping of chemically induced mutations with the avenue of whole genome sequencing technology is a method of the past. Yet, in our lab, we have tested numerous mapping strategies, and the easiest mapping method accessible to all *Drosophila* researchers that can be performed in any laboratory set up for fly research is to map mutations with defined *P* element insertions (60) (Fig. 4). To map a mutation located on

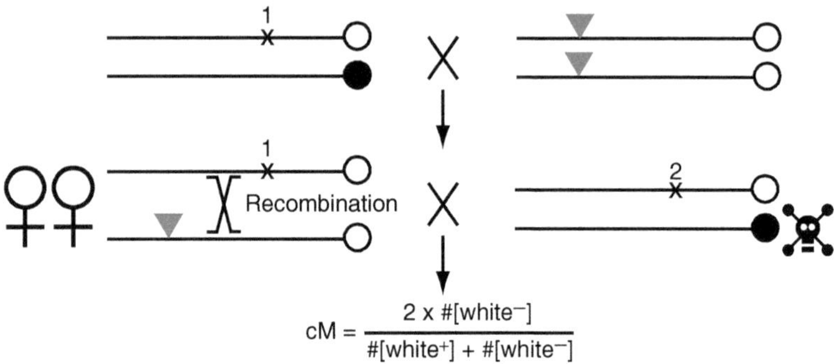

Fig. 4. Mapping of lethal mutations with transposons. An unmapped lethal mutation (*indicated by 1*) maintained over a balancer chromosome (*indicated in black*), is crossed to a mapped viable *P* element insertion (*gray triangle*) marked by the eye dominant marker *white*$^{+}$. Female progeny containing the mutant chromosome are crossed to males containing a second lethal mutation belonging to the same complementation group (*indicated by 2*) and maintained over a conditionally lethal balancer chromosome (*indicated in black and with skull*). A subsequent heat shock removes all unwanted progeny containing the conditionally lethal balancer chromosome, as well as all homozygous mutant lethal flies. Since recombination between the unmapped lethal mutation 1 and the mapped viable *P* element insertion can occur in the female germ line during the second genetic cross, the ratio of *white*$^{-}$ flies compared to the total number of flies in the resulting progeny is indicative of the genetic distance between the unmapped mutation and the mapped *P* element insertion. Performing this crossing scheme with several mapped *P* elements allows estimating the location of the mutation within a certain genetic interval.

a known chromosome arm, two rounds of genetic mapping are performed. In a first rough mapping round, the locus can be mapped within one centiMorgan using defined *P* elements whose mapping position is known. This requires only 30 vials and typically allows mapping a mutation to an approximate 400 kb interval. Stocks for rough mapping of mutations on the second and third chromosomes have been carefully selected so that they show strong expression of mini-*white*$^{+}$, resulting in a red adult eye, facilitating screening when scoring large populations of flies. They are available from stock centers, such as the BDSC: http://flystocks.bio.indiana.edu/Browse/misc-browse/Baylor-kits.htm. In the second round of mapping, we use a set of *P* elements that are located near the mapped area and carefully chosen from publicly available lines. This allows mapping to a less than 100 kb physical interval, depending on the number of flies scored. Such a region is small enough to be investigated directly by standard molecular biology techniques using Sanger sequencing (60) or to use targeted whole genome sequencing approaches (61). The advantage of whole genome sequencing is that the method is rapidly becoming cheaper and that it is less labor intensive, especially if one only needs to focus the analysis of sequencing data to a 100 kb interval. Positives in these intervals tend to be genuine.

2.2. Genome-Wide Deletion Collections Generated by FRT Containing Transposons and Flp Recombinase

Several strategies have been designed to generate deletions in the fly genome (3). The most useful methodology is based on the Flp recombinase and *FRT* containing transposon. Two genome-wide transposon collections have been independently generated that contain *FRT* sites and can therefore be used for genome engineering (54). When two nearby mapped transposable elements, each containing an *FRT* site in the same orientation, are brought in *trans*, a precise deletion can be generated when Flp is provided (Fig. 5). One set of *FRT* sites containing transposons was obtained by hopping a previously tested *P* element (54). Several thousand of new insertions were generated and used to create hundreds of defined deficiencies, commonly known as the DrosDel project (http://www.drosdel.org.uk/) (14, 15). An independent collection of insertions was generated with both *P* element and *piggyBac* transposons containing *FRT* sites (48), allowing the creation of hundreds of deficiencies (http://flystocks.bio.indiana.edu/Browse/df/dfextract.php?num=all&symbol=exeldef) (13). Both insertion collections were used to push coverage to 98% of all euchromatic genes (62). These three deficiency sets formed the basis of a collection that was compiled at the BDSC (http://flystocks.bio.indiana.edu/Browse/df/dftop.htm) (62). These deficiencies cover 98% of all fly genes.

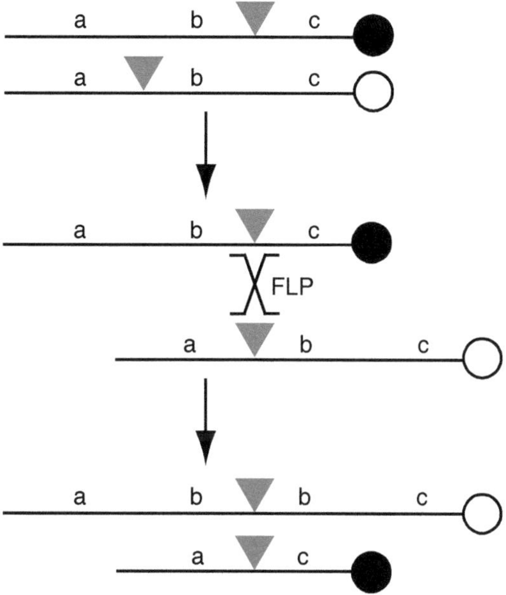

Fig. 5. Generation of deletions with transposable elements containing *FRT* recombination sites. Two mapped transposon insertions (*gray triangles*) with *FRT* recombination sites in the same orientation are brought in *trans* through a genetic cross. Recombination between both *FRT* sites catalyzed by Flp recombinase in the germ line results in a deletion and a reciprocal duplication for which the extents are exactly known. Genome-wide availability of transposons containing an *FRT* site allowed the generation of deletions covering majority of the *Drosophila* genome with this simple strategy.

These sets of stocks are the most used stocks in the fly community as they permit mapping of mutations through complementation tests and represent true null alleles of almost all genes.

2.3. Genome-Wide Mutagenic Mosaic Screens Using Centromeric FRT Insertions

One strategy that has been very useful to identify novel genes for specific biological processes is based on *FRT* containing transposons that are located at the base of each chromosome arm near the centromere (63). Flp-mediated mitotic recombination results in the exchange of entire chromosome arms during cell division (Fig. 6). This allows the generation of homozygous mutant tissue in an otherwise heterozygous animal, therefore limiting the effect of a potentially detrimental mutant phenotype at an earlier developmental state. Conveniently, screens can be designed as F1 screens, meaning that the progeny of mutagenized flies can be directly screened and mutations isolated and balanced to generate stable stocks.

These screens are typically based on organ or cell-specific Flp expression driven by regulatory elements or the binary GAL4/UAS system. The mutation is made homozygous in a tissue of

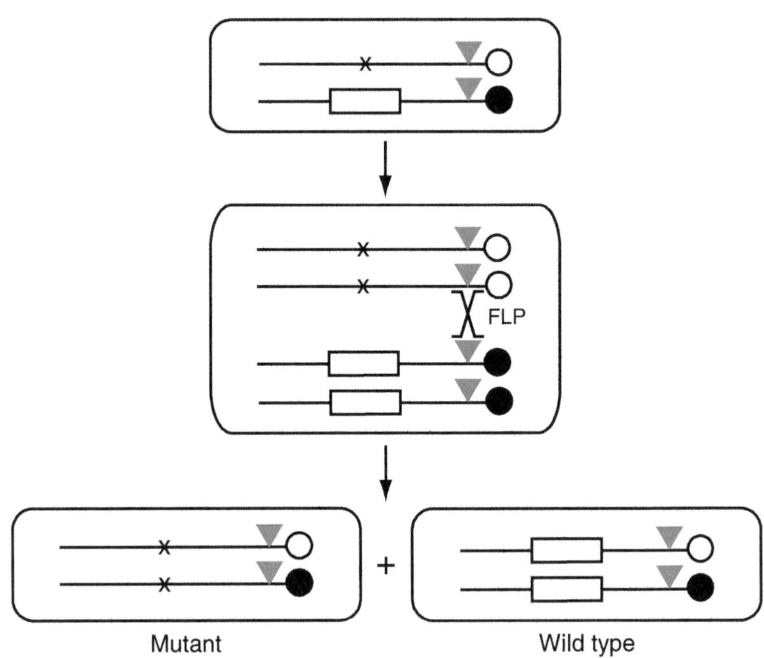

Fig. 6. Mitotic analysis mediated by Flp recombinase. Both chromosomes in the progenitor cell have a *P* element insertion with an *FRT* site (*triangle*) proximal to the centromere (*circle*). The cell is heterozygous with one of the chromosomes having a mutation (*X*) while the other chromosome bears a marker (*rectangle*). After Flp-mediated recombination between both *FRT* sites on sister chromatids during mitosis, one cell becomes homozygous marked and wild type (also known as the twin spot), while the other cell becomes homozygous mutant resulting in a negatively marked mutant clone that can be analyzed and compared to the nonmutant tissue in the same animal.

interest and the phenotype is scored. The most widely used tissue so far is the eye. The main reason is that there is an excellent eye driver available, *eyeless*, which is directly fused to the Flp protein (64). Alternatively, a similar driver is fused to the GAL4 transcriptional activator driving UAS-Flp (65). To obtain clones that are large enough, it is important to use a driver that is expressed early in development and the *eyeless* regulatory element fulfills this role since it is already expressed in embryos. The size of the mutant clone is often enlarged by using a cell lethal mutation (66) or a *Minute* (67) on the homologous chromosome of the one that has been mutagenized so that wild type clones are unable to propagate equally well. The large clones in the eye allow to screen for morphological defects of the eye cells, such as aberrant photoreceptor cells, pigment cells, lens cells, interommatidial bristles (64), as well as head cuticle (68) and bristles on the head cuticle (67), failure of pupae to eclose (69), as well as defects that affect axonal targeting from the retina toward the fly brain (70) or the electrophysiological function of photoreceptors using electroretinograms (71). Other tissues can be analyzed using this technology as well. Obviously, these screens can be combined with mosaic analysis with a repressible cell marker (MARCM), in which the mutant cells are labeled with a fluorescent marker (72) as well as more sophisticated labeling strategies (73, 98). Flp recombinase can also be induced spatially by the binary GAL4/UAS system (74) or temporally in all tissues using heat shocks and the hs-Flp transgene (49).

2.4. Transgene Manipulations Through Recombineering and ΦC31-Mediated Site-Specific Integration

Recombineering was first introduced to the fly field in a versatile transgenic platform called P[acman] (P/ΦC31 artificial chromosome for manipulation) (18) (http://www.pacmanfly.org/). P[acman] was generated because of the need to integrate large transgenes into the fly genome. P[acman] was possible because of the availability of genome-wide annotated BAC libraries (http://bacpac.chori.org/) for *D. melanogaster* (75) (Fig. 7). P[acman] combines three powerful technologies: recombineering (76), a site-specific integrase ΦC31 (16) and a conditionally amplifiable BAC (77). Recombineering allows the incorporation of virtually any change into DNA constructs, and was used to retrieve 102 kb of DNA from existing libraries via gap-repair (Fig. 8). A double gap-repair procedure, in which two pieces of overlapping BACs were joined together allowed the generation of a 133 kb fragment that reconstituted the *Tenascin major* gene. The ΦC31 integrase allows the integration of *attB* containing constructs into defined docking sites in the fly genome that contain *attP* sites (16–19). Moreover, this study illustrated that the ΦC31 integrase is able to integrate plasmids up to at least 146 kb in vivo. The conditionally amplifiable BAC maintains the plasmids at low-copy number to ensure stability of large constructs under normal conditions, but

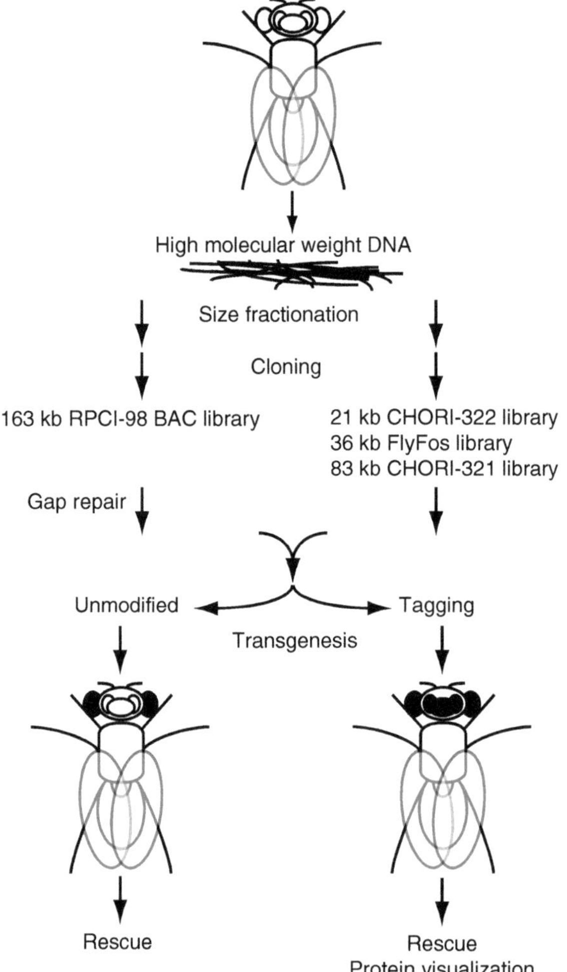

Fig. 7. Gene manipulations with genome-wide genomic DNA libraries and recombineering paradigms. High molecular weight DNA is isolated from flies, fractionated and subcloned. This resulted in genome-wide genomic DNA BAC libraries that require recombineering-mediated gap-repair into a transgenesis competent vector (*left*). Alternatively, DNA is subcloned directly into transgenesis competent plasmids, resulting in 3 libraries: the 21 kb CH322 P[acman] library, the 36 kb FlyFos library, and the 83 kb CH321 P[acman] library (*right*). In both cases, unmodified clones can be integrated into docking sites, transgenic events can be isolated by the eye markers *white*$^+$ or *3xP3-DsRed*, and transgenic flies can be used for rescue experiments of mutations. Alternatively, clones can be modified through recombineering to incorporate protein tags for protein visualization that for simplicity illustrates an expression domain in the adult brain.

that copy number can be artificially induced with a simple sugar solution to high copy number facilitating cloning purposes and DNA preparation tremendously (18, 77). The low-copy condition is also very helpful when performing recombineering since changes are introduced at low-copy number and therefore fixed and amplified within the bacterial colony. This is next to impossible to achieve

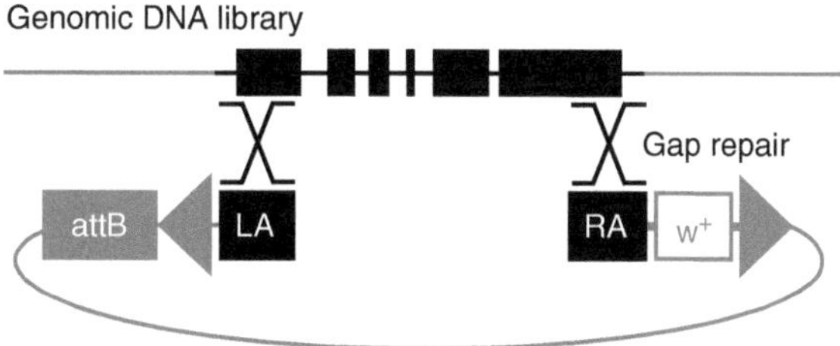

Fig. 8. Recombineering-mediated gap-repair. Two homology arms, left (LA) and right (RA), flanking the DNA region of interest contained within a clone of an available genomic DNA library are subcloned into a plasmid compatible with ΦC31 (*attB*) and *P* transposase-mediated transgenesis (*gray triangles*). The plasmid is linearized between both homology arms and transformed into bacterial cells containing the available genomic DNA library clone and recombineering functions. Recombineering-mediated gap-repair retrieves specifically the DNA region of interest into the transgenesis compatible plasmid which can be modified further with recombineering or directly used for transgenesis.

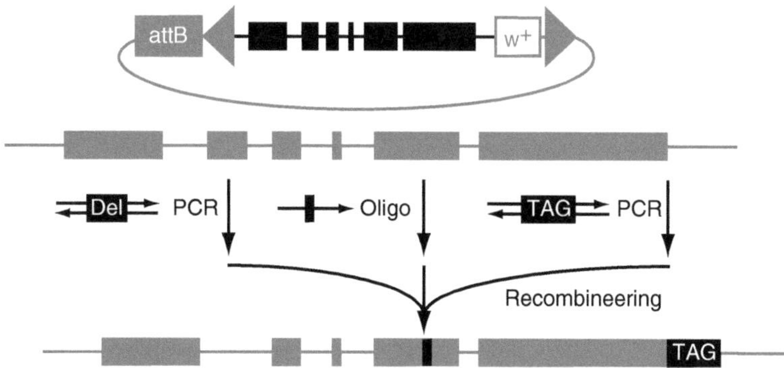

Fig. 9. Recombineering-mediated gene tagging and manipulation. A plasmid containing the gene of interest is transformed into bacterial cells containing recombineering functions. Recombineering can then be used to incorporate a variety of changes for structure-function analysis of transgenes, such as deletions generated by PCR or point mutations encompassed within oligonucleotides. Alternatively, protein tags can be generated by PCR and integrated at the N terminus (not shown) or C terminus for protein visualization.

when performing recombineering in high-copy number plasmids. Subsequently, recombineering was used to introduce deletions in genomic loci for structure/function analysis (78) and different protein tags for visualization of protein expression or acute protein inactivation (79) (Fig. 9).

In a next step, the P[acman] methodology was extrapolated toward the creation of genome-wide *Drosophila* libraries (22, 23). Two genomic *D. melanogaster* libraries were generated, end-sequenced, and annotated onto the reference fly genome: the CHORI-322 library with an average insert size of 21 kb, and the CHORI-321 library with an average insert size of 83 kb in P[acman] (23) (http://www.pacmanfly.org/). The CHORI-321 and CHORI-322 libraries represent a 12-fold coverage of the genome and allow rescue of more than 95% of annotated genes of the fly genome. Another group generated a FlyFos library with an average insert size of 36 kb in a fosmid instead of the P[acman] backbone (22) (http://transgeneome.mpi-cbg.de/). The FlyFos library has a 3.3-fold coverage. These libraries eliminate the need for gap-repair except for genes that are not contained within the libraries, for genes that require tailored customization, or for genes that are too large to be in any of the libraries. Hence, the gap-repair approach and library approach have their advantages and disadvantages. The gap-repair procedure requires additional cloning steps but allows defined genomic fragments to be analyzed while the libraries eliminate cloning but the clones were randomly generated potentially resulting in suboptimal coverage of the gene of interest. The functionality of all three libraries was tested by integrating many clones of each into *attP* docking sites and showing that they rescue the mutant phenotypes associated with known mutations in the region of interest. Moreover, genes in numerous clones were tagged by recombineering using several strategies and shown to faithfully recapitulate the expression pattern of their endogenous counterparts (22, 23). Hence, tagged clones should be very useful to determine the expression patterns of numerous uncharacterized genes and their corresponding proteins.

2.5. X Chromosome Duplications

The X chromosome of *D. melanogaster* contains 22 Mb of euchromatic DNA encompassing approximately 2,300 protein-coding genes (1). Mutations in essential and male fertility genes on the X chromosome are tedious to map, unless X duplications present on other chromosomes are available. Two recent studies generated extensive sets of duplications to allow mapping of X chromosome mutations. One study was focused on generating relatively large duplications through genetic means to allow quick but rough mapping (62). As shown in Fig. 10, in a first set of crosses, the goal is to recombine on an attached XY chromosome two transposons that contain *FRT* sites that point toward each other (Fig. 10c). The attached-XY chromosome contains a complete X chromosome without centromere fused to a complete Y chromosome that compensates for the loss of the centromere of the X chromosome. This attached XY chromosome is then subjected to Flp recombinase to mediate an inversion between both inverted *FRT* sites (Fig. 10d). This inverted attached XY is then exposed to X-rays to induce random deletions. This irradiation typically breaks in the heterochromatin and somewhere in the euchromatin (Fig. 10e). Upon resealing, the resulting deletions form a nested set of overlapping duplications on a Y chromosome as most of the duplicated X material has been deleted (62) (Fig. 10f). Hence, the resulting chromosomes consist of a small tip of the X chromosome, an interchromosomal duplicated portion of the X chromosome near the tip, and a full Y chromosome. Independent deletion events

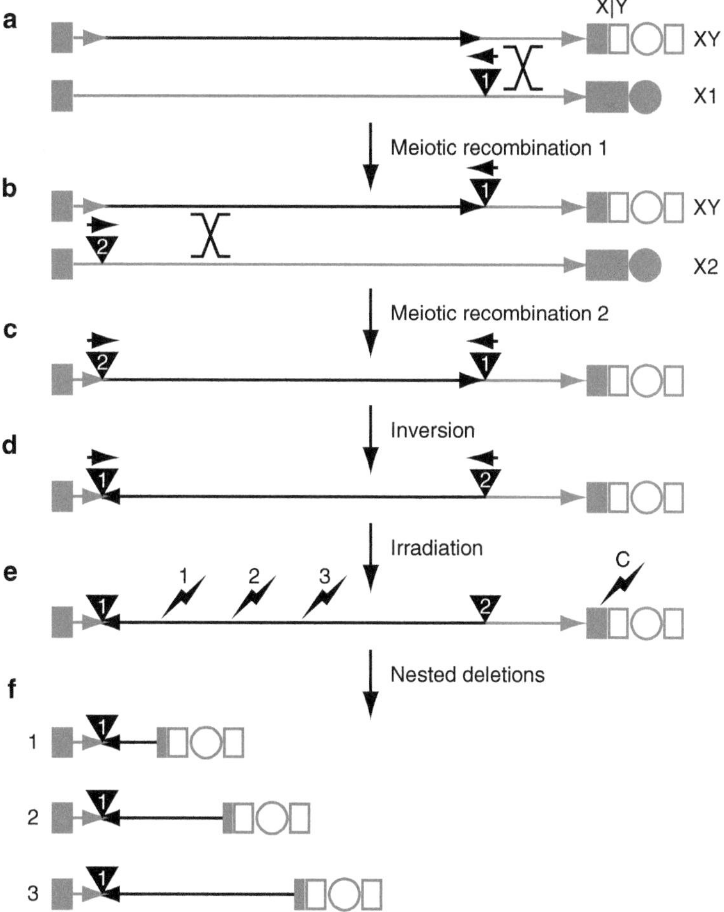

Fig. 10. The X chromosome duplication project generated by in vivo recombination and X-ray irradiation. (**a–c**) Two *FRT* containing *P* elements (*triangles*) are recombined sequentially over two generations (meiotic recombination 1 and 2) from two separate X chromosomes (X1 and X2) onto an attached-XY chromosome (XY). This is followed by an inversion of the chromosome section between both *FRT* sites (**d**). Next, irradiation creates double stranded breaks at a proximal random site near the X/Y boundary (C) and a distal random site (for example 1) (**e**). This results in a large internal deletion of that section of the X chromosome between breakpoints C and 1. The isolation of independent events after irradiation results in several deletions of the same inverted section between breakpoints C and 1, 2, or 3. As such, a nested set of medial duplicated segments is generated from each inversion progenitor (1, 2, or 3) (**f**). Repeating this procedure for several such progenitors resulted in covering majority of the X chromosome with sets of nested duplications that cover 78% of the X chromosome and can be used for rescue experiments of X chromosome genes not yet mapped.

using the same inversion precursor results in nested chromosomal duplicated segments of the X chromosome portion contained within the inversion, and proximal to the first transposon insertion. Hence, mutations mapping within the largest duplication of the set can be fine mapped using the nested set of duplications. By repeating the overall procedure and creating many different inversions

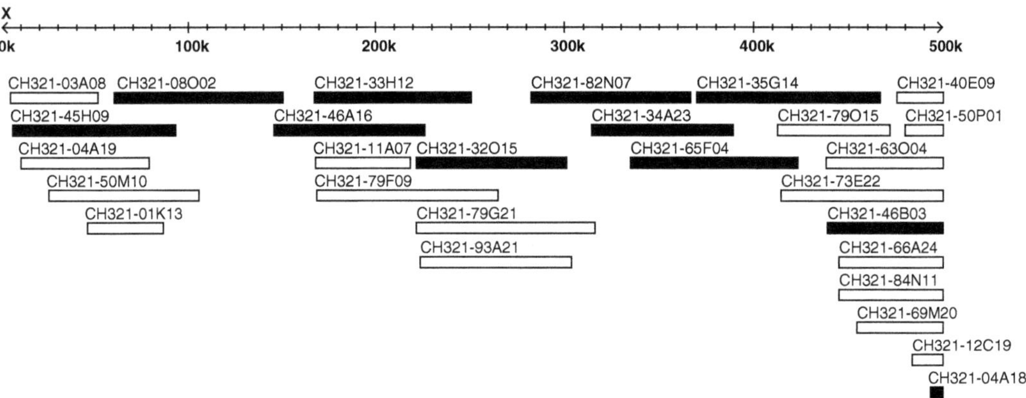

Fig. 11. The X chromosome duplication project generated by transgenesis. The CHORI-321 P[acman] library contains DNA clones (CH321) that have an average insert size of 83 kb. A minimum tiling path (*indicated in black*) of 408 overlapping P[acman] clones for the entire 22 megabase X chromosome of *Drosophila* was selected and integrated in the same docking site, VK33 on chromosome arm 3 L, using the ΦC31 integrase. The resulting transgenic duplication kit contains 382 fly lines covering 96% of the entire X chromosome and can be used to perform rescue experiments of X chromosome genes not yet mapped. Illustrated is part of the tiling path encompassing the first half megabase of the X chromosome.

and repeating the irradiation, numerous sets of nested duplications can be generated. These duplications currently cover 90% of the X chromosome. Since X-ray irradiation creates molecularly undefined breakpoints, comparative genome hybridization microarrays and PCR were used to establish the extents of the different duplications. The entire collection is available at the BDSC: http://flystocks.bio.indiana.edu/Browse/dp/BDSC-Dps.php.

The second study used the availability of the P[acman] libraries (23). The CHORI-321 library clones contain an average insert size of 83 kb. A minimal tiling path of 408 overlapping clones predominantly from the CHORI-321 library was integrated into a single *attP* docking site, VK33 on 3 L, using the ΦC31 integrase system (80) (Fig. 11). Currently, 96% of the X chromosome is covered with the P[acman] duplications contained in 382 fly strains. About 100 clones were tested for rescue experiments and 92% rescued existing mutations in genes as well as entire deficiencies previously generated with the Flp/*FRT* system (13–15). Interestingly, this study also illustrated that most genes are tolerated at twice the normal dosage, and allowed the more precise mapping of two regions that were previously involved in diplo-lethality. The entire collection is available at the BDSC: http://flystocks.bio.indiana.edu/Browse/dp/DC-Dps.php.

Both duplication collections complement each other well. Unmapped mutations can be mapped with two rounds of crosses. In the first round of crosses using the large duplications (62), a mutation can be mapped to an interval of a 100 kb. In a second set of crosses using the P[acman] duplications (80), the mutation can be mapped within a 20–30 kb region typically encompassing two to three genes. This can be followed by Sanger sequencing of overlapping PCR products, as described before (60). This strategy has proven to be very efficient in mapping numerous loci derived from a large-scale X chromosome mutagenesis effort ongoing in the Bellen lab.

2.6. Genome-Wide In Vivo RNA Libraries and RNAi Rescue

RNA interference or RNAi has had a tremendous impact on biology (81). The transcript levels of genes can be knocked down using RNAi, leading to phenotypes that are typically associated with hypomorphic mutations. In *Drosophila*, RNAi was initially performed through embryonic microinjections (82). RNAi was subsequently demonstrated to be possible via transgenesis using the GAL4/UAS system allowing the tissue-specific knockdown of transcripts of specific genes (83–85). This work was eventually expanded into genome-wide RNAi libraries encompassing all fly genes. Three libraries have been generated. The first library was generated in a *P* element backbone (86). An impressive 22,270 transgenic lines were generated covering 88% of all the predicted protein-coding genes. The lines are available from the Vienna Drosophila RNAi Center (http://www.vdrc.at/) and have been used for several genome-wide screens for immunity, Notch signaling, heart function, obesity and triglycerides levels, muscle development, and heat nociception (87, 88). Since the integration site of *P* elements cannot be controlled, *P* element transgenes are under the influence of position effects and therefore the different RNAi lines of this library cause variable knock-down levels. This problem was circumvented by using the ΦC31 integrase system in the remaining libraries (20, 86) (http://www.vdrc.at/, http://www.flyrnai.org/TRiP-HOME.html, http://flystocks.bio.indiana.edu/Browse/RNAi.php, http://flystocks.bio.indiana.edu/Browse/TRiPtb.htm and http://flystocks.bio.indiana.edu/Browse/VDRCtb.htm).

RNAi experiments sometimes result in unwanted phenotypes due to off-target knockdown of genes other than the desired one. Hence, rescue of RNAi phenotypes is an often overlooked critical proof that the RNAi experiment is specifically directed toward the target gene. Three strategies are currently available to perform RNAi rescue (Fig. 12). Two of the strategies are based on the availability of genome-wide libraries of a related species, *Drosophila pseudoobscura* (22, 89, 90). Genes and their regulatory regions of *D. pseudoobscura* are similar enough to rescue genes of *D. melanogaster*, but divergent enough to resist the RNAi machinery. In one of the strategies, the authors used an existing *D. pseudoobscura* fosmid library (91) that was upgraded with Cre-assisted BAC modification (92) to incorporate the conditionally amplifiable BAC system, an *attP* site and the mini-*white*$^+$ marker, so the clones were compatible with the ΦC31 integrase system (89). To demonstrate cross-species functionality, the retrofitted *D. pseudoobscura* fosmids

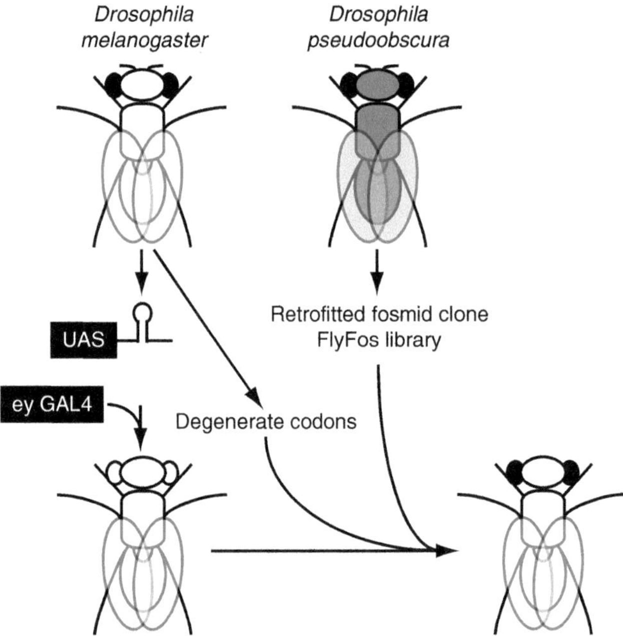

Fig. 12. In vivo RNAi libraries and RNAi rescue. Genome-wide in vivo RNA libraries of *Drosophila melanogaster* were generated in plasmids containing UAS binding sites that can be used with GAL4 drivers to knock down target transcripts tissue specifically, for example in the eye as illustrated for a hairpin loop directed against the *white*+ marker. In addition, genomic DNA libraries of *Drosophila pseudoobscura*, such as retrofitted fosmid or FlyFos libraries can be used to rescue the RNAi phenotype since that species is both similar enough to rescue the phenotype yet divergent enough to resist the RNAi machinery. Alternatively, the codon usage of *D. melanogaster* can be changed significantly in engineered cDNA rescue constructs to resist the RNAi machinery as well.

rescued existing mutations for *D. melanogaster* genes in four cases. In the second strategy, a completely new FlyFos fosmid library was generated for *D. pseudoobscura* that was immediately compatible with fly transgenesis (22, 90). The final strategy did not require a genome-wide library but is more labor intensive. An RNAi-resistant transgene was engineered by changing several codons in the rescue transgene (93). Due to the degeneracy of the genetic code, codons can be changed without changing amino acid composition. The advantage of this technique is that rescue is performed by restoring the expression of the same gene of *D. melanogaster*. All three strategies have been proven useful in rescuing RNAi-induced phenotypes. However, in a few cases a fosmid of *D. pseudoobscura* was not able to rescue the RNAi phenotype (89, 90).

2.7. Genome-Wide Enhancer Detection Libraries

The GAL4/UAS system has been extremely useful to drive expression of numerous different constructs in specific cell populations. Many of these drivers were isolated by hopping *P* element enhancer detector elements (28, 29) that carry the binary GAL4

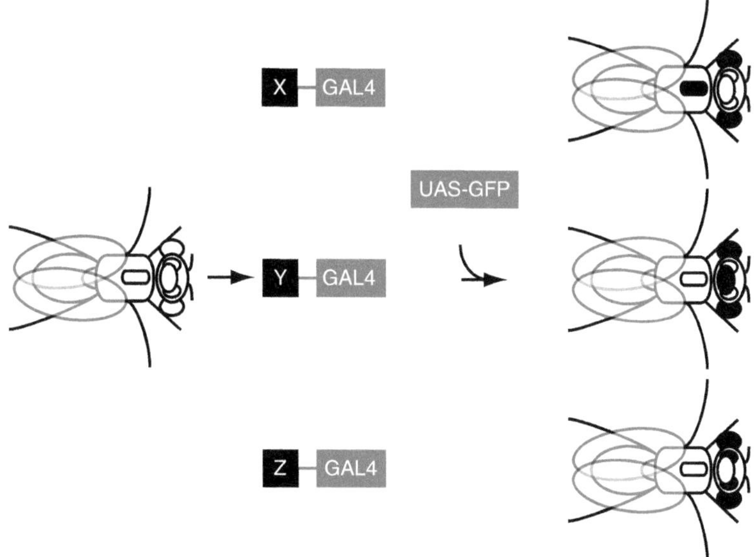

Fig. 13. Enhancer detection libraries. Different regulatory elements (X, Y, and Z) are subcloned in front of a GAL4 driver and all are integrated into the same genomic docking site using the *white*+ eye marker and the ΦC31 system, thereby neutralizing position effects. A fluorescent reporter (here, UAS-GFP) is used to assess their expression domain. Indicated here are different brain regions highlighted by specific regulatory elements that include for simplicity, the entire ventral nerve cord, the main brain or the optic lobes.

transactivator (94). By integrating in the regulatory regions of genes, the GAL4 comes under the control of the endogenous enhancers. However, relatively few enhancer detector insertion strains have truly highly specific expression patterns and can be used to manipulate-restricted cell populations. This lack of specificity is a problem when one desires to remove specific cells or functionally inactivate or manipulate, for example a few neurons in a complex circuit. To isolate individual components from the complex regulatory environment of most genes, it typically suffices to isolate smaller pieces and bring them in front of a neutral promoter, the so-called enhancer bashing strategy. To neutralize position effects that may influence the analysis of different regulatory regions, it is advantageous to integrate the different constructs in the same "neutral" *attP* docking site using the ΦC31 integrase system. Hence, a large set of docking sites were analyzed for expression levels (19) and one, *attP2*, was used to integrate a genome-wide library of regulatory elements (21) (Fig. 13). Many fragments were analyzed for nervous system-specific expression with the ultimate goal of generating a brain expression atlas. These libraries are currently being integrated into the BDSC collection and will allow one to selectively manipulate numerous cell populations with much greater precision than was previously possible (21).

3. Concluding Remarks and Future Directions

Here, we provided an overview of how transposons, recombinases, and integrases can be incorporated into genome-wide studies of the fruit fly *D. melanogaster*. The development of new techniques and methods that become simpler allow us to spend more time to answer biological questions. Transposons have been instrumental in transgenesis and mutagenesis techniques in flies. Optimizing transposons for downstream modifications by including recombination or attachment sites was a significant step forward. A genome-wide set of transposons containing two attachment sites, such as the *Minos* derivative MiMIC, will allow virtually any change of the genome through regular site-specific transgenesis or RMCE. It will also allow the generation of duplications of genes that are modified. These can then be used for gene-targeting strategy. Flp recombinase will be further critical in manipulating and interrogating the genome.

The ΦC31 integrase system has opened many new avenues for the fly community. Since its birth in 2004 (16), it has contributed to a wealth of novel applications in the fly community, and we are most likely only seeing the tip of the iceberg. Robust, simple, and efficient as it is, it will promote fly biology to new heights. An additional key to its success is the virtual neutralization of position effects, therefore allowing the comparison of different transgenes at the same genomic location. This was previously impossible. The ΦC31 integrase system has allowed us to generate genome-wide libraries of many great reagents. These include genome-wide genomic libraries, genome-wide RNAi libraries, genome-wide regulatory element libraries, and others. Needless to say, these genome-encompassing technological achievements will move the fly field forward and attract numerous young biologists to the fly field.

Acknowledgments

We apologize to those whose work we could not cite due to space limitations. We thank Karen Schulze and Kevin Cook for critical comments on the manuscript.

References

1. Adams, M. D., Celniker, S. E., Holt, R. A., Evans, C. A., Gocayne, J. D., Amanatides, P. G., Scherer, S. E., Li, P. W., Hoskins, R. A., Galle, R. F., George, R. A., Lewis, S. E., Richards, S., Ashburner, M., Henderson, S. N., Sutton, G. G., Wortman, J. R., Yandell, M. D., Zhang, Q., Chen, L. X., Brandon, R. C., Rogers, Y. H., Blazej, R. G., Champe, M., Pfeiffer, B. D., Wan, K. H., Doyle, C., Baxter, E. G., Helt, G., Nelson, C. R., Gabor, G. L., Abril, J. F., Agbayani, A., An, H. J., Andrews-Pfannkoch, C., Baldwin, D., Ballew, R. M., Basu, A., Baxendale, J., Bayraktaroglu, L., Beasley, E. M., Beeson, K. Y., Benos, P. V., Berman, B. P., Bhandari, D., Bolshakov, S., Borkova, D., Botchan, M. R., Bouck, J., Brokstein, P., Brottier, P., Burtis, K. C., Busam, D. A., Butler, H., Cadieu, E., Center, A., Chandra, I., Cherry, J. M., Cawley, S., Dahlke, C., Davenport, L. B., Davies, P., de, P. B., Delcher, A., Deng, Z., Mays, A. D., Dew, I., Dietz, S. M., Dodson, K., Doup, L. E., Downes, M., Dugan-Rocha, S., Dunkov, B. C., Dunn, P., Durbin, K. J., Evangelista, C. C., Ferraz, C., Ferriera, S., Fleischmann, W., Fosler, C., Gabrielian, A. E., Garg, N. S., Gelbart, W. M., Glasser, K., Glodek, A., Gong, F., Gorrell, J. H., Gu, Z., Guan, P., Harris, M., Harris, N. L., Harvey, D., Heiman, T. J., Hernandez, J. R., Houck, J., Hostin, D., Houston, K. A., Howland, T. J., Wei, M. H., Ibegwam, C., Jalali, M., Kalush, F., Karpen, G. H., Ke, Z., Kennison, J. A., Ketchum, K. A., Kimmel, B. E., Kodira, C. D., Kraft, C., Kravitz, S., Kulp, D., Lai, Z., Lasko, P., Lei, Y., Levitsky, A. A., Li, J., Li, Z., Liang, Y., Lin, X., Liu, X., Mattei, B., McIntosh, T. C., McLeod, M. P., McPherson, D., Merkulov, G., Milshina, N. V., Mobarry, C., Morris, J., Moshrefi, A., Mount, S. M., Moy, M., Murphy, B., Murphy, L., Muzny, D. M., Nelson, D. L., Nelson, D. R., Nelson, K. A., Nixon, K., Nusskern, D. R., Pacleb, J. M., Palazzolo, M., Pittman, G. S., Pan, S., Pollard, J., Puri, V., Reese, M. G., Reinert, K., Remington, K., Saunders, R. D., Scheeler, F., Shen, H., Shue, B. C., Siden-Kiamos, I., Simpson, M., Skupski, M. P., Smith, T., Spier, E., Spradling, A. C., Stapleton, M., Strong, R., Sun, E., Svirskas, R., Tector, C., Turner, R., Venter, E., Wang, A. H., Wang, X., Wang, Z. Y., Wassarman, D. A., Weinstock, G. M., Weissenbach, J., Williams, S. M., WoodageT, Worley, K. C., Wu, D., Yang, S., Yao, Q. A., Ye, J., Yeh, R. F., Zaveri, J. S., Zhan, M., Zhang, G., Zhao, Q., Zheng, L., Zheng, X. H., Zhong, F. N., Zhong, W., Zhou, X., Zhu, S., Zhu, X., Smith, H. O., Gibbs, R. A., Myers, E. W., Rubin, G. M., and Venter, J. C. (2000) *Science 287*, 2185–2195.
2. Venken, K. J. and Bellen, H. J. (2007) *Development 134*, 3571–3584.
3. Venken, K. J. and Bellen, H. J. (2005) *Nat. Rev. Genet. 6*, 167–178.
4. Rubin, G. M. and Spradling, A. C. (1982) *Science 218*, 348–353.
5. Bellen, H. J., Levis, R. W., Liao, G., He, Y., Carlson, J. W., Tsang, G., Evans-Holm, M., Hiesinger, P. R., Schulze, K. L., Rubin, G. M., Hoskins, R. A., and Spradling, A. C. (2004) *Genetics 167*, 761–781.
6. Handler, A. M. and Harrell, R. A. (1999) *Insect Mol. Biol. 8*, 449–457.
7. Loukeris, T. G., Arca, B., Livadaras, I., Dialektaki, G., and Savakis, C. (1995) *Proc. Natl. Acad. Sci. U. S. A 92*, 9485–9489.
8. Cook, K. R., Parks, A. L., Jacobus, L. M., Kaufman, T. C., and Matthews, K. A. (2010) *Fly. (Austin.) 4*, 88–91.
9. Yamamoto, M. T. (2010) *Exp. Anim 59*, 125–138.
10. Bischof, J. and Basler, K. (2008) *Methods Mol. Biol. 420*, 175–195.
11. Gong, W. J. and Golic, K. G. (2003) *Proc. Natl. Acad. Sci. U. S. A 100*, 2556–2561.
12. Rong, Y. S. and Golic, K. G. (2000) *Science 288*, 2013–2018.
13. Parks, A. L., Cook, K. R., Belvin, M., Dompe, N. A., Fawcett, R., Huppert, K., Tan, L. R., Winter, C. G., Bogart, K. P., Deal, J. E., alHerr, M. E., Grant, D., Marcinko, M., Miyazaki, W. Y., Robertson, S., Shaw, K. J., Tabios, M., Vysotskaia, V., Zhao, L., Andrade, R. S., Edgar, K. A., Howie, E., Killpack, K., Milash, B., Norton, A., Thao, D., Whittaker, K., Winner, M. A., Friedman, L., Margolis, J., Singer, M. A., Kopczynski, C., Curtis, D., Kaufman, T. C., Plowman, G. D., Duyk, G., and Francis-Lang, H. L. (2004) *Nat. Genet. 36*, 288–292.
14. Ryder, E., Blows, F., Ashburner, M., Bautista-Llacer, R., Coulson, D., Drummond, J., Webster, J., Gubb, D., Gunton, N., Johnson, G., O'Kane, C. J., Huen, D., Sharma, P., Asztalos, Z., Baisch, H., Schulze, J., Kube, M., Kittlaus, K., Reuter, G., Maroy, P., Szidonya, J., Rasmuson-Lestander, A., Ekstrom, K., Dickson, B., Hugentobler, C., Stocker, H., Hafen, E., Lepesant, J. A., Pflugfelder, G., Heisenberg, M., Mechler, B., Serras, F., Corominas, M., Schneuwly, S., Preat, T., Roote, J., and Russell, S. (2004) *Genetics 167*, 797–813.

15. Ryder, E., Ashburner, M., Bautista-Llacer, R., Drummond, J., Webster, J., Johnson, G., Morley, T., Chan, Y. S., Blows, F., Coulson, D., Reuter, G., Baisch, H., Apelt, C., Kauk, A., Rudolph, T., Kube, M., Klimm, M., Nickel, C., Szidonya, J., Maroy, P., Pal, M., Rasmuson-Lestander, A., Ekstrom, K., Stocker, H., Hugentobler, C., Hafen, E., Gubb, D., Pflugfelder, G., Dorner, C., Mechler, B., Schenkel, H., Marhold, J., Serras, F., Corominas, M., Punset, A., Roote, J., and Russell, S. (2007) *Genetics 177*, 615–629.
16. Groth, A. C., Fish, M., Nusse, R., and Calos, M. P. (2004) *Genetics 166*, 1775–1782.
17. Bischof, J., Maeda, R. K., Hediger, M., Karch, F., and Basler, K. (2007) *Proc. Natl. Acad. Sci. U. S. A 104*, 3312–3317.
18. Venken, K. J., He, Y., Hoskins, R. A., and Bellen, H. J. (2006) *Science 314*, 1747–1751.
19. Markstein, M., Pitsouli, C., Villalta, C., Celniker, S. E., and Perrimon, N. (2008) *Nat. Genet. 40*, 476–483.
20. Ni, J. Q., Liu, L. P., Binari, R., Hardy, R., Shim, H. S., Cavallaro, A., Booker, M., Pfeiffer, B. D., Markstein, M., Wang, H., Villalta, C., Laverty, T. R., Perkins, L. A., and Perrimon, N. (2009) *Genetics 182*, 1089–1100.
21. Pfeiffer, B. D., Jenett, A., Hammonds, A. S., Ngo, T. T., Misra, S., Murphy, C., Scully, A., Carlson, J. W., Wan, K. H., Laverty, T. R., Mungall, C., Svirskas, R., Kadonaga, J. T., Doe, C. Q., Eisen, M. B., Celniker, S. E., and Rubin, G. M. (2008) *Proc. Natl. Acad. Sci. U. S. A 105*, 9715–9720.
22. Ejsmont, R. K., Sarov, M., Winkler, S., Lipinsk, K. A., and Tomancak, P. (2009) *Nat. Methods 6*, 435–437.
23. Venken, K. J., Carlson, J. W., Schulze, K. L., Pan, H., He, Y., Spokony, R., Wan, K. H., Koriabine, M., de Jong, P. J., White, K. P., Bellen, H. J., and Hoskins, R. A. (2009) *Nat. Methods 6*, 431–434.
24. Ryder, E. and Russell, S. (2003) *Brief. Funct. Genomic. Proteomic. 2*, 57–71.
25. Castro, J. P. and Carareto, C. M. (2004) *Genetica 121*, 107–118.
26. Cooley, L., Kelley, R., and Spradling, A. (1988) *Science 239*, 1121–1128.
27. Robertson, H. M., Preston, C. R., Phillis, R. W., Johnson-Schlitz, D. M., Benz, W. K., and Engels, W. R. (1988) *Genetics 118*, 461–470.
28. Bellen, H. J., O'Kane, C. J., Wilson, C., Grossniklaus, U., Pearson, R. K., and Gehring, W. J. (1989) *Genes Dev. 3*, 1288–1300.
29. Bier, E., Vaessin, H., Shepherd, S., Lee, K., McCall, K., Barbel, S., Ackerman, L., Carretto, R., Uemura, T., and Grell, E. (1989) *Genes Dev. 3*, 1273–1287.
30. Spradling, A. C., Stern, D., Beaton, A., Rhem, E. J., Laverty, T., Mozden, N., Misra, S., and Rubin, G. M. (1999) *Genetics 153*, 135–177.
31. Ochman, H., Gerber, A. S., and Hartl, D. L. (1988) *Genetics 120*, 621–623.
32. Hui, E. K., Wang, P. C., and Lo, S. J. (1998) *Cell Mol. Life Sci. 54*, 1403–1411.
33. Voelker, R. A., Greenleaf, A. L., Gyurkovics, H., Wisely, G. B., Huang, S. M., and Searles, L. L. (1984) *Genetics 107*, 279–294.
34. Gloor, G. B., Nassif, N. A., Johnson-Schlitz, D. M., Preston, C. R., and Engels, W. R. (1991) *Science 253*, 1110–1117.
35. Sepp, K. J. and Auld, V. J. (1999) *Genetics 151*, 1093–1101.
36. Norga, K. K., Gurganus, M. C., Dilda, C. L., Yamamoto, A., Lyman, R. F., Patel, P. H., Rubin, G. M., Hoskins, R. A., Mackay, T. F., and Bellen, H. J. (2003) *Curr. Biol. 13*, 1388–1396.
37. Buszczak, M., Paterno, S., Lighthouse, D., Bachman, J., Planck, J., Owen, S., Skora, A. D., Nystul, T. G., Ohlstein, B., Allen, A., Wilhelm, J. E., Murphy, T. D., Levis, R. W., Matunis, E., Srivali, N., Hoskins, R. A., and Spradling, A. C. (2007) *Genetics 175*, 1505–1531.
38. Quinones-Coello, A. T., Petrella, L. N., Ayers, K., Melillo, A., Mazzalupo, S., Hudson, A. M., Wang, S., Castiblanco, C., Buszczak, M., Hoskins, R. A., and Cooley, L. (2007) *Genetics 175*, 1089–1104.
39. Handler, A. M., Gomez, S. P., and O'Brochta, D. A. (1993) *Arch. Insect Biochem. Physiol 22*, 373–384.
40. Pirrotta, V. (1988) *Biotechnology 10*, 437–456.
41. Patton, J. S., Gomes, X. V., and Geyer, P. K. (1992) *Nucleic Acids Res. 20*, 5859–5860.
42. Berghammer, A. J., Klingler, M., and Wimmer, E. A. (1999) *Nature 402*, 370–371.
43. Horn, C. and Wimmer, E. A. (2000) *Dev. Genes Evol. 210*, 630–637.
44. Rorth, P., Szabo, K., Bailey, A., Laverty, T., Rehm, J., Rubin, G. M., Weigmann, K., Milan, M., Benes, V., Ansorge, W., and Cohen, S. M. (1998) *Development 125*, 1049–1057.
45. Metaxakis, A., Oehler, S., Klinakis, A., and Savakis, C. (2005) *Genetics 171*, 571–581.
46. Hacker, U., Nystedt, S., Barmchi, M. P., Horn, C., and Wimmer, E. A. (2003) *Proc. Natl. Acad. Sci. U. S. A 100*, 7720–7725.
47. Horn, C., Offen, N., Nystedt, S., Hacker, U., and Wimmer, E. A. (2003) *Genetics 163*, 647–661.

48. Thibault, S. T., Singer, M. A., Miyazaki, W. Y., Milash, B., Dompe, N. A., Singh, C. M., Buchholz, R., Demsky, M., Fawcett, R., Francis-Lang, H. L., Ryner, L., Cheung, L. M., Chong, A., Erickson, C., Fisher, W. W., Greer, K., Hartouni, S. R., Howie, E., Jakkula, L., Joo, D., Killpack, K., Laufer, A., Mazzotta, J., Smith, R. D., Stevens, L. M., Stuber, C., Tan, L. R., Ventura, R., Woo, A., Zakrajsek, I., Zhao, L., Chen, F., Swimmer, C., Kopczynski, C., Duyk, G., Winberg, M. L., and Margolis, J. (2004) *Nat. Genet. 36*, 283–287.
49. Golic, K. G. and Lindquist, S. (1989) *Cell 59*, 499–509.
50. Heidmann, D. and Lehner, C. F. (2001) *Dev. Genes Evol. 211*, 458–465.
51. McLeod, M., Craft, S., and Broach, J. R. (1986) *Mol. Cell Biol. 6*, 3357–3367.
52. Groth, A. C., Olivares, E. C., Thyagarajan, B., and Calos, M. P. (2000) *Proc. Natl. Acad. Sci. U. S. A 97*, 5995–6000.
53. Golic, M. M., Rong, Y. S., Petersen, R. B., Lindquist, S. L., and Golic, K. G. (1997) *Nucleic Acids Res. 25*, 3665–3671.
54. Golic, K. G. and Golic, M. M. (1996) *Genetics 144*, 1693–1711.
55. Schlake, T. and Bode, J. (1994) *Biochemistry 33*, 12746–12751.
56. Bateman, J. R., Lee, A. M., and Wu, C. T. (2006) *Genetics 173*, 769–777.
57. Horn, C. and Handler, A. M. (2005) *Proc. Natl. Acad. Sci. U. S. A 102*, 12483–12488.
58. Oberstein, A., Pare, A., Kaplan, L., and Small, S. (2005) *Nat. Methods 2*, 583–585.
59. St Johnston, D. (2002) *Nat. Rev. Genet. 3*, 176–188.
60. Zhai, R. G., Hiesinger, P. R., Koh, T. W., Verstreken, P., Schulze, K. L., Cao, Y., Jafar-Nejad, H., Norga, K. K., Pan, H., Bayat, V., Greenbaum, M. P., and Bellen, H. J. (2003) *Proc. Natl. Acad. Sci. U. S. A 100*, 10860–10865.
61. Mamanova, L., Coffey, A. J., Scott, C. E., Kozarewa, I., Turner, E. H., Kumar, A., Howard, E., Shendure, J., and Turner, D. J. (2010) *Nat. Methods 7*, 111–118.
62. Cook, R. K., Deal, M. E., Deal, J. A., Garton, R. D., Brown, C. A., Ward, M. E., Andrade, R. S., Spana, E. P., Kaufman, T. C., and Cook, K. R. (2010) *Genetics 186*, 1095–1109.
63. Xu, T. and Rubin, G. M. (1993) *Development 117*, 1223–1237.
64. Newsome, T. P., Asling, B., and Dickson, B. J. (2000) *Development 127*, 851–860.
65. Stowers, R. S. and Schwarz, T. L. (1999) *Genetics 152*, 1631–1639.
66. Verstreken, P., Koh, T. W., Schulze, K. L., Zhai, R. G., Hiesinger, P. R., Zhou, Y., Mehta, S. Q., Cao, Y., Roos, J., and Bellen, H. J. (2003) *Neuron 40*, 733–748.
67. Tien, A. C., Rajan, A., Schulze, K. L., Ryoo, H. D., Acar, M., Steller, H., and Bellen, H. J. (2008) *J. Cell Biol. 182*, 1113–1125.
68. Kango-Singh, M., Nolo, R., Tao, C., Verstreken, P., Hiesinger, P. R., Bellen, H. J., and Halder, G. (2002) *Development 129*, 5719–5730.
69. Menut, L., Vaccari, T., Dionne, H., Hill, J., Wu, G., and Bilder, D. (2007) *Genetics 177*, 1667–1677.
70. Mehta, S. Q., Hiesinger, P. R., Beronja, S., Zhai, R. G., Schulze, K. L., Verstreken, P., Cao, Y., Zhou, Y., Tepass, U., Crair, M. C., and Bellen, H. J. (2005) *Neuron 46*, 219–232.
71. Hiesinger, P. R., Fayyazuddin, A., Mehta, S. Q., Rosenmund, T., Schulze, K. L., Zhai, R. G., Verstreken, P., Cao, Y., Zhou, Y., Kunz, J., and Bellen, H. J. (2005) *Cell 121*, 607–620.
72. Lee, T. and Luo, L. (1999) *Neuron 22*, 451–461.
73. Awasaki, T. and Lee, T. (2011) *Glia 59*, 1377–1386.
74. Duffy, J. B., Harrison, D. A., and Perrimon, N. (1998) *Development 125*, 2263–2271.
75. Hoskins, R. A., Carlson, J. W., Kennedy, C., Acevedo, D., Evans-Holm, M., Frise, E., Wan, K. H., Park, S., Mendez-Lago, M., Rossi, F., Villasante, A., Dimitri, P., Karpen, G. H., and Celniker, S. E. (2007) *Science 316*, 1625–1628.
76. Sharan, S. K., Thomason, L. C., Kuznetsov, S. G., and Court DL (2009) *Nat. Protoc. 4*, 206–223.
77. Wild, J., Hradecna, Z., and Szybalski, W. (2002) *Genome Res. 12*, 1434–1444.
78. Pepple, K. L., Atkins, M., Venken, K., Wellnitz, K., Harding, M., Frankfort, B., and Mardon, G. (2008) *Development 135*, 4071–4079.
79. Venken, K. J., Kasprowicz, J., Kuenen, S., Yan, J., Hassan, B. A., and Verstreken, P. (2008) *Nucleic Acids Res. 36*, e114.
80. Venken, K. J., Popodi, E., Holtzman, S. L., Schulze, K. L., Park, S., Carlson, J. W., Hoskins, R. A., Bellen, H. J., and Kaufman, T. C. (2010) *Genetics 186*, 1111–1125.
81. Fire, A., Xu, S., Montgomery, M. K., Kostas, S. A., Driver, S. E., and Mello, C. C. (1998) *Nature 391*, 806–811.
82. Kennerdell, J. R. and Carthew, R. W. (1998) *Cell 95*, 1017–1026.
83. Fortier, E. and Belote, J. M. (2000) *Genesis. 26*, 240–244.
84. Lam, G. and Thummel, C. S. (2000) *Curr. Biol. 10*, 957–963.

85. Kennerdell, J. R. and Carthew, R. W. (2000) *Nat. Biotechnol. 18*, 896–898.
86. Dietzl, G., Chen, D., Schnorrer, F., Su, K. C., Barinova, Y., Fellner, M., Gasser, B., Kinsey, K., Oppel, S., Scheiblauer, S., Couto, A., Marra, V., Keleman, K., and Dickson, B. J. (2007) *Nature 448*, 151–156.
87. Neumuller, R. A. and Perrimon, N. (2011) *Wiley. Interdiscip. Rev. Syst. Biol. Med. 3*, 471–478.
88. Neely, G. G., Hess, A., Costigan, M., Keene, A. C., Goulas, S., Langeslag, M., Griffin, R. S., Belfer, I., Dai, F., Smith, S. B., Diatchenko, L., Gupta, V., Xia, C. P., Amann, S., Kreitz, S., Heindl-Erdmann, C., Wolz, S., Ly, C. V., Arora, S., Sarangi, R., Dan, D., Novatchkova, M., Rosenzweig, M., Gibson, D. G., Truong, D., Schramek, D., Zoranovic, T., Cronin, S. J., Angjeli, B., Brune, K., Dietzl, G., Maixner, W., Meixner, A., Thomas, W., Pospisilik, J. A., Alenius, M., Kress, M., Subramaniam, S., Garrity, P. A., Bellen, H. J., Woolf, C. J., and Penninger, J. M. (2010) *Cell 143*, 628–638.
89. Kondo, S., Booker, M., and Perrimon, N. (2009) *Genetics 183*, 1165–1173.
90. Langer, C. C., Ejsmont, R. K., Schonbauer, C., Schnorrer, F., and Tomancak, P. (2010) *PLoS. ONE. 5*, e8928.
91. Richards, S., Liu, Y., Bettencourt, B. R., Hradecky, P., Letovsky, S., Nielsen, R., Thornton, K., Hubisz, M. J., Chen, R., Meisel, R. P., Couronne, O., Hua, S., Smith, M. A., Zhang, P., Liu, J., Bussemaker, H. J., van Batenburg, M. F., Howells, S. L., Scherer, S. E., Sodergren, E., Matthews, B. B., Crosby, M. A., Schroeder, A. J., Ortiz-Barrientos, D., Rives, C. M., Metzker, M. L., Muzny, D. M., Scott, G., Steffen, D., Wheeler, D. A., Worley, K. C., Havlak, P., Durbin, K. J., Egan, A., Gill, R., Hume, J., Morgan, M. B., Miner, G., Hamilton, C., Huang, Y., Waldron, L., Verduzco, D., Clerc-Blankenburg, K. P., Dubchak, I., Noor, M. A., Anderson, W., White, K. P., Clark, A. G., Schaeffer, S. W., Gelbart, W., Weinstock, G. M., and Gibbs, R. A. (2005) *Genome Res. 15*, 1–18.
92. Wang, Z., Engler, P., Longacre, A., and Storb, U. (2001) *Genome Res. 11*, 137–142.
93. Schulz, J. G., David, G., and Hassan, B. A. (2009) *Nucleic Acids Res. 37*, e93.
94. Brand, A. H. and Perrimon, N. (1993) *Development 118*, 401–415.
95. Bellen, H. J., Levis, R. W., He, Y., Carlson, J. W., Evans-Holm, M., Bae, E., Kim, J., Metaxakis, A., Savakis, C., Schulze, K. L., Hoskins, R. A., and Spradling, A. C. (2011) The Drosophila gene disruption project: progress using transposons with distinctive site specificities. *Genetics 188*, 731–743.
96. Spradling, A. C., Bellen, H. J., and Hoskins, R. A. (2011) Drosophila P elements preferentially transpose to replication origins. *Proc. Natl. Acad. Sci. U. S. A. 108*, 15948–15953.
97. Venken, K. J., Schulze, K. L., Haelterman, N. A., Pan, H., He, Y., Evans-Holm, M., Carlson, J. W., Levis, R. W., Spradling, A. C., Hoskins, R. A., and Bellen, H. J. (2011) MiMIC: a highly versatile transposon insertion resource for engineering Drosophila melanogaster genes. *Nat. Methods 8*, 737–743.
98. Venken, K. J., Simpson, J. H., and Bellen, H. J. (2011) Genetic manipulation of genes and cells in the nervous system of the fruit fly. *Neuron 72*, 202–230.

Chapter 13

The *Sleeping Beauty* Transposon Toolbox

Ismahen Ammar, Zsuzsanna Izsvák, and Zoltán Ivics

Abstract

The mobility of class II transposable elements (DNA transposons) can be experimentally controlled by separating the two functional components of the transposon: the terminal inverted repeat sequences that flank a gene of interest to be mobilized and the transposase protein that can be conditionally supplied to drive the transposition reaction. Thus, a DNA molecule of interest (e.g., a fluorescent marker, an shRNA expression cassette, a mutagenic gene trap or a therapeutic gene construct) cloned between the inverted repeat sequences of a transposon-based vector can be stably integrated into the genome in a regulated and highly efficient manner. *Sleeping Beauty* (*SB*) was the first transposon ever shown capable of gene transfer in vertebrate cells, and recent results confirm that *SB* supports a full spectrum of genetic engineering in vertebrate species, including transgenesis, insertional mutagenesis, and therapeutic somatic gene, transfer both *ex vivo* and *in vivo*. This methodological paradigm opened up a number of avenues for genome manipulations for basic and applied research. This review highlights the state-of-the-art in *SB* transposon technology in diverse genetic applications with special emphasis on the transposon as well as transposase vectors currently available in the *SB* transposon toolbox.

Key words: Mutagenesis, Gene transfer, Transgenesis, Gene therapy, Transposase, Inverted repeats, Gene vectors

1. Introduction

Transposable elements are discrete DNA segments with the distinctive ability to move from one genomic location to another. DNA transposons in particular can be considered as natural gene delivery vehicles that are capable of efficient genomic insertion. A typical DNA transposon consists of a single gene encoding the transposase protein, the enzymatic factor needed for transposition, flanked by two inverted repeat (IR) sequences. DNA transposons move via a cut-and-paste mechanism. The transposase binds to specific DNA sequences within the IRs and catalyzes the excision and reintegration of the transposon. DNA transposons have been successfully used as

Yves Bigot (ed.), *Mobile Genetic Elements: Protocols and Genomic Applications*, Methods in Molecular Biology, vol. 859, DOI 10.1007/978-1-61779-603-6_13, © Springer Science+Business Media, LLC 2012

tools for transgenesis and insertional mutagenesis in invertebrate animal models, including *Drosophila* (1) and *Caenorhabditis elegans* (2). In vertebrates, however, there was no known transposon that was sufficiently active to be used as a tool for such purposes. The situation changed when the synthetic transposon named *Sleeping Beauty* (*SB*) was resurrected from multiple inactive *Tc1/mariner* element fossils found in fish genomes (3). The reconstructed *SB* transposon showed appreciable transposition efficiencies in vertebrate cells (4), and opened up a variety of new, safe, simple, and efficient technologies using DNA transposons for functional genomics, insertional mutagenesis, transgenesis, and somatic gene therapy (reviewed in refs. (5, 6)). In its natural configuration, the *SB* element is about 1,600-bp long comprising the transposase coding sequence flanked by two 230-bp long IRs. The IRs with the proper transposase binding sites are sufficient for DNA mobilization. Therefore, it is possible to physically separate the transposase gene from the IRs. This represents the basis for utilizing the *SB* transposon as a transgene vector. Virtually, any gene of interest can be cloned between the IRs and mobilized by supplying the transposase function in *trans*. The two components of the *SB* system can be maintained and delivered on separate plasmids or as a *cis*-vector, in which the transposon and transposase expression cassette are present on the same plasmid (Fig. 1a–c).

Both components of the *SB* system, transposon and transposase, have undergone extensive engineering. Significant efforts have been made to enhance the transpositional activity of the originally

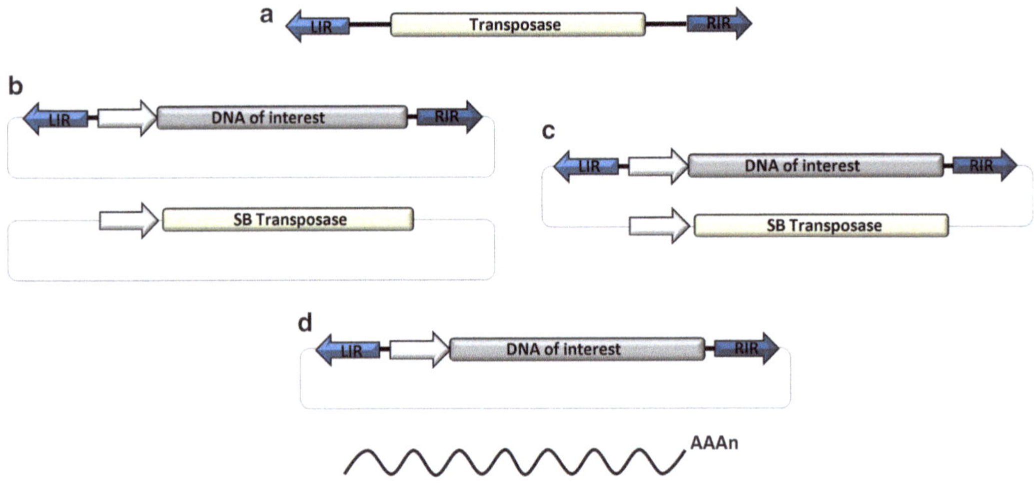

Fig. 1. The *Sleeping Beauty* transposon system. (**a**) Natural arrangement of the *Sleeping Beauty* (*SB*) transposon. The transposase coding sequence is flanked by inverted repeats (LIR and RIR for left and right IR). (**b**) Two-component system, in which the transposon carrying a DNA of interest is maintained on one plasmid and the transposase is provided in *trans* in the form of a second plasmid vector under the control of a suitable promoter (*white arrow*). (**c**) *Cis*-vector configuration, in which the transposon carrying a DNA of interest and the transposase gene are maintained on the same plasmid. (**d**) The transposon containing the DNA of interest is supplied with in vitro-transcribed transposase mRNA.

Table 1
Outline of currently available *SB* transposon vectors

pT	– First-generation *SB* transposon – Salmonid-type *Tc1*-like element isolated from *Tanichthys albonubes* (3)
pT2	– Improved pT transposon – Site-specific mutations in the IR elements accounted for optimized *SB* transposase binding sites and thus higher transposition activity – Addition of TATA-dinucleotides flanking the transposon sequence in order to promote increased excision of the element (7)
pT3	– Hyperactive *SB* transposon – Has an extra TA-dinucleotide flanking the 3′ end of the transposon increasing the excision activity (11)
pT2B	– pT transposon with two left IRs (LIR) – Showed a threefold higher transpositional activity than the original pT transposon (8)
Sandwich transposon	– Two complete SB transposons flank a transgene in an inverted orientation – Transposon vector with superior ability to transpose large transgenes (>10-kb) (12)

resurrected *SB* transposon. For example, base pair changes within the IR sequences were identified that enhanced the transposition activity (pT2 transposons in Table 1, (7)) as compared to the first-generation transposon vectors (pT transposons (3)). It was also found that doubling the left IR on both ends of the transposon increases the efficiency of transposition (Table 1), possibly due to a transpositional enhancer sequence situated in the left IR (8). Constant improvement of the originally resurrected *SB* transposase led to the development of a set of hyperactive transposase enzymes (Table 2) (9–12). By far, the most active of these is the recently developed variant *SB100X*, an *SB* transposase that is 100-fold more potent in chromosomal insertion of a transgene than the originally reconstructed protein (Table 2) (13).

The design of an *SB* transposon vector depends on the particular application. Several practical considerations, including the cargo capacity of the *SB* transposon and the integration site preference, need to be kept in mind. A transgene construct, including coding regions of genes with all the necessary transcriptional regulatory elements, can exceed several kilobases in size. *SB* transposition frequency decreases with increasing length of the transposon, thereby reducing the cloning capacity of *SB* transposon vectors (14). The generation of "sandwich" transposons that have two complete *SB* elements flanking a transgene to be mobilized enhances transposition of large (>10 kb) transgene constructs (Table 1) (12). Hence, sandwich *SB* vectors are the vector of choice for large transgene constructs that would otherwise poorly transpose (Fig. 2c).

Table 2
Outline of all currently available *SB* transposases

SB10	– The originally resurrected *SB* transposase enzyme – Constructed from inactive *Tc1-mariner* transposon elements found in fish genomes (3)
SB11	– Created by amino acid substitutions based on phylogenetic comparison with active *Tc1* transposases – Has a threefold higher transpositional activity than *SB10* (10)
SB12	– Generated by a combination of four single amino acid substitutions – Resulted in an almost fourfold enhancement of transpositional activity as compared to *SB10* (12)
HSB1–HSB5	– A series of hyperactive transposase proteins generated by alanine scanning mutagenesis of the N-terminal DNA-binding domain of *SB10* – Hyperactive mutants with up to 10-fold higher transpositional activity than *SB10* (11)
HSB13–HSB17	– Created through combination of novel mutants generated by phylogenetic-based mutagenesis approaches and previously identified hyperactive *SB* transposase mutants – The most active transposase, *HSB17* showed a 17-fold higher transpositional activity as compared to *SB10* (9)
SB100X	– Generated by a high throughput, PCR-based, DNA-shuffling strategy followed by large-scale genetic screen in mammalian cells – Hyperactive transposase version which is 100-fold more potent in chromosomal insertion of a transgene than *SB10* (13)

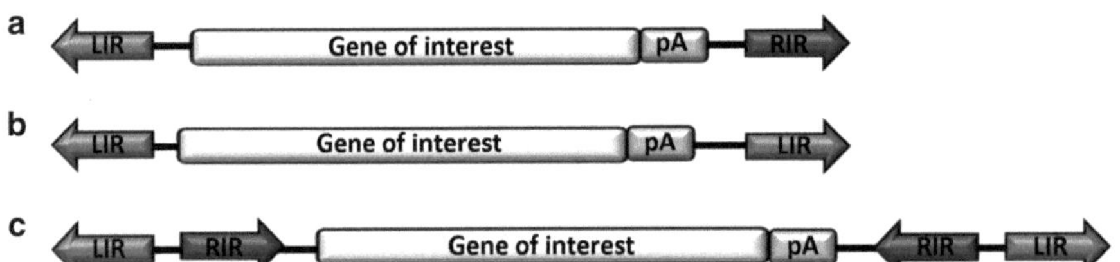

Fig. 2. Types of *SB* transposon vectors. (**a**) Standard *SB* vector. The gene of interest is flanked by one left (LIR) and one right (RIR) inverted repeat element. (**b**) *SB* pT2B transposon vector. The gene of interest is flanked by two LIRs. (**c**) *SB* sandwich transposon vector. The transgene cassette is flanked by two complete *SB* elements in an inverted orientation. Due to mutations in the transposase binding sites of the two inner IRs, only the full composite element can be mobilized by the transposase.

2. Transgenesis

SB-mediated gene delivery is a powerful tool for efficient gene transfer in cultured cells and in primary cell types, including stem cells. Compared to conventional, transfection-based methods for delivering foreign genes into cultured cells, *SB* transposon-mediated gene delivery increases the efficiency of chromosomal integration and facilitates single-copy insertion events. Single-copy insertions are less prone to transgene silencing than are concatemeric insertions created by classical methods (15). The introduction of foreign DNA elements into the host chromosomes tend to invoke epigenetic changes, such as DNA methylation, which may induce transcriptional silencing by heterochromatin formation. Potential silencing of *SB* transposon-carried transgenes was found to be primary determined by the cargo transgene construct, and not by the transposon vector sequence itself (16). Notably, the promoter within the cargo transgene construct plays a major role in triggering epigenetic modifications of the integrating transposon (17). With careful promoter choice, *SB*-mediated transposition can provide long-term transgene expression *in vivo*, as seen in mice after gene delivery in the liver (18–21), lung (22, 23), brain (24), and blood after hematopoietic reconstitution *in vivo* (25).

SB can efficiently transpose from a plasmid to chromosomes of vertebrate cells (3). The generation of stable cell lines using the *SB* transposon system is based on co-delivery of a donor plasmid carrying the *cis*-regulatory IR sequences flanking the transgene cassette together with a source for the transposase (Fig. 2a). Efficient *SB*-mediated gene transfer in somatic cells and tissues has been achieved by supplying *SB*-encoding plasmid DNA as the source of the transposase. The amounts of the delivered plasmids can be adjusted to obtain the desired insertion frequencies per cell (15). In order to reduce genotoxic risks associated with repeated rounds of genomic transposition events evoked by continued expression of the transposase, *in vitro*-transcribed mRNA can be used as a transient transposase source (Fig. 1d) (26). Cell- or tissue-specific expression of the transgene is ensured by a specific promoter inside the *SB* transposon driving transgene expression (Fig. 4a). Efficient and long-term *SB*-mediated transgene integration and expression was shown in cultured cells from fish, mouse, human, frog, chick, sheep, cow, dog, rabbit, hamster, and monkey (14). The *SB* transposon system was also shown to support efficient transposition in mouse (27) and human (28) embryonic stem cells.

The combination of *SB*-mediated gene transfer and RNA interference technology provides an alternative to silence gene expression in mammalian cells. The *SB* transposon system can be used to integrate plasmid-based short hairpin RNA (shRNA)-expression cassettes into chromosomes (Fig. 3b). Following transcription, the shRNA are cleaved by the DICER enzyme into active small interfering RNAs (siRNA), which activate the RNA-induced silencing complex (RISC), and thus promote the sequence-specific degradation and translational repression of the targeted gene. Transposition-based stable delivery of shRNA expression cassettes allows the continuous production of siRNAs and a stable knockdown of target gene expression in the cells. *SB* transposon-based RNAi delivery

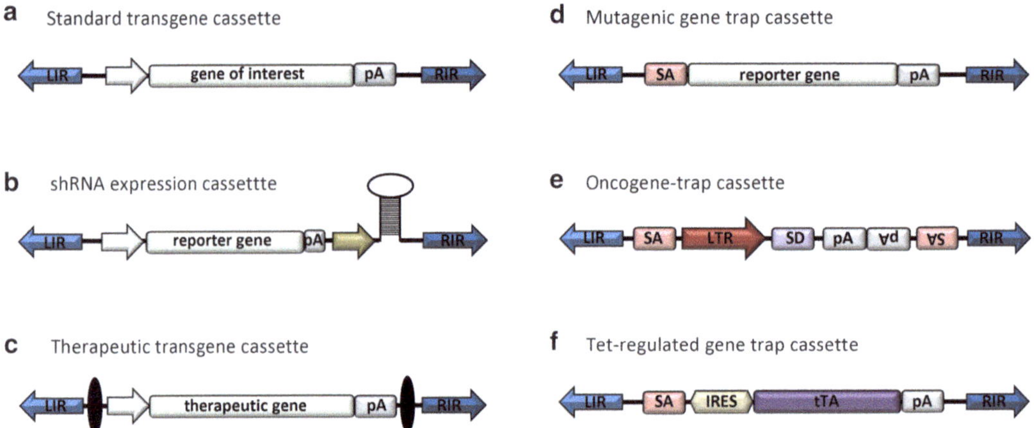

Fig. 3. Types of expression cassettes delivered by *SB* transposon vectors. (**a**) Standard transgene cassette. A gene of interest equipped with a suitable promoter (*white arrow*) and a poly-A sequence (pA) is flanked by the inverted repeat elements (LIR and RIR) (**b**) shRNA expression cassette. The expression cassette includes a polymerase II promoter (*white arrow*) that drives expression of a reporter gene and a polymerase III promoter (*green arrow*) that drives the expression of a short hairpin (sh) RNA. (**c**) Therapeutic transgene cassette. The expression of the therapeutic transgene is driven by a ubiquitous or tissue-specific enhancer/promoter. To further improve the safety of *SB*-based gene therapy vectors, the expression cassette might be flanked by insulator elements (*black disks*) that will block transactivation of endogenous promoters by the transposon insertion, and simultaneously protect the expression of the therapeutic transgene from position effects. (**d**) Mutagenic gene trap cassette. The gene trap cassette contains a splice acceptor sequence (SA) followed by a promoterless reporter gene and a pA. Expression of the promoter gene occurs only if its transcription is initiated from a promoter of a disrupted, actively transcribed endogenous transcription unit. (**e**) Oncogene-trap transposon vector used in *SB*-induced models of cancer. The oncogene-trap cassette includes a strong viral promoter/enhancer long terminal repeat (LTR) sequence followed by a splice donor (SD) element designed to cause overexpression of downstream oncogenes (gain-of-function mutations). The SA pA sequences allow the vector to function as a gene trap and to disrupt expression of tumor suppressor genes (loss-of-function mutations). (**f**) Tet-regulated gene trap cassette for the conditional expression of genes that are under the control of a Tet response element promoter. The cassette contains an SA, an internal ribosomal entry site (IRES), the tetracycline transactivator protein coding sequence (tTA) and a pA. The gene trap vector inserts into genes, traps the splicing of the gene, express the tTA transcription factor, which then activates the expression of tetracycline response element promoter-regulated transgenes in a doxycycline-dependent manner.

vectors were used to down-regulate genes in human cell lines (29). Such technology represents a potential approach to the therapy of acquired immunodeficiency syndrome by stable RNA interference with *SB* vectors knocking down the CCR5 and CXCR4 cell surface co-receptors that are required for viral entry as a first step to confer resistance to HIV (30).

The *SB* transposon system offers a powerful tool for the generation of transgenic animals. Efficient germ line transgenesis is achieved by co-injection of an *SB* transposon carrying a transgene cassette of interest together with an in vitro synthesized mRNA as a transposase source. This technique offers elevated germ line transmission rates of the transgene due to the fast availability of the transposase in the injected oocytes or eggs and reduces transgene mosaicism in the embryo. *In vivo* co-injection of *SB* transposons with transposase mRNA has been employed to generate transgenic zebrafish (31), *Xenopus laevis* (32) and mice (33).

3. Insertional Mutagenesis

The ability of the *SB* transposon system to mediate efficient integration of versatile transgene cassettes into chromosomes of embryonic stem cells as well as into the genome of somatic and germ line tissues of animal models together with the random integration pattern of *SB* transposons provide the basis for a very useful tool for insertional mutagenesis.

The *SB* transposon system has been a useful tool for insertional mutagenesis in embryonic stem cells (34, 35), in somatic tissues (36, 37) and germ line tissues (4, 20, 26, 38, 39) in animal models. *SB*-based insertional mutagenesis screens in mice and rats have been accomplished using the "jump-starter" and "mutator" scheme (40–42). The mutator transgenic lines carry concatemeric *SB* transposon-based gene-trapping vectors, whereas the jump-starter line expresses the transposase. Upon crossing of both lines, the expressed transposase in the jump-starter line can efficiently mobilize the transposon-based gene-trapping cassette in the mutator line. A gene-trap transposon contains a splice acceptor (SA) sequence followed by a promoterless reporter gene, such as a *neo-lacZ* fusion (β-geo) and a poly-A (pA) signal cloned between the IRs of the transposon (Fig. 3d). The β-geo reporter gene is only expressed when its transcription is initiated from the promoter of a disrupted, actively transcribed endogenous transcription unit. Therefore, the expression pattern of the transposon reporter reflects that of the endogenous gene harboring the transposon insertion. Conditional expression of a gene of interest in a tissue- or temporal-specific manner was achieved by using a tetracycline-inducible system. The *SB* transposon-based gene trap cassette comprised a doxycycline-repressible Tet-Off system that is capable to activate the expression of a gene under the control of a Tet response element promoter (Fig. 3f) (43).

The jump-starter and mutator approach does not allow for a preselection for transposition events that disrupt gene expression in the germ line before producing mutant progeny. However, in germ line stem cells it is possible to select in tissue culture for transposition events that disrupt gene expression before being used to produce transgenic animals. Applying insertional mutagenesis to embryonic and spermatogonial stem cells offer the advantage to perform a preselection of the modified stem cell clones before generating mutant animals. *SB* transposon-mediated insertional mutagenesis can help to simultaneously mutate thousands of genes in rat spermatogonial stem cells. Knockout mutations were generated by co-delivering an *SB* transposon gene-trap selection cassette and an *SB100X* expression plasmid into primary rat spermatogonial cells. The culturing of the rat spermatogonial stem cells allows for screening of a large number of transposition events and the generation of gene knockout libraries in spermatogonial stem cells. Selected spermatogonial stem cell clones can be then transplanted to repopulate the testes of sterilized, wild-type recipient male rats. Upon crossing of the recipient males with wild-type females the stem cell genomes can be passed to the transgenic offspring (44, 45).

SB-mediated insertional mutagenesis has been used to identify new cancer genes. A general forward genetics strategy for identifying cancer genes in animal models involves the induction of tumors through insertional mutagenesis. In the mouse, the most common animal model to study cancer, insertional mutagenesis was mainly achieved by using retroviruses, such as the mouse mammary tumor virus (MMTV) or the murine leukemia virus (MuLV). Although retrovirus-induced insertional mutagenesis enabled the identification of hundreds of candidate cancer genes in mice (46) one major limitation is that these retroviruses only produce mammary and hematopoietic cancer (47). Therefore, an insertional mutagen for use in other mouse somatic cells was needed in order to identify genes involved in the tumor formation of a wider variety of tissues. The *SB* system transposes efficiently in numerous mouse somatic tissues, and was successfully used as somatic insertional mutagen to identify genes involved in solid tumor formation in the mouse model system (36). In order to enhance the mutagenicity as well as reporting capabilities of *SB*-based insertional vector technologies, gene trapping vectors have been established. Mutagenic *SB* transposons were constructed that are capable of inducing tumors by mutating both tumor suppressors and proto-oncogenes. In its most common form, an *SB*-based gene trapping construct consists of a splice acceptor element, a reporter gene (usually, an antibiotic resistance, a fluorescent protein or ß-galactosidase) and a poly-A (pA) signal flanked by two IR sequences (Fig. 3d). More sophisticated vectors contain a dual tagging system that combines both gene trap- and poly-A trap elements (Fig. 3e). The T2/Onc vector (36) contains SAs and poly-A sequences on both strands that function as a bidirectional gene trap generating loss-of-function mutations when integrating and thereby disrupting the expression of a tumor suppressor gene. Gain-of-function mutations are achieved by the incorporated sequences from the 5′ long terminal repeat (LTR) of the murine stem cell virus (MSCV) that serve as promoter and enhancer elements. The MSCV 5′LTR can promote gene expression when integrated upstream or within a gene. The splice donor downstream of the MSCV 5′LTR sequence ensures splicing of a transcript initiated from the LTR into downstream exons of endogenous genes (36). Expression of a truncated oncogene is induced if the transposon integrates within the gene itself. Based on the T2/Onc vector the mutagenic transposon T2/Onc2 was constructed. T2/Onc2 contains a larger fragment for one of the splice acceptors and is flanked by two optimized *SB* transposase binding sites (37). A further development was the design of a new mutagenic transposon called T2/Onc3. In this transgene cassette, the MSCV promoter was replaced with the CMV enhancer/chicken beta-actin promoter (CAG). The CAG promoter has been shown to be only weakly expressed in hematopoietic cell lineages, but at high levels in epithelial cells and resulted in the removal of the bias toward inducing mostly lymphomas (48). The choice of the promoter within the mutagenic transposon can have profound effects on the tumor type induced and thus broaden the tumor spectrum that can be produced by *SB* transposon-mediated insertional mutagenesis.

Tissue-specific screens to identify candidate cancer genes can be designed by controlling the spatial expression of the *SB* transposase. The expression of the *SB* transposase from a tissue-specific promoter ensures that transposon mutagenesis is occurring only in sites where the *SB* transposase is expressed (Fig. 4a). Tissue-specific expression of the *SB* transposase can also be achieved by generating a *Cre*-inducible transposase allele (Fig. 4b), in which transposase expression from a ubiquitous promoter is interrupted by a transcriptional stop cassette flanked by loxP sites (lox-stop-lox). This design ensures that the *SB* transposase is only expressed when the lox-stop-lox cassette is removed by *Cre*-recombination. The numerous available tissue-specific *Cre*-transgenic mouse strains allow controlled activation of the *SB* transposase in specific tissues through targeted expression of the *Cre*-recombinase. Activation of SB transposase expression by *Cre*-mediated recombination in germinal center

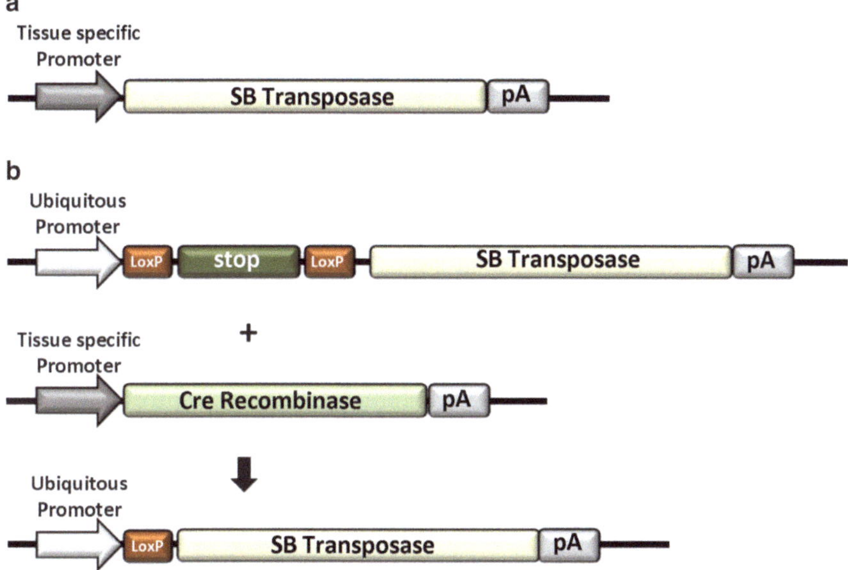

Fig. 4. Tissue-specific expression of the *SB* transposase. (**a**) A tissue specific promoter (*gray arrow*) regulates the expression of the *SB* transposase gene (*light yellow box*). (**b**) Tissue-specific expression of the *SB* transposase is achieved by inserting a lox-stop-lox cassette (*orange and green boxes*) between a ubiquitously active promoter (*white arrow*) and the *SB* transposase gene. The lox-stop-lox cassette can be removed by a tissue-specifically expressed *Cre*-recombinase (*light green box*). The *Cre*-recombinase excises the lox-stop-lox cassette in the tissue of choice and thus enables tissue-specific expression of the *SB* transposase gene.

B-cells generated lymphoma specifically in this tissue (48), whereas hepatocyte-specific *Cre* expression was shown to limit *SB* transposition to the liver (49).

4. Gene Therapy

SB transposon-mediated gene transfer offers an alternative to the use of viral-based delivery systems. The *SB* transposon system offers stable and long-term transgene expression combined with an integration profile that is likely safer than that of retroviruses, and is thus particularly suitable to be adapted for human gene therapy applications. *SB* transposons exhibit a fairly random genomic integration pattern with no preference for integration into genes or upstream regulatory units (50). The use of the *SB100X* hyperactive transposon system enables efficient gene transfer and long-term transgene expression in therapeutically relevant primary cell types, including hematopoietic stem cells (13), mesenchymal stem cells, myoblasts, and iPS cells (51). *SB* transposon vectors have an inherently low enhancer/promoter activity (52). However, the efficient expression of the therapeutic transgene may require the incorporation of strong enhancer/promoter sequences. The safety profile of the *SB* transposon system can be further increased by flanking the transcription units of the cargo with insulator sequences that prevent accidental transactivation of promoters of neighboring genes, and simultaneously protect the expression of the therapeutic transgene from position effects (Fig. 3c) (52).

SB transposon-mediated gene delivery has been used for the correction of hemophilia A (19, 21, 22) and hemophilia B (53), junctional epidermolysis bullosa (54), Huntington disease (29), tyrosinemia type I (55), diabetes (56), and mucopolysaccharidosis I and VII (38) in animal models. The *SB* transposon system has been also used to treat a xenograft model for glioblastoma (19, 24). In the first in-human application, the *SB* transposon system is used to genetically modify human T cells. The *SB* transposon to be introduced carriers a chimeric antigen receptor to render the T cells specifically cytotoxic toward $CD19^+$ B-lineage tumors (25).

Acknowledgments

Work in the authors' laboratories was supported by EU FP6 (INTHER) and EU FP7 (PERSIST and InduStem), grants from the Deutsche Forschungsgemeinschaft SPP1230 "Mechanisms of gene vector entry and persistence," and from the Bundesministerium für Bildung und Forschung (NGFN-2, NGFNplus, iGene, InTherGD, and ReGene).

References

1. Cooley L, Kelley R, Spradling A (1988) Insertional mutagenesis of the Drosophila genome with single P elements. Science 239:1121–1128
2. Bessereau JL, et al. (2001) Mobilization of a Drosophila transposon in the Caenorhabditis elegans germ line. Nature 413:70–74
3. Ivics Z, et al. (1997) Molecular reconstruction of Sleeping Beauty, a Tc1-like transposon from fish, and its transposition in human cells. Cell 91:501–510
4. Fischer SE, Wienholds E, Plasterk RH (2001) Regulated transposition of a fish transposon in the mouse germ line. Proc Natl Acad Sci USA 98:6759–64
5. Ivics Z, Izsvak Z (2010) The expanding universe of transposon technologies for gene and cell engineering. Mob DNA 1:25
6. Ivics Z, et al. (2009) Transposon-mediated genome manipulation in vertebrates. Nat Methods. 6:415–422
7. Cui Z, et al. (2002) Structure-function analysis of the inverted terminal repeats of the sleeping beauty transposon. J Mol Biol 318:1221–1235
8. Izsvak Z, et al. (2002) Involvement of a bifunctional, paired-like DNA-binding domain and a transpositional enhancer in Sleeping Beauty transposition. J Biol Chem 277: 34581–34588
9. Baus J, et al. (2005) Hyperactive transposase mutants of the Sleeping Beauty transposon. Mol Ther 12:1148–1156
10. Geurts AM, et al. (2003) Gene transfer into genomes of human cells by the sleeping beauty transposon system. Mol Ther 8:108–117
11. Yant SR, et al. (2004) Mutational analysis of the N-terminal DNA-binding domain of sleeping beauty transposase: critical residues for DNA binding and hyperactivity in mammalian cells Mol Cell Biol 24:9239–9247
12. Zayed H, et al. (2004) Development of hyperactive sleeping beauty transposon vectors by mutational analysis. Mol Ther 9:292–304
13. Mates L, et al. (2009) Molecular evolution of a novel hyperactive Sleeping Beauty transposase enables robust stable gene transfer in vertebrates. Nat Genet 41:753–761
14. Izsvak Z, Ivics Z, Plasterk RH (2000) Sleeping Beauty, a wide host-range transposon vector for genetic transformation in vertebrates. J Mol Biol 302:93–102
15. Grabundzija I, et al. (2010) Comparative analysis of transposable element vector systems in human cells. Mol Ther 18:1200–1209
16. Park CW, et al. (2005) DNA methylation of Sleeping Beauty with transposition into the mouse genome. Genes Cells 10:763–776
17. Garrison BS, et al. (2007) Postintegrative gene silencing within the Sleeping Beauty transposition system. Mol Cell Biol 27:8824–8833
18. Yant SR, et al. (2000) Somatic integration and long-term transgene expression in normal and haemophilic mice using a DNA transposon system. Nat Genet 25:35–41
19. Ohlfest JR, et al. (2005) Phenotypic correction and long-term expression of factor VIII in hemophilic mice by immunotolerization and nonviral gene transfer using the Sleeping Beauty transposon system. Blood 105:2691–2698
20. Aronovich EL, et al. (2009) Systemic correction of storage disease in MPS I NOD/SCID mice using the sleeping beauty transposon system. Mol Ther 17:1136–1144
21. Kren BT, et al. (2009) Nanocapsule-delivered Sleeping Beauty mediates therapeutic Factor VIII expression in liver sinusoidal endothelial cells of hemophilia A mice. J Clin Invest 119:2086–99
22. Liu L, Mah C, Fletcher BS (2006) Sustained FVIII expression and phenotypic correction of hemophilia A in neonatal mice using an endothelial-targeted sleeping beauty transposon. Mol Ther 13:1006–1015
23. Belur LR, et al. (2003) Gene insertion and long-term expression in lung mediated by the Sleeping Beauty transposon system. Mol Ther 8:501–507
24. Ohlfest JR, et al. (2005) Combinatorial antiangiogenic gene therapy by nonviral gene transfer using the sleeping beauty transposon causes tumor regression and improves survival in mice bearing intracranial human glioblastoma. Mol Ther 12:778–788
25. Xue X, et al. (2009) Stable gene transfer and expression in cord blood-derived CD34+ hematopoietic stem and progenitor cells by a hyperactive Sleeping Beauty transposon system. Blood 114:1319–30
26. Wilber A, et al. (2006) RNA as a source of transposase for Sleeping Beauty-mediated gene insertion and expression in somatic cells and tissues. Mol Ther 13:625–630
27. Luo G, et al. (1998) Chromosomal transposition of a Tc1/mariner-like element in mouse embryonic stem cells. Proc Natl Acad Sci USA 95:10769–10773
28. Orban TI, et al. (2009) Applying a "double-feature" promoter to identify cardiomyocytes differentiated from human embryonic stem

cells following transposon-based gene delivery. Stem Cells 27:1077–1087

29. Chen ZJ, et al. (2005) Sleeping Beauty-mediated down-regulation of huntingtin expression by RNA interference. Biochem Biophys Res Commun 329:646–652
30. Tamhane M, Akkina R (2008) Stable gene transfer of CCR5 and CXCR4 siRNAs by sleeping beauty transposon system to confer HIV-1 resistance. AIDS Res Ther 5:16
31. Nasevicius A, Ekker SC (2000) Effective targeted gene 'knockdown' in zebrafish. Nat Genet 26:216–220
32. Sinzelle L, et al. (2006) Generation of trangenic Xenopus laevis using the Sleeping Beauty transposon system. Transgenic Res 15:751–760
33. Dupuy AJ, et al. (2002) Mammalian germ-line transgenesis by transposition. Proc Natl Acad Sci USA 99:4495–4499
34. Liang Q, et al. (2009) Chromosomal mobilization and reintegration of Sleeping Beauty and PiggyBac transposons. Genesis 47:404–408
35. Kokubu C, et al. (2009) A transposon-based chromosomal engineering method to survey a large cis-regulatory landscape in mice. Nat Genet 41:946–952
36. Collier LS, et al. (2005) Cancer gene discovery in solid tumours using transposon-based somatic mutagenesis in the mouse. Nature 436:272–276
37. Dupuy AJ, et al. (2005) Mammalian mutagenesis using a highly mobile somatic Sleeping Beauty transposon system. Nature 436:221–226
38. Aronovich EL, et al. (2007) Prolonged expression of a lysosomal enzyme in mouse liver after Sleeping Beauty transposon-mediated gene delivery: implications for non-viral gene therapy of mucopolysaccharidoses. J Gene Med 9:403–415
39. Keng VW, et al. (2005) Region-specific saturation germline mutagenesis in mice using the Sleeping Beauty transposon system. Nat Methods 2:763–769
40. Carlson CM, et al. (2003) Transposon mutagenesis of the mouse germline. Genetics 165:243–256
41. Dupuy AJ, Fritz S, Largaespada DA (2001) Transposition and gene disruption in the male germline of the mouse. Genesis 30:82–88
42. Horie K, et al. (2001) Efficient chromosomal transposition of a Tc1/mariner- like transposon Sleeping Beauty in mice. Proc Natl Acad Sci USA 98:9191–9196
43. Geurts AM, et al. (2006) Conditional gene expression in the mouse using a Sleeping Beauty gene-trap transposon. BMC Biotechnol 6:30
44. Izsvak Z, et al. (2010) Generating knockout rats by transposon mutagenesis in spermatogonial stem cells. Nat Methods 7:443–445
45. Ivics Z, et al. (2011) Sleeping Beauty transposon mutagenesis of the rat genome in spermatogonial stem cells. Methods 53:356–65
46. Mikkers H, Berns A (2003) Retroviral insertional mutagenesis: tagging cancer pathways. Adv Cancer Res 88:53–99
47. Dupuy AJ, Jenkins NA, Copeland NG (2006) Sleeping beauty: a novel cancer gene discovery tool. Hum Mol Genet 15 Spec N°1: R75-9
48. Dupuy AJ, et al. (2009) A modified sleeping beauty transposon system that can be used to model a wide variety of human cancers in mice. Cancer Res 69:8150–8156
49. Keng VW, et al. (2009) A conditional transposon-based insertional mutagenesis screen for genes associated with mouse hepatocellular carcinoma. Nat Biotechnol 27:264–274
50. Yant SR, et al. (2005) High-resolution genome-wide mapping of transposon integration in mammals Mol Cell Biol 25:2085–2094
51. Belay E, et al. (2011) Novel hyperactive transposons for genetic modification of induced pluripotent and adult stem cells: a nonviral paradigm for coaxed differentiation. Stem Cells 28:1760–1771
52. Walisko O, et al. (2008) Transcriptional activities of the Sleeping Beauty transposon and shielding its genetic cargo with insulators. Mol Ther 16:359–369
53. Yant SR, et al. (2002) Transposition from a gutless adeno-transposon vector stabilizes transgene expression in vivo. Nat Biotechnol 20:999–1005
54. Ortiz-Urda S, et al. (2003) Sustainable correction of junctional epidermolysis bullosa via transposon-mediated nonviral gene transfer. Gene Ther 10:1099–1104
55. Montini E, et al. (2002) In vivo correction of murine tyrosinemia type I by DNA-mediated transposition. Mol Ther 6:759–69
56. Heggestad AD, Notterpek L, Fletcher BS (2004) Transposon-based RNAi delivery system for generating knockdown cell lines. Biochem Biophys Res Commun 316:643–650

Chapter 14

PiggyBac Toolbox

Mario Di Matteo, Janka Mátrai, Eyayu Belay, Tewodros Firdissa, Thierry VandenDriessche, and Marinee K.L. Chuah

Abstract

The PiggyBac (PB) transposon system was originally derived from the cabbage looper moth *Trichoplusia ni* and represents one of the most promising transposon systems to date. Engineering of the PB transposase enzyme (PBase) and its cognate transposon DNA elements resulted in a substantial increase in transposition activities. Consequently, this has greatly enhanced the versatility of the PB toolbox. It is now widely used for stable gene delivery into a broad variety of cell types from different species, including mammalian cells. This opened up new perspectives for potential therapeutic applications in the fields of gene therapy and regenerative medicine. In particular, we have recently demonstrated that PB transposons could be used to stably deliver genes into human $CD34^+$ hematopoietic stem cells (HSCs) resulting in sustained transgene expression in its differentiated progeny. The PB transposon system is particularly attractive for the generation of induced pluripotent stem cells (iPS). Typically, this can be accomplished by stable gene transfer of genes encoding one or more reprogramming factors (i.e., c-MYC, KLF-4, OCT-4, and/or SOX-2). We have generated a PB-based nonviral reprogramming toolbox that contains different combinations of these reprogramming genes. The main advantage of using this PB toolbox for iPS generation is that the reprogramming cassette can be excised by *de novo* transposase expression, without leaving any molecular trace in the target cell genome. This "traceless excision" paradigm obviates potential risks associated with inadvertent re-expression of reprogramming factors in the iPS progeny. These various applications in gene therapy, stem cell engineering, and regenerative medicine underscore the emerging versatility of the PB toolbox.

Key words: PiggyBac, Gene therapy, Induced pluripotent stem cell, Reprogramming, Hematopoietic stem cells

Abbreviations

BSA	Bovine serum albumin
CAG	CMV early enhancer/chicken β-actin promoter
CFU-E	Erythroid colony forming unit
CFU-GM	Granulocyte/monocyte/macrophage colony forming unit
ePB	Enhanced PB system (14)
FACS	Fluorescence-activated Cell Sorter

Yves Bigot (ed.), *Mobile Genetic Elements: Protocols and Genomic Applications*, Methods in Molecular Biology, vol. 859, DOI 10.1007/978-1-61779-603-6_14,

GFP	Green fluorescence protein
hFlt3-L	Human FMS-like tyrosine kinase 3-ligand
hIL-3	Human interleukin-3
hIL-6	Human interleukin-6
HSCs	Hematopoietic stem cells
hSCF	Human stem cell factor
hTPO	Human thrombopoietin
hyPB	Hyperactive PB (10)
iPS	Induced pluripotent stem cells
IRs	Inverted terminal repeats
mPB	Codon-optimized mouse PB (19)
PB	PiggyBac
PBase	PB transposase enzyme
$PBase_{CO\text{-}MT}$	Human or mouse codon-optimized PB transposase (Chuah-VandenDriessche lab, unpublished)

1. Introduction

1.1. PiggyBac Technology

The PiggyBac (PB) transposon was first identified when it jumped from the cabbage looper moth *Trichoplusia ni*, into the genome of a baculovirus (1). It is a DNA transposon that can mobilize genetic elements from one location to another in the host genome. Essentially, the PB transposon system consists of two components: a DNA element flanked by two inverted terminal repeats (IRs) and a transposase enzyme that catalyzes the transposon's mobilization (Fig. 1a). This transposition occurs by a "cut-and-paste" mechanism (Fig. 1b). Genomic integration is essentially random but requires a minimal tetrad sequence corresponding to a TTAA nucleotide sequence (2). This mechanism of PB transposition has several features that make them particularly attractive as gene delivery vectors. Importantly, only one protein (i.e., PB transposase) is required for in vitro or in vivo transposition and secondly, the PB transposase can act efficiently in trans on virtually any DNA substrate that contains the cis sequences required for transposase recognition and cleavage (3). Consequently, PB transposition in the desired target cell can be achieved by co-transfection of an expression plasmid encoding the transposase and a donor plasmid containing the DNA to be integrated flanked cis-acting sequences required for transposition (Fig.1a). This underlying concept has been exploited to develop a versatile and powerful PB toolbox that is widely used for many different applications, including forward and reverse genetic screens (4, 5), loss-of-function studies (6), functional analysis of parasites (7–9), gene therapy (10) and generation of induced pluripotent stem cells (iPS; (11–16)). This review focuses primarily on the use of the PB toolbox for gene transfer into clinically relevant target cells, particularly hematopoietic stem cells (HSCs). Furthermore, a diverse PB toolbox for iPS generation is discussed.

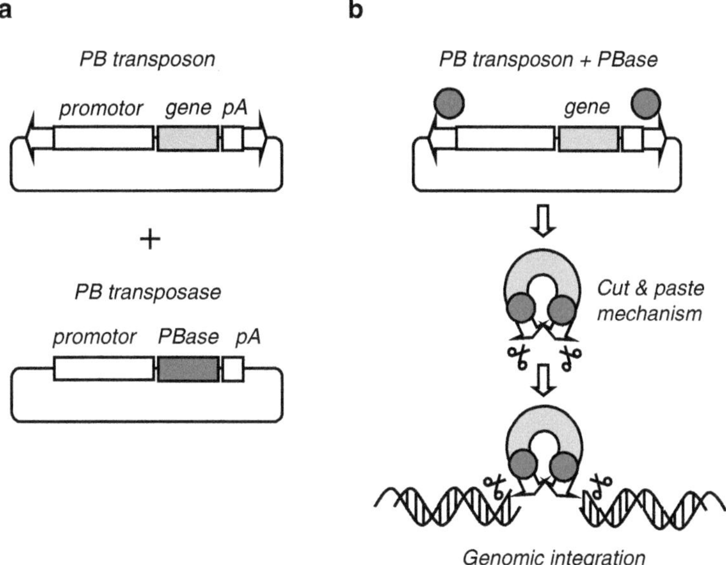

Fig. 1. Schematic representation of the PB two component system and the transposition reaction. (**a**) The PB two component system consists of (1) a donor plasmid containing the gene to be integrated (transposon), (2) an expression plasmid that encodes the transposase. (**b**) The transposase (PBase) binds the terminal repeats (*arrows*) flanking the gene of interest and catalyzes the excision of the transposon and its subsequent genomic integration.

1.2. PiggyBac for Gene Therapy

Since PB transposons integrate into the target cell genome, they have the potential for long-term gene expression, which makes them particularly attractive for gene therapy applications. Moreover, PB transposons have several advantages over viral vector systems for gene therapy. As a nonviral vector, they can be delivered into primary cells by commonly used conventional transfection techniques, like electroporation (17). In addition, since there are no viral antigens present in the delivery vehicle, they have the potential to be less immunogenic than viral vectors. Finally, the use of PB transposons may overcome some of the manufacturing and regulatory hurdles intrinsic to the production of viral vectors. This may ultimately facilitate clinical translation and result in relatively low-cost nonviral vector production. Another advantage of using PB transposons for gene therapy is that relatively large transgenes can easily be accommodated in these vectors. Though transposition efficiency may decrease with greater insert size, it has been shown that PB can still efficiently transpose inserts of up to 14 kb (18). This would open new perspectives for gene delivery of large therapeutic transgenes (>10 kb), which cannot be packaged in most viral vectors due to their intrinsic packaging constraints. Several strategies have recently been devised to further boost the efficiency of PB transposition. This could be achieved by increasing the expression levels of the PB transposase. Since PB is derived from insects, the codon usage is not optimal to ensure high expression in mammalian cells. To overcome this limitation, the PB transposase was therefore subjected to a codon-optimization strategy resulting in a codon usage that is more typical for mouse cells (designated as mPB; (19)). This could be achieved by incorporating silent mutations that improved translation of the PB transcript. The improved transposase enhanced transposition activity in mouse embryonic stem (ES) cells due to the higher translation rate and transposase production (6). Alternatively, the transposon IR themselves can be engineered to augment transposition efficiencies. More specifically, using random PCR mutagenesis, mutants were identified carrying two substitutions in the 5′IR (T53C/C136T) that resulted in an increased transposition efficiency. Combining the mutant PB transposon carrying these T53C/C136T substitutions with a human codon-optimized PB transposase, resulted in a transposition efficiency of about 90% in transfected human ES cells. This enhanced PB system (ePB) (14) also increased the cargo capacity up to 18 kb. Finally, transposition could be increased by mutagenesis of PB transposase using error-prone PCR and selection of hyperactive mutants in yeast which were incorporated in the mPB codon-optimized version. The resulting hyperactive PB (hyPB; (10)) contained seven amino acid changes and yielded a 17-fold enhanced excision and ninefold enhanced integration activity compared to mPB.

PB transposons have been explored for both in vivo and ex vivo gene delivery approaches. In particular, high-pressure hydrodynamic cotransfection of PB transposon and transposase-encoding constructs (18, 20). Though hydrodynamic transfection is not readily applicable to a clinical setting, these studies establish proof-of-principle that the PB system can potentially be useful for in vivo gene therapy. This warrants the development of alternative in vivo gene delivery approaches, such as those based on nanoparticles.

Over the past 20 years, it has been particularly challenging to develop an efficient, nonviral gene delivery system to genetically modify HSCs. Only retroviral, lentiviral, and foamy viral vectors were sufficiently robust to achieve stable gene transfer into HSCs at efficiencies that were clinically relevant (21–25). However, we have recently demonstrated that efficient and stable nonviral gene transfer into bona fide HSCs or progenitor stem cells could be achieved by nonviral means. To achieve this, $CD34^+$ HSCs were enriched from cord blood and transfected by nucleofection with a Sleeping Beauty SB100X transposon encoding a marker gene (17). We now report gene transfer into cord blood derived $CD34^+$ cells using a PB encoding GFP (PB-CAG-GFP-pA) and human or mouse codon-optimized PB vectors ($PBase_{CO\text{-}MT}$; Fig. 2a) that were generated in our laboratory. Up to 15% of the hematopoietic colonies (CFU-E) expressed the reporter gene (GFP) encoded by the PB transposon from a CMV/β-actin promoter (CAG; Fig. 2b). Since the transfection efficiency falls around 50%, we hereby

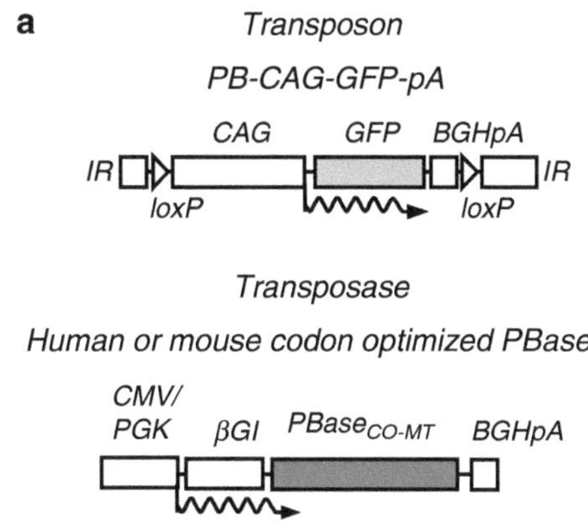

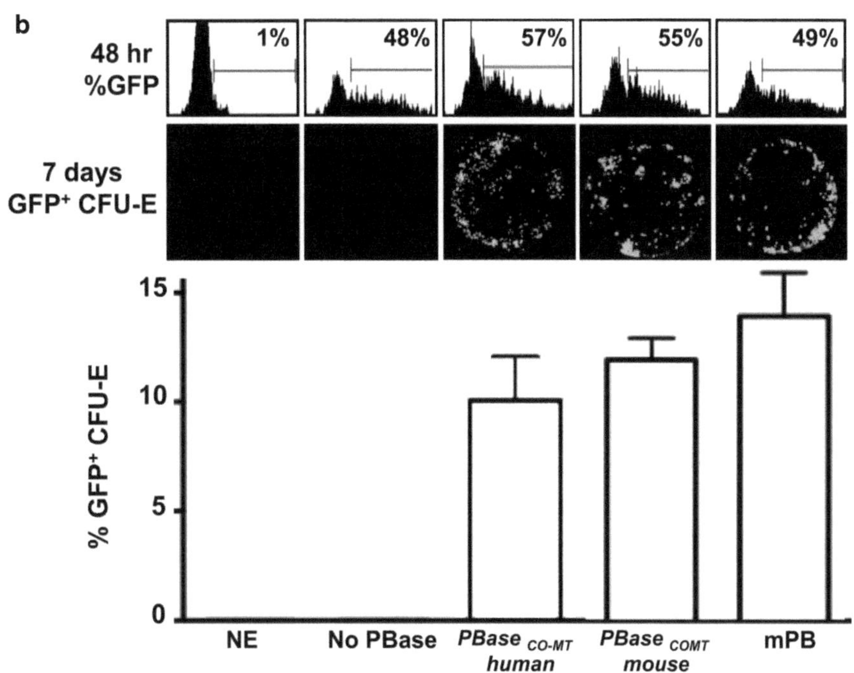

Fig. 2. Schematic representation of PB gene transfer into clinically relevant HSC. (**a**) The PB transposon constructs contain the marker gene (GFP) driven from the CAG promoter and flanked between two loxP sites, whereas the transposase constructs encode either the human or mouse codon-optimized PB transposes ($PBase_{CO-MT}$) driven from the CMV or PGK promoter cloned upstream of a β-globin intron. (**b**) $CD34^+$ cells isolated from umbilical cord blood were nucleofected with the *PB-CAG-GFP plasmid* and the $PBase_{CO-MT.}$ plasmids. Nucleofection method is described in details in Subheadings 2 and 3. Transfection efficiency was determined by FACS analysis 48 h after nucleofection. Nucleofected cells were subjected to an erythroid colony assay. Colony forming units in the erythroid lineage (CFU-E) were counted and microscopic analysis was performed at day 7 post-nucleofection. Three different transposase constructs used in this HSC experiment were the PBase codon-optimized mouse or human generated in our lab designated as $PBase_{CO-MT}$ mouse or human (described above) and were tested side by side with another mouse codon-optimized transposase version called mPB (19). Transfection efficiency of these different transposases falls in the similar range of about 50%. 7 days after transfection, about 13–15% of GFP + CFU-E were scored indicating no significant difference in transposition efficiency between the mouse or human codon-optimized constructs in HSC. The PB system successfully transposed the $CD34^+$ cells giving rise to about 30% transposition efficiency.

demonstrated that the PB vector led to about 30% transposition efficiency in human HSC. The transfected $CD34^+$ cells retained their ability to differentiate in vitro along distinct lymphohematopoietic lineages using conventional clonogenic assays (Fig. 2b). These results are consistent with our previously published work (26). The high efficiency of sustainable expression after high cell proliferation as well as the promising safety features of the treated stem cells are expected to facilitate clinical implementation of ex vivo gene therapies. Hence, PB can offer a safer alternative to viral vectors in gene therapy. The availability of PB may greatly facilitate clinical implementation of ex vivo gene therapy based on the nonviral genetic modification of HSCs for the treatment of hematopoietic disorders and cancer. A detailed description of the use of PB for HSC engineering is presented in the subsequent Subheadings 2 and 3.

1.3. PiggyBac for iPS Generation

iPS are promising adult stem cells for regenerative medicine. Typically, they are derived from autologous somatic cells after genetic reprogramming and have first been described by Yamanaka and colleagues (27, 28). Genetic reprogramming of mouse and human fibroblasts could be achieved after ectopic expression of a defined combination of four transcription factors, namely *c-MYC*, *KLF-4*, *OCT-4*, and *SOX-2*. The main advantage of iPS cells is their remarkable pluripotency, which resembles that of embryonic stem cells. iPS cells can be obtained from autologous somatic cells, obviating the need for prolonged immunosuppressive therapy in the context of cell transplantation. iPS cells can be genetically modified and can differentiate into the three germ layer endodermal, mesodermal, and ectodermal. The differentiated progeny of these iPS could potentially be used for transplantation to treat degenerative or genetic diseases. Though retroviral and lentiviral vectors had been used for iPS generation, there are several reasons why an alternative approach would still be desirable. The untoward expression of the delivered transcription factors, such as *c-MYC* and *KLF-4*, which are known oncogenes, their expression or reactivation in iPS-derived mice may cause tumors (29). In addition, insertional mutagenesis mediated by retroviral or lentiviral vectors may contribute to oncogenesis in the iPS-derived progeny. Lastly, although iPS cells have been generated by transient transfection of the reprogramming genes with nonviral vectors, the efficiency of iPS cell induction with nonviral vectors is rather modest and was unsuccessful in many primary human cell types (30, 31).

Several recent studies provide an alternative, more efficient, and safer strategy that involves viral vector-free integration of reprogramming genes, followed by their removal ((11–13); Fig. 3). This was achieved by incorporating all four reprogramming genes into a single PB vector. Typically, the PB transposons contain a single polycistronic transcript that encodes the c-Myc, Klf-4, Oct-4, and Sox-2 reprogramming factors separated by viral 2A peptides. This allows for the subsequent post-translational cleavage of the cognate polyprotein into its individual constituents. The efficiency

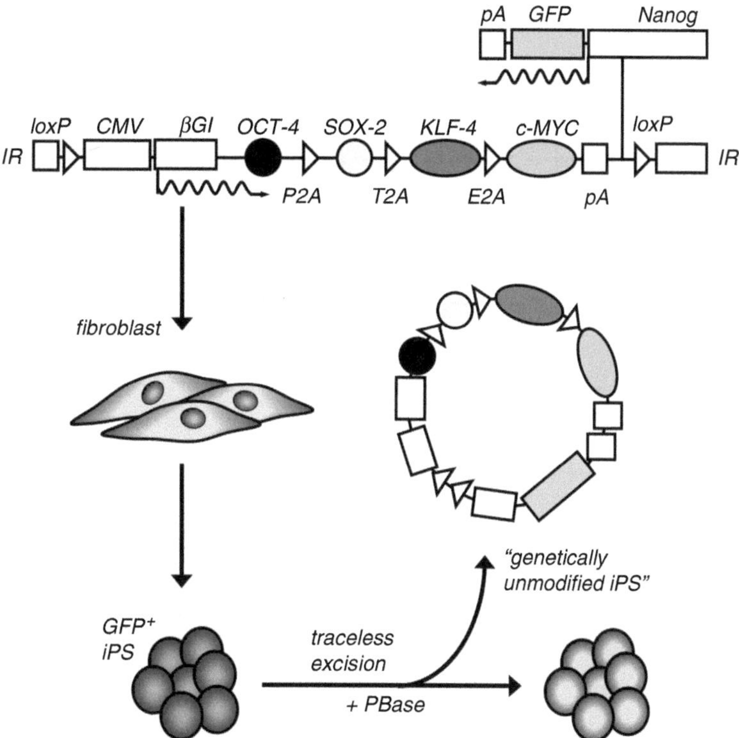

Fig. 3. PB-based reprogramming with traceless excision as a nonviral paradigm for iPS generation. The PB transposon contains a single, polycistronic transcript that encodes Oct-4, Sox-2, Klf-4, and c-Myc, respectively, and flanked between the two PB IRs. Each of these factors is separated by a viral 2A peptide. These four reprogramming factors are driven from the CMV promoter cloned upstream of the β-globin intron. A nanog promoter-driven GFP expression cassette is incorporated downstream of the four factors. This entire reprogramming cassette is flanked with loxP sites allowing for its subsequent excision in IPS cells by CRE recombinase. Alternatively, the entire reprogramming cassette, can be excised following re-expression of the PBase leaving no residual trace of the reprogramming transposon. See Subheading 1.3 for reprogramming.

of iPS generation could be increased eightfold when the hyPB system was used (10) compared to mPB (19). Similarly, excision of the transposon encoding the reprogramming factors was increased 20-fold with hyPB compared to when mPB was used. The expression of the reprogramming factors can be driven either from a viral promoter or an inducible promoter that relies on the rtTA system to achieve regulated expression of the reprogramming factors in response to doxycycline. The most important and unique feature of this approach is that reintroduction of the PB transposase by transient transfection resulted in the excision of the reprogramming cassette from the iPS cell (11). Consequently, there was no residual genetic remnant of the transposon in the iPS. This "traceless excision" paradigm using PB technology obviates most of the aforementioned concerns or limitations associated with the use of viral vectors for iPS induction, while maintaining a relatively robust reprogramming efficiency. Alternatively, the reprogramming

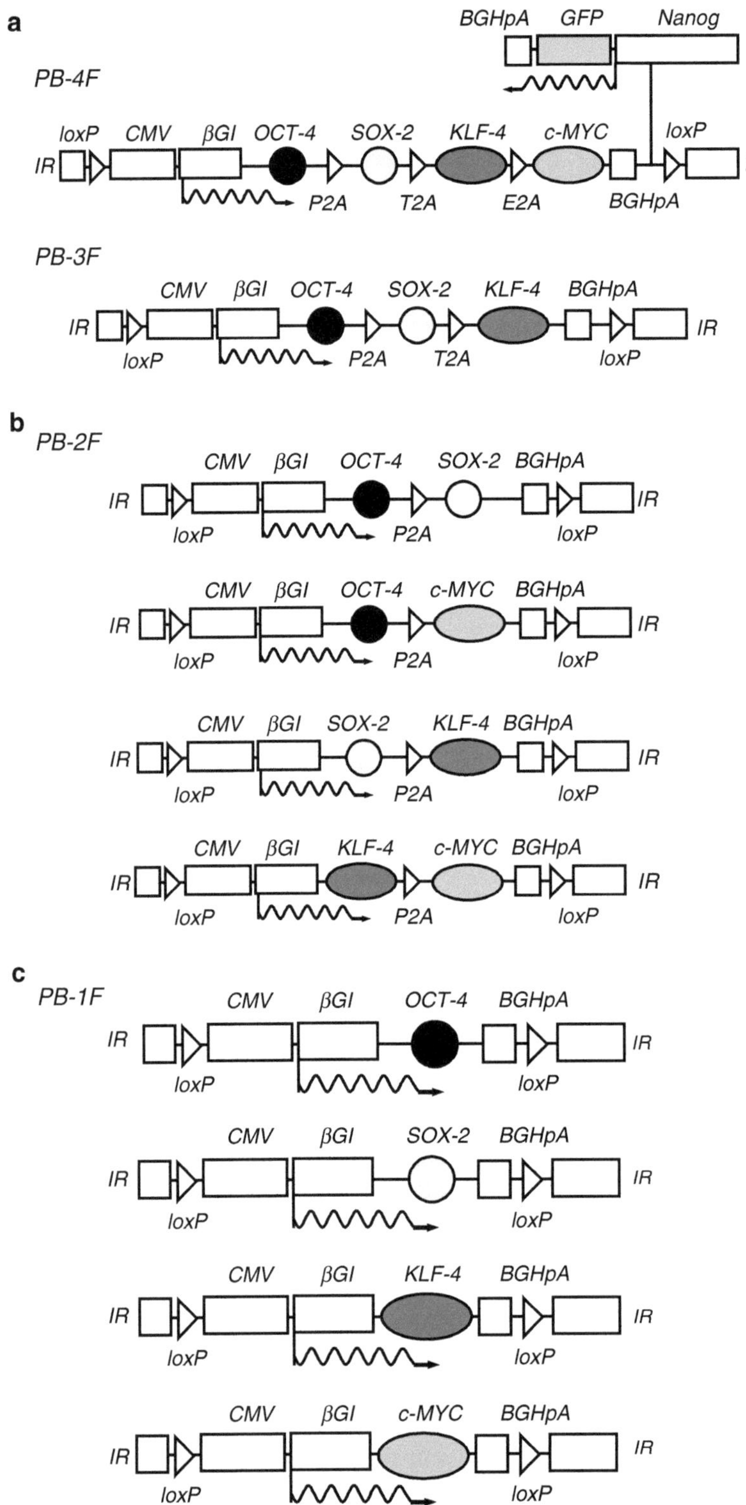

Fig. 4. PiggyBac toolbox for iPS generation. The PiggyBac toolbox for iPS generation consists of ten different constructs generated in our laboratory. The general design of the reprogramming cassettes is similar to the PB construct depicted and described in Fig. 3. This extended PB toolbox encodes 4, 3, 2, or 1 reprogramming factor(s) in different combination. (**a**) PB reprogramming cassette with 4 (with or without nanog-GFP) and 3 factors (**b**) PB reprogramming cassette with 2 factors (**c**) PB reprogramming cassette with 1 factor. The 3, 2, or 1 factor constructs do not encode the nanog-GFP cassette.

cassette could be flanked with lox P sites allowing for its subsequent excision in iPS cells by CRE recombinase. This approach leaves a minimal genetic trace in the iPS genome corresponding to a residual loxP site with the IRs.

The PB toolbox paves the way toward the generation of patient- or disease-specific iPS cells devoid of reprogramming factors for basic stem cell studies and ultimate regenerative medicine applications. Depending on the cell type used to generate iPS, not all reprogramming factors may be required. In particular, only Oct4 is needed to reprogram adult mouse neural stem cells into iPS (32). Similarly, *KLF-4* and *OCT-4* is sufficient to generate iPS from dermal papilla cells (33). In view of these developments, we have therefore generated an extended PB toolbox encoding 4, 3, 2, or 1 reprogramming factor(s) in different combinations. The most optimal PB configuration can then be selected from this toolbox, depending on the cell type used to generate the iPS. A detailed description of our PB toolbox for iPS generation is presented in Fig. 4.

In this chapter, we now specifically focus on the detailed description of an optimized protocol for using the PB system in cord blood-derived CD34$^+$ enriched HSCs. In this protocol, the transfection of the PB transposon encoding GFP (PB-CAG-GFP-pA) and the codon-optimized PBase (PBase$_{\text{CO-MT}}$) into human CD34$^+$ cells results in robust and stable gene expression, reaching up to 60% of the cells that express the GFP transgene (Fig. 2b).

2. Materials

2.1. CD34$^+$ Cell Isolation and Nucleofection

2.1.1. CD34$^+$ Cell Isolation

1. Dilution buffer: PBS (Invitrogen, 14190-094) supplemented with 2 mM EDTA, filter sterilize, and store at 4°C.
2. Lymphoprep (Axis-Shield, Lys 3773). Should be stored at room temperature away from light.
3. Human CD34$^+$ magnetic cell separation kit (Milteny) composed of CD34 magnetic MicroBead kit (130-046-702) and LS separation column (130-042-401) which fit to the MidiMacs separation magnet. The MicroBead kit contains the anti-human-CD34 antibodies conjugated magnetic microbeads (120-000-267) and the FcR human blocking reagent (120-000-442), stored at 4°C.
4. MACS pre-separation filter, sterile packed (130-041-407).
5. MACS buffer: PBS + 2 mM EDTA buffer is supplemented with BSA (Sigma, A9418) at 0.5% final concentration, filter sterilized, and stored at 4°C.
6. Stimulation medium. It is prepared from Stemline II HSC medium (Sigma–Aldrich, MO, USA, S0192) and cytokine cocktail supplement. The cytokines are human stem cell factor (hSCF), human interleukin-6 (hIL-6), human interleukin-3

(hIL-3), human FMS-like tyrosine kinase 3-ligand (hFlt3-L), human thrombopoietin (hTPO) (Peprotech). The supplemented medium with cytokines should be stored at 4°C and not longer than 48 h. After 48 h medium should be prepared fresh.

2.1.2. Nucleofection of Cord Blood-Derived CD34+ Cells

1. Nucleofection kit for CD34+ cell containing the nucleofection buffer and the cuvettes and plastic Pasteur pipettes (Lonza, VPA-1003).
2. Cord blood-derived CD34+ HSC obtained from independent donors. Constructs: The PB transposon contains the green fluorescent protein (GFP) driven by the CAG promoter (PB-CAG-GFP-pA). The PB transposase constructs ($PBase_{CO-MT}$). *All the vectors are available from the author upon request.* The plasmid DNA should have a transfection grade quality (Invitrogen Cat. # K210017). It is critical to determine the A260:A280 ratio. It should be at least 1.8. The DNA can be stored as a stock at −20°C at a concentration of 1 μg/μl in MQ water.
3. The PB-CAG-GFP-pA transposon donor plasmid and the second plasmid, encoding for the PB transposase ($PBase_{CO-MT}$) are mixed prior to nucleofection. In our protocol, the amount is 10 μg PB-CAG-GFP-pA while the amount of transposase coding plasmids is 5 μg.
4. Fluorescence-activated Cell Sorter (FACS) machine to assess marker gene expression.

2.1.3. Clonogenic Assays

Semi-solid Methocult GF (Stem Cell Technologies, H4534) was used for CFU-GM and Methocult SF (Stem Cell Technologies, H4436) for CFU-E and for CFU-GM assays.

3. Methods

3.1. Protocol for CD34+ Cell Isolation

3.1.1. Ficoll Density Gradient Separation

1. Keep the cord blood sample on room temperature. Sample should not be older than 1 day.
2. Dilute cord blood with equal volume of dilution buffer and mix well.
3. Pipette 15 ml of Ficoll-Paque into 50 ml falcon tube.
4. Overlay 35 ml of diluted cord blood slowly over the Ficoll (see Notes 1 and 2).
5. Centrifuge 2,400 rpm (800 × *g*) 30 min at room temperature, with the lowest brake force (see Note 3).
6. Collect the interphase consisting of PBMCs into a 15 ml falcon tube with a Pasteur pipette.
7. Centrifuge at 1,200 rpm (300 × *g*), for 5 min at RT. Brakes can be set onto the maximum level from now on in every centrifugation step.

8. Aspirate off the supernatant and resuspend the pellet in 10 ml dilution buffer. Count cells in hemocytometer and proceed to magnetic labeling.

3.1.2. Magnetic Labeling

1. Centrifuge at 1,200 rpm (300 × *g*), for 5 min at RT. Aspirate off the supernatant and resuspend the cell pellet in 300 μl MACS buffer per 1×10^8 cells.
2. Add 100 μl of FcR blocking reagent per 1×10^8 cells, then add 100 μl of CD34 microbeads per 1×10^8 cells. Scale up according to cell number.
3. Mix well and incubate at +4°C for 30 min.
4. Add an equal volume of MACS buffer and proceed to magnetic separation.

3.1.3. Magnetic Separation

1. Place an LS column with a filter in the magnetic separation unit and calibrate the LS column with 3 ml of MACS buffer.
2. After calibration apply the cell suspension on the column.
3. Wash the column 3× with 3 ml MACS buffer.
4. Take the column out of the separation unit and transfer it to a new collection tube.
5. Elute the cells with 5 ml wash buffer with the piston provided. Repeat this elution step one more time. Your $CD34^+$ cells are now in the 15 ml tube in the total volume of 10 ml MACS buffer.
6. Apply the 10 ml cell suspension obtained into a new calibrated LS column and then repeat step 5.
7. Count the cells and take aliquots for FACS analysis (to check purity of cells).
8. Centrifuge the cells 5 min at 1,200 rpm (300 × *g*) and proceed to nucleofection.

3.2. Nucleofection

Nucleofection was performed using the human $CD34^+$ cells nucleofection kit and by the manufacturer-optimized protocol. We routinely carry out at least three independent experiments using cells isolated from three to five independent donors.

1. MACS enriched $CD34^+$ cells were centrifuged at 1,200 rpm (300 × *g*) (4°C) for 5 min.
2. 1×10^6 human $CD34^+$ cells are resuspended in 100 μl Nucleofector buffer at room temperature.
3. 100 μl cell suspension is transferred to Nucleofector cuvette.
4. Transposon/transposase plasmid mixture (our protocol is 10 μg transposon plasmid + 5 μg transposase plasmid = 15 μg) is added to the cell suspension, mixed gently.

5. As a control the non-nucleofected cells without plasmids or a PB plasmid without PBase is used.
6. Electroporate the cells with purified plasmids containing the transposon (10 μg) and transposase (5 μg; see Note 4). Use program U-08 for nucleofactor I and U-008 for nucleofactor II devices.
7. At the end of the nucleofection transfer the cells into 500 μl pre-equilibrated Stem line medium to the cuvette and immediately transfer cells to desired culture condition.
8. After 48 h of incubation, measure transfection efficiency by FACS (GFP+). Cells should be kept in Stem line medium, supplemented with 100 ng/ml SCF, 20 ng/ml IL-6, 100 ng/ml IL-3, 20 ng/ml Flt3-L, and 100 ng/ml TPO.
9. Stable gene transfer can be determined in differentiated cells by a clonogenic assay (see below).

3.3. Clonogenic Assay

3.3.1. Granulocyte/Monocyte/Macrophage and Erythroid Differentiation Medium

1. Semi-solid Methocult GF H4534 is supplemented with 1% methylcellulose (4,000 cps), 30% fetal bovine serum, 1% BSA, 10^{-4} M 2-mercaptoethanol, 2 mM L-glutamine, 50 ng/ml rhSCF, 10 ng/ml rhGM-CSF, 10 ng/ml rhIL-3 in Iscove's modified Dulbecco's medium (MDM).
2. Methocult SF H4436 is a methylcellulose-based medium in Iscove's MDM supplemented with BSA, 2-mercaptoethanol, L-glutamine, rh Insulin, Human Transferrin (iron saturated), rh Stem Cell Factor, rh GM-CSF, rh IL-3, rh IL-6, rh G-CSF, and rh erythropoietin.
3. Thaw complete Methocult GF H4534 for (CFU-GM) and Methocult SF H4436 for (CFU-E) medium overnight under refrigeration (2–8°C) or at room temperature. (Not at 37°C). The medium is ready to use for differentiation of transfected cells to granulocyte/monocyte/macrophage (CFU-GM) and erythroid (CFU-E) lineages (Fig. 3).
4. Thawed medium should be vigorously shaken for 30–60 s (let it stand to allow bubbles to dissipate).
5. Dispense MethoCult medium into tubes.
6. Add nucleofected cells at 1:10 (v/v) ratio, i.e., nucleofected cell to MethoCult medium.
7. Usually, 100 μl of nucleofected cell suspension is added to 1 ml of the semisolid medium in a single 6-well plate. This could be scaled down to 24-well plate accordingly.
8. Mix by vortexing or pipetting and dispense.
9. Incubate for 7 days at 37°C for the erythroid (CFU-E) and 14 days for granulocyte/monocyte/macrophage lineages (CFU-GM) and count colonies.

4. Notes

1. Diluted blood should be applied gently and slowly onto the surface of the Lymphoprep. Tilting the falcon tube at an angle of 45° allows better separation of the blood from the liquid phase.
2. Upon completion of loading the blood onto the Ficoll, tilt the falcon tube slowly back to vertical position and gently put the tube into the centrifuge to avoid mixing of the two layers.
3. Apply minimum break force of the centrifuge to avoid turbulence occurring on the mononuclear cell layer when the centrifugation is spinning to a stop.
4. Nucleofected cells should be put into the respective medium as quickly as possible and then into the incubator to avoid subjecting the cells to further stress. Any delay in the handling of the cells, can negatively influence cell viability and nucleofection efficiency. To decrease cell mortality, repeated pipetting should be avoided.

Acknowledgments

This work was supported by the 7th EU framework programme (grant agreement no 222878, PERSIST) FWO, GOA EPIGEN (VUB), EHA, and AFM.

References

1. Fraser MJ, Smith GE, Summers MD (1983) Acquisition of Host Cell DNA Sequences by Baculoviruses: Relationship Between Host DNA Insertions and FP Mutants of Autographa californica and Galleria mellonella Nuclear Polyhedrosis Viruses. J Virol 47:287–300
2. Fraser MJ, et al. (1996) Precise excision of TTAA-specific lepidopteran transposons piggyBac (IFP2) and tagalong (TFP3) from the baculovirus genome in cell lines from two species of Lepidoptera. Insect Mol Biol 5:141–151
3. Craig NL, Eickbush TH, Voytas DF (2010) Welcome to mobile DNA. Mob DNA 1:1
4. Rad R, et al. (2010) PiggyBac transposon mutagenesis: a tool for cancer gene discovery in mice. Science 330:1104–1107
5. Chew SK, Rad R, Futreal PA, Bradley A, Liu P (2011) Genetic screens using the piggyBac transposon. Methods 53:366–371
6. Bjork BC, et al. (2010) A transient transgenic RNAi strategy for rapid characterization of gene function during embryonic development. PLoS One 5:e14375
7. Balu B, et al. (2010) A genetic screen for attenuated growth identifies genes crucial for intraerythrocytic development of *Plasmodium falciparum*. PLoS One 5:e13282
8. Balu B, et al. (2009) piggyBac is an effective tool for functional analysis of the Plasmodium falciparum genome. BMC Microbiol 9:83
9. Labbe GM, Nimmo DD, Alphey L (2010) piggybac- and PhiC31-mediated genetic transformation of the Asian tiger mosquito, Aedes albopictus (Skuse). PLoS Negl Trop Dis 4:e788
10. Yusa K, et al. (2011) A hyperactive piggyBac transposase for mammalian applications. Proc Natl Acad Sci USA 108:1531–1536

11. Yusa K, et al. (2009) Generation of transgene-free induced pluripotent mouse stem cells by the piggyBac transposon. Nat Methods 6:363–369
12. Woltjen K, et al. (2009) piggyBac transposition reprograms fibroblasts to induced pluripotent stem cells. Nature 458:766–70
13. Kaji K, et al. (2009) Virus-free induction of pluripotency and subsequent excision of reprogramming factors. Nature 458:771–775
14. Lacoste A, Berenshteyn F, Brivanlou AH (2009) An efficient and reversible transposable system for gene delivery and lineage-specific differentiation in human embryonic stem cells. Cell Stem Cell 5:332–342
15. Wang W, et al. (2008) Chromosomal transposition of PiggyBac in mouse embryonic stem cells. Proc Natl Acad Sci USA 105:9290–9295
16. Saridey SK, et al. (2009) PiggyBac transposon-based inducible gene expression in vivo after somatic cell gene transfer. Mol Ther 17:2115–20
17. Mates L, et al. (2009) Molecular evolution of a novel hyperactive Sleeping Beauty transposase enables robust stable gene transfer in vertebrates. Nat Genet 41:753–761
18. Ding S, et al. (2005) Efficient transposition of the piggyBac (PB) transposon in mammalian cells and mice. Cell 122:473–483
19. Cadinanos J, Bradley A (2007) Generation of an inducible and optimized piggyBac transposon system. Nucleic Acids Res 35:e87
20. Nakanishi H, et al. (2010) piggyBac transposon-mediated long-term gene expression in mice. Mol Ther 18:707–714
21. Cavazzana-Calvo M, et al. (2000) Gene therapy of human severe combined immunodeficiency (SCID)-X1 disease. Science 288:669–72
22. Aiuti A, et al. (2009) Gene therapy for immunodeficiency due to adenosine deaminase deficiency. N Engl J Med 360:447–458
23. Rethwilm A (2007) Foamy virus vectors: an awaited alternative to gammaretro- and lentiviral vectors. Curr Gene Ther 7:261–71
24. Bauer TR Jr, et al. (2008) Successful treatment of canine leukocyte adhesion deficiency by foamy virus vectors. Nat Med 14:93–97
25. Kohn DB (2008) Gene therapy for childhood immunological diseases. Bone Marrow Transplant 41:199–205
26. Grabundzija I, Irgang M, Mates L, Belay E, Mátrai J, Gogol-Doring A, Kawakami K, Chen W, Ruiz P, Chuah MK, VandenDriessche T, Izsvak Z, Ivics Z (2010). Comparative analysis of transposable element vector systems in human cells. Mol Ther 18: 1200–1209
27. Takahashi K, Yamanaka S (2006) Induction of pluripotent stem cells from mouse embryonic and adult fibroblast cultures by defined factors. Cell 126:663–676
28. Takahashi K, T, et al. (2007) Induction of pluripotent stem cells from adult human fibroblasts by defined factors. Cell 131, 861–872
29. Nakagawa M, et al. (2008) Generation of induced pluripotent stem cells without Myc from mouse and human fibroblasts. Nat Biotechnol 26:101–106
30. Okita K, et al. (2008) Generation of mouse induced pluripotent stem cells without viral vectors. Science 322:949–953
31. Stadtfeld M, et al. (2008) Induced pluripotent stem cells generated without viral integration. Science 322:945–949
32. Kim JB, et al. (2009) Direct reprogramming of human neural stem cells by OCT4. Nature 461:649–643
33. Tsai SY, et al. (2010) Oct4 and klf4 reprogram dermal papilla cells into induced pluripotent stem cells. Stem Cells 28:221–8

Chapter 15

Insertion Site Pattern: Global Approach by Linear Amplification-Mediated PCR and Mass Sequencing

Cynthia C. Bartholomae, Hanno Glimm, Christof von Kalle, and Manfred Schmidt

Abstract

In gene therapy, viral or nonviral integrating vectors are used to deliver a corrected gene to replace the corresponding defective cellular gene. As vector delivery is (yet) commonly not targeted to a specific site in the host genome, and vector integration may lead to unwanted cellular gene deregulation, the comprehensive analysis of vector locations is a crucial approach to assess vector biosafety and to follow the fate of the gene corrected cells in vivo. The retrieved vector integration sites are unique for each transduced cell clone, thereby serving as a molecular marker and allowing to track distinct cell clones in various samples. Today, several PCR-based methods are available for the identification and characterization of unknown flanking DNA sequences (Mueller and Wold Science 246:780–786, 1989; Paruzynski et al. Nat Protoc 5:1379–1395, 2010; Schmidt et al. Nat Methods 4:1051–1057, 2007; Silver and Keerikatte J Virol 63:1924–1928, 1989). Thereof, the linear amplification-mediated PCR (LAM-PCR) proved to exhibit the highest sensitivity, allowing the detection of miscellaneous vector integration sites in one sample. The broad application spectrum and robustness of LAM-PCR has been approved by its application as a tool for the molecular follow up of gene-modified cells in preclinical and clinical gene therapy trials (Li et al. Science 296:497, 2002; Cartier et al. Science 326:818–823, 2009; Ott et al. Nat Med 12:401–409, 2006; Deichmann et al. J Clin Invest 117:2225–2232, 2007). The combination of LAM-PCR and next-generation sequencing (NGS) platforms offers the opportunity to study the clonal inventory and pharmacokinetics in clinical gene therapy studies.

Key words: Gene therapy, LAM-PCR, Clonality, Integration sites, Safety, In vivo monitoring, Next-generation sequencing, Clonal inventory

1. Introduction

Integrating retroviral vectors allow the long-term expression of transgenes in transduced cells and have already been used in various clinical trials for the treatment of monogenetic disorders (1–4). As viral vector integration sites are representing molecular markers for individual cell clones, the monitoring of the

Yves Bigot (ed.), *Mobile Genetic Elements: Protocols and Genomic Applications*, Methods in Molecular Biology, vol. 859, DOI 10.1007/978-1-61779-603-6_15,

cellular fate also in a complex background is enabled. Optimized transduction protocols have led to higher efficiency and successful treatment in gene therapy. However, the increased efficiency has been accompanied by side effects, ranging from cell immortalization in in vitro experiments (5) to clonal dominance and even leukemia in preclinical (6, 7) and clinical trials (8–10). Thus, the intense study of integration profiles of distinct vector systems became a crucial determinant to assess vector-related genotoxicity.

Linear amplification-mediated PCR (LAM-PCR) has proven to be the superior method for analyzing and sequencing unknown DNA sequences which are flanked by known sequences, i.e., nonviral and viral vector integration sites (11). However, LAM-PCR can be easily applied to any other scientific projects, aiming to characterize unknown localization of "donor" DNA in any host genome. Limitation of LAM-PCR, due to a bias introduced by a restriction digest being a crucial step in this methodology, has been comprehensively addressed and highlighted the necessity of a careful experimental design a priori. This challenge can be largely solved by a combination of restriction enzymes with different recognition motifs, thereby avoiding enzymes with "CG" motifs by favoring "Non-CG" motifs (12). Furthermore, our recent development of a nonrestrictive (nr)LAM-PCR circumvents the need of restriction digestion (12, 13). Since its lower sensitivity, attributed to the less efficient single strand ligation step carried out by RNA ligase, nrLAM-PCR is advisable if starting material is not limited (>1 μg). As nrLAM-PCR amplicons of various lengths are generated for each vector–genome junctions, one should take into consideration that the illustration of the clonal composition of the samples is strictly dependent on sequencing.

LAM-PCR is a multiple step procedure. In a first step, the preamplification of vector–genome junctions present in the template DNA is carried out by biotinylated primers hybridizing to the vector, comparable to a repeated primer extension or linear PCR. A selective amplification of vector–genome junctions is secured by magnetic separation of the biotinylated DNA strands via paramagnetic beads. After synthesis of DNA double strands by hexanucleotide random priming, a restriction digest and the ligation of an arbitrary linker cassette on the genomic part of the amplified fragments are completing the enzymatic reactions on the semisolid streptavidin phase. Two exponential PCRs with nested arranged vector- and linker cassette-specific primers are finalizing the LAM-PCR, which results in DNA fragments consisting of vector-, genomic- and linker cassette sequences. The generated fragments can either be shotgun cloned and sequenced by Sanger sequencing technology, or by high-throughput NGS platforms like "FLX Titanium" (454/Roche) or "Genome Analyzer/HiSeq" (Illumina). NGS platforms comprise an additional PCR step adding a molecular barcode and the NGS amplification and sequencing specific oligonucleotides. After sequencing, integration sites can be identified. In the following, the protocol for LAM-PCR combined with NGS is outlined exemplarily for transposon integration site analysis.

2. Material

2.1. Linear PCR

1. Taq DNA Polymerase.
2. dNTPs.
3. Aqua ad injectabilia (B.Braun, Melsungen, Germany).
4. Human genomic DNA (Roche Diagnostics, Mannheim, Germany).
5. Primer (MWG Biotech, Ebersberg, Germany): 5′ biotin-modified primers are marked with (B) IRDRL I1 (B) 5′>GACTT GTGTCATGCACAAAGTAGATGTCC<3′.

2.2. Magnetic Capture

1. Magnetic particles: Dynabeads M-280 Streptavidin (Invitrogen, Carlsbad, USA).
2. Magnetic separation units: Single Place Magnetic Stand, DynaMag™-96 Side Skirted (Invitrogen, Carlsbad, USA).
3. PBS/0.1% BSA (pH 7.5): For 1 L of PBS/0.1% BSA solution, dissolve 1 g of BSA in PBS. Store at room temperature for up to 3 months.
4. 3 M/6 M LiCl solution (Carl Roth, Karlsruhe, Germany): For 50 ml 3 M (6 M) solution, dissolve 6.36 (12.72) g of LiCl in 0.5 ml 1 M Tris–HCl (pH 7.5) and 0.1 ml 0.5 M EDTA (pH 8.0) and adjust the volume with distilled water to 50 ml. Filtrate the solution using a 0.45 μm filter. Solutions can be stored for several months at room temperature (see Note 1).
5. Aqua ad injectabilia.

2.3. Construction of the Linker Cassette

1. 250 mM Tris–HCl, pH 7.5.
2. 100 μM $MgCl_2$.
3. Oligonucleotides:

Oligonucleotide I:	5′>GACCCGGGAGATCTGAATTCAGTG GCACAGCAGTTAGG<3′
Oligonucleotide II:	5′>AATTCCTAACTGCTGTGCCACTGAA TTCAGATC<3′.

4. Microcon-30 (Millipore, Bedford, USA).
5. Aqua ad injectabilia.

2.4. Double Strand Synthesis (Hexanucleotide Priming)

1. Klenow Polymerase.
2. Hexanucleotide mixture.
3. dNTPs.
4. Aqua ad injectabilia.

2.5. Restriction Digest

1. Enzyme Tsp509I (New England Biolabs, Frankfurt am Main, Germany, see Note 2).
2. Aqua ad injectabilia.

2.6. Ligation of the Linker Cassette

1. Fast-Link DNA Ligation Kit (Epicentre Biotechnologies, Madison, USA).
2. Aqua ad injectabilia.

2.7. Denaturation

1. 0.1 N NaOH.

2.8. Exponential PCRs

1. Taq DNA Polymerase.
2. dNTPs.
3. Aqua ad injectabilia.
4. Primers: 5′ biotin-modified primers are marked with (B):

First exponential PCR:	
IRDRL II (B)	5′ > GACTTGCCAAAACTATTGTTTG < 3′
LC1	5′ > GACCCGGGAGATCTGAATTC < 3′
Second exponential PCR:	
IRDRL III	5′ > ACGAGTTTTAATGACTCCAAC < 3′
LC2	5′ > GATCTGAATTCAGTGGCACAG < 3′

2.9. Separation of the LAM-PCR Product on a Spreadex High-Resolution Gel

1. Spreadex high-resolution gel (Elchrom Scientific, Cham, Switzerland).
2. Gel electrophoresis apparatus SAE 2000 (Elchrom Scientific, Cham, Switzerland).
3. 40× TAE buffer.
4. 5× Blue run loading buffer: 25 mM Tris–HCl, pH 7.0, 150 mM ethylenediaminetetraacetic acid (EDTA), pH 8.0, 0.05% bromophenol blue, 25% glycerol.
5. Ethidium bromide.
6. 100 bp ladder.

2.10. Sample Preparation (Barcoding) for NGS Using 454/Roche Technology

1. QIAquick PCR Purification Kit (Qiagen, Hilden, Germany).
2. Taq DNA Polymerase.
3. dNTPs.
4. Aqua ad injectabilia.
5. Guanidinhydrochlorid.

6. Primers (see Note 3):

5′ biotin-modified primers are marked with (B)	
Mega-Linker:	5′>GCCTTGCCAGCCCGCTCAGAGTGGCACAGCAGTTAGG<3′
Barcoded Mega-IRDRL:	5′>GCCTCCCTCGCGCCATCAG(N)$_6$ACGAGTTTTAATGACTCCAAC<3′

7. Agarose LE.
8. 10× TBE buffer.
9. 5× Blue run loading buffer.
10. Ethidium bromide.
11. 100 bp ladder.

3. Methods

LAM-PCR is the most sensitive method to allow the identification of unknown DNA sequences which are flanked by known sequences, as exemplarily described for Sleeping Beauty (SB) transposons integration sites in gene therapy studies in this chapter. As explained in the introduction LAM-PCR is a multiple step procedure, which we recently have adapted for NGS platforms to allow in depth analyses.

3.1. Linear PCR

The first step of the LAM-PCR is a linear PCR accomplished with a 5′-biotinylated (B) primer hybridizing at the 5′-end of SB transposon left inverted repeat direct repeat (IRDRL). The primer sequence is given in the material section (see Subheading 2.1).

1. Put 0.01 ng to 1 μg of transduced DNA together with 1× concentrated PCR buffer, 200 μM dNTPs each, 0.0835 μM Primer (83.5 nM), and 1.25 U Taq polymerase into a PCR tube and fill up the reaction with distilled water to a final volume of 50 μl. The same amount of untransduced DNA is used as negative control. The initial denaturation is carried out for 2 min at 95°C followed by 50 PCR cycles composed of the denaturation for 45 s at 95°C, the annealing for 45 s at 60°C, and the extension for 60 s at 72°C. Perform the final extension for 5 min at 72°C. After completion of the PCR add another 2.5 U Taq polymerase to each PCR reaction and repeat the 50 cycle PCR.

3.2. Magnetic Capture

1. Expose 20 μl of magnetic particles (10 μg/μl) to a magnetic field for 60 s and discard the supernatant in the presence of the magnetic field using the "Single Place Magnetic Stand"

magnet device for steps 1–3, all subsequent steps of this protocol are carried out using the "DynaMag™-96 Side Skirted" magnetic device (see Note 4).

2. Resuspend the particles in 40 μl PBS/0.1% BSA (pH 7.5) and discard the supernatant in the presence of the magnetic field. Repeat this step once.
3. Wash the particles once in 20 μl 3 M LiCl, resuspend them in 50 μl 6 M LiCl.
4. Incubate each linear PCR product with 50 μl of the magnetic particle solution on a shaker at 300 rpm at room temperature over night (see Note 5).
5. Expose the sample for 60 s to a magnetic field, discard the supernatant in the presence of the magnetic field and wash the beads once in 100 μl distilled water.

3.3. Construction of the Linker Cassette

1. Blend 40 μl of oligonucleotide I (100 pmol/μl), 40 μl of oligonucleotide II (100 pmol/μl), 110 μl 250 mM Tris–HCl, pH 7.5, 10 μl 100 mM $MgCl_2$, and incubate the reaction for 5 min at 95°C in a heat block. Switch off the heat block and let the sample cool down slowly over night within the heat block. The sequences of oligonucleotide I and oligonucleotide II are given in the material section (see Subheading 2.5). Excess of linker cassette may be stored as aliquots of 5 μl, 10 μl, and 50 μl at −20°C.
2. Add 300 μl of distilled water and put the sample on a Microcon-30 column.
3. Centrifuge the sample for 10 min at room temperature and 14,000 × *g*.
4. Place the column reversed onto a fresh tube and centrifuge the sample for 2 min at room temperature and 1,000 × *g*.
5. Fill the concentrated sample up with distilled water to a final volume of 80 μl.
6. Aliquot the sample and freeze it at −20°C (see Note 6).

3.4. Double Strand Synthesis (Hexanucleotide Priming)

1. Insert 1× concentrated hexanucleotide mixture, 200 μM dNTPs each, 1 U Klenow-Polymerase, and fill the mixture up with distilled water to a final volume of 10 μl.
2. Resuspend the particles with 10 μl of the reaction mixture.
3. Incubate the sample for exactly 1 h at 37°C.
4. Add 90 μl of distilled water and expose the sample to a magnetic field for 60 s.
5. Discard the supernatant in the presence of the magnetic field and wash the particles once with 100 μl of distilled water.

3.5. Restriction Digest

1. Blend 1× concentrated restriction buffer, 2 U of the enzyme Tsp509 I and fill the reaction up with distilled water to a final volume of 10 μl (see Note 7).
2. Incubate the DNA-particle complex with 10 μl of the restriction mixture for 1 h at 65°C (see Note 8).
3. Add 90 μl of distilled water to the reaction, expose the sample for 60 s to a magnetic field, discard the supernatant in the presence of the magnetic field, and wash the DNA-particle complex once with 100 μl of distilled water.

3.6. Ligation of the Linker Cassette

1. Blend 2 μl of the linker cassette, 1× concentrated fast-link ligation buffer, 10 mM ATP, 2 U fast-link ligase, and fill the reaction up to 10 μl with distilled water.
2. Add 10 μl of the ligation mixture to the DNA-particle complex and incubate the reaction for 5 min at room temperature.
3. Add 90 μl of distilled water, expose the sample for 60 s to a magnetic field and discard the supernatant in the presence of the magnetic field.
4. Wash the DNA-particle complex once with 100 μl of distilled water.

3.7. Denaturation

1. Incubate the DNA-particle complex with 5 μl of freshly prepared 0.1 N NaOH at room temperature on a shaker (300 rpm) for 10 min.
2. Expose the sample for 60 s to a magnetic field and collect the supernatant which contains the DNA in the presence of the magnetic field.

3.8. Exponential PCRs and Magnetic Capture (Optional)

Primer sequences for the first and second exponential PCR are listed in the material section (see Subheading 2.8). An additional magnetic capture step after the first exponential PCR is optional (step 3, see Note 7).

1. Use 1 μl of the denaturation product as template for the first exponential PCR. Blend 1× concentrated PCR Buffer, 200 μM dNTPs each, 8.3 μM (8,350 nM) of each primer, 1.25 U Taq polymerase, and fill the reaction up with distilled water to a final volume of 25 μl.
2. Carry out an initial denaturation of 95°C for 2 min followed by 35 cycles composed of 95°C for 45 s, 60°C for 45 s, and 72°C for 1 min. As final extension choose 72°C for 5 min.
3. Optional, see Note 7: Expose 20 μl of the magnetic particles (10 μg/μl) to a magnetic field for 60 s and discard the supernatant in the presence of the magnetic field. Resuspend the particles in 40 μl PBS/0.1% BSA (pH 7.5) and discard the

supernatant in the presence of the magnetic field. Repeat this step once. Wash the particles once in 20 μl 3 M LiCl and resuspend them in 25 μl 6 M LiCl. Incubate each PCR product of the first exponential PCR with 25 μl of the magnetic particle solution for at least 1 h on a shaker at 300 rpm at room temperature. Expose the sample for 60 s to a magnetic field, discard the supernatant in the presence of the magnetic field, and wash the beads once in 100 μl distilled water. For denaturation, incubate the DNA-particle complex with 20 μl of fresh 0.1 N NaOH for 10 min at room temperature and at 300 rpm on a shaker. Expose the sample for 60 s to a magnetic field and collect the supernatant which contains the DNA in the presence of the magnetic field.

4. Use 0.1–5% of the first exponential product or 1 μl of denaturation product as template for second exponential PCR. Blend 1× concentrated PCR Buffer, 200 μM dNTPs each, 8.3 μM (8,350 nM) of each primer, 2.5 U Taq polymerase, and fill the reaction up with distilled water to a final volume of 50 μl.
5. Carry out an initial denaturation of 95°C for 2 min followed by 35 cycles composed of 95°C for 45 s, 60°C for 45 s, and 72°C for 1 min. As final extension choose 72°C for 5 min.

3.9. Separation of the LAM-PCR Product on a Spreadex High-Resolution Gel

1. Fill the electrophoresis tank with 1.9 L of 1× concentrated TAE buffer and fix a Spreadex gel within the electrophoresis tank using an appropriate catamaran.
2. Load 10 μl of each LAM-PCR product with 2 μl of 5× concentrated blue run loading buffer. Include 1 well for a molecular weight marker.
3. Let the gel run at 10 V/cm electrode gap.
4. After 5 min switch on the buffer pump.
5. As soon as the dye front reaches the lower edge of the gel, detach the gel from the plastic matrix using a nylon fiber and stain the gel for 20 min in ethidium bromide solution (0.5 μg ethidium bromide/ml distilled water) on a shaker (50 rpm) at room temperature.
6. Apply the gel on a gel documentation system.

3.10. Sample Preparation (Barcoding) for NGS Using 454/Roche Technology

1. Use QIAquick PCR Purification Kit to purify second exponential PCR products. Add 5 volumes of buffer PB to 1 volume of the PCR sample and mix. Place a QIAquick spin column in a provided 2 ml collection tube. Apply the sample to the QIAquick column and centrifuge for 30–60 s. Discard flow-through. Add 750 μl 35% Guanidinhydrochlorid and centrifuge for 30–60 s. Wash two times with 35% Guanidinhydrochlorid. Discard flow-through. Add 750 μl PE buffer and centrifuge for 30–60 s. Discard flow-through and place the QIAquick

column back in the same tube. Centrifuge the column for an additional 1 min. Place QIAquick column in a clean 1.5 ml microcentrifuge tube. To elute DNA, add 50 μl water to the center of the QIAquick membrane and centrifuge the column for 1 min.

2. Use 40 ng purified PCR product as template DNA, insert 1× concentrated PCR buffer, 200 μM dNTPs each, 0.2 μM (200 nM) each primer (see Subheading 2.10), and 2.5 U Taq polymerase into the PCR and fill the reaction up with distilled water to a final volume of 50 μl. Use the same amount of untransduced DNA as a negative control. Carry out the initial denaturation for 2 min at 95°C followed by 15 PCR cycles composed of the denaturation for 45 s at 95°C, the annealing for 45 s at 58°C, and the extension for 60 s at 72°C. Conduct the final extension for 5 min at 72°C.
3. Prepare a 2% agarose gel using 2 g agarose per 100 ml 1× concentrated TBE buffer.
4. Boil the mixture until the agarose is completely dissolved.
5. Add 0.5 μg/ml ethidium bromide after cooling down of the gel for 5 min at room temperature (see Note 9).
6. Insert a comb into a sledge and cast the gel.
7. Remove the comb after polymerization of the gel and transfer the sledge into an electrophoresis tank filled with 1× concentrate TBE buffer.
8. Load 10 μl of the PCR product mixed with 2 μl of 5× concentrated blue run loading buffer in each well. Include 1 well for a molecular weight marker.
9. Connect the gel unit to a power supply and run the gel at 10 V/cm electrode gap until the dye front runs 7 cm below the upper edge of the gel.
10. Apply the gel on a gel documentation system to verify whether the PCR reaction was successful.
11. Measure the DNA concentration of barcoded PCR products.
12. Pool the PCR products with different barcodes which will be sequenced in one sequencing chamber together into one vial (see Note 10). The amount of DNA of each sample within the pool is proportional to the number of retrieved sequencing reads.

3.11. Bioinformatics/Sequence Analyses

The retrieved sequences can be processed by public available bioinformatics tools like Seqmap 2.0 (14), "QuickMap" (15) or own developed programs, such as noted in ref. (13).

4. Notes

1. Alternatively binding solution provided by the manufacturer of the magnetic beads can be used. LiCl in our hand performs in a comparable way and is cost-effective.
2. The ratio of PCR product and LiCl solution must always be 1:1. Drying and freezing of magnetic beads needs to be avoided.
3. To distinguish between different PCR products after pooling them together for NGS, we introduce a six to ten nucleotide long barcode into a joint oligonucleotide consisting out of the adapter A sequence of FLX platform (454/Roche) and LTR-specific sequence.
4. This capturing step needs to be carried out at least for 8 h.
5. After thawing one aliquot do not refreeze the linker cassette.
6. The use of different "Non-CG" motif restriction enzymes is highly recommended to assure a sophisticated analysis. If two different enzymes are used, the linear PCR product might be divided into two reactions following Subheading 3.3.
7. The restriction time may be extended to several hours.
8. This magnetic capture step is performed to increase the sensitivity and specificity.
9. Since ethidium bromide is mutagenic, we accomplish the insertion of the ethidium bromide beneath a flue.
10. For NGS preparation of LAM-PCR amplicons, the emulsion PCR, it is crucial to estimate the DNA concentration properly; DNA amount of at least 500 ng/pool is advisable to ensure accurate quality control.

References

1. Cartier N, et al. (2009) Hematopoietic stem cell gene therapy with a lentiviral vector in X-linked adrenoleukodystrophy. Science 326:818–823
2. Ott MG, et al. (2006) Correction of X-linked chronic granulomatous disease by gene therapy, augmented by insertional activation of MDS1-EVI1, PRDM16 or SETBP1. Nat Med 12:401–409
3. Hacein-Bey-Abina S, et al. (2003) LMO2-associated clonal T cell proliferation in two patients after gene therapy for SCID-X1. Science 302:415–419
4. Aiuti A, et al. (2007) Multilineage hematopoietic reconstitution without clonal selection in ADA-SCID patients treated with stem cell gene therapy. J Clin Invest 117:2233–2240
5. Du Y, Jenkins NA, Copeland NG (2005) Insertional mutagenesis identifies genes that promote the immortalization of primary bone marrow progenitor cells. Blood 106:3932–3939
6. Calmels B, et al. (2005) Recurrent retroviral vector integration at the Mds1/Evi1 locus in nonhuman primate hematopoietic cells. Blood 106:2530–2533
7. Modlich U, et al. (2008) Leukemia induction after a single retroviral vector insertion in Evi1 or Prdm16. Leukemia 22:1519–1528
8. Hacein-Bey-Abina S, et al. (2008) Insertional oncogenesis in 4 patients after retrovirus-

mediated gene therapy of SCID-X1. J Clin Invest 118:3132–3142

9. Howe SJ, et al. (2008) Insertional mutagenesis combined with acquired somatic mutations causes leukemogenesis following gene therapy of SCID-X1 patients. J Clin Invest 118:3143–3150
10. Stein S, et al. (2010) Genomic instability and myelodysplasia with monosomy 7 consequent to EVI1 activation after gene therapy for chronic granulomatous disease. Nat Med 16:198–204
11. Schmidt M, et al. (2007) High-resolution insertion-site analysis by linear amplification-mediated PCR (LAM-PCR). Nat Methods 4:1051–1057
12. Gabriel R, et al. (2009) Comprehensive genomic access to vector integration in clinical gene therapy. Nat Med 15:1431–1436
13. Paruzynski A, et al. (2010) Genome-wide high-throughput integrome analyses by nrLAM-PCR and next-generation sequencing. Nat Protoc 5:1379–1395
14. Hawkins TB, et al. (2011) Identifying viral integration sites using SeqMap 2.0. Bioinformatics. 27:720–722
15. Appelt JU, et al. (2009) QuickMap: a public tool for large-scale gene therapy vector insertion site mapping and analysis. Gene Ther 16:885–893

Chapter 16

Comprehensive DNA Methylation Profiling of Human Repetitive DNA Elements Using an MeDIP-on-RepArray Assay

Eric Gilson and Béatrice Horard

Abstract

Hypomethylation of repetitive DNA elements is a common epigenetics event in cancer. Although it is believed that this hypomethylation impacts chromosomal and transcriptional stability of the genome, the extent of repetitive sequences contribution to the development and progression of human cancers remains to be clarified. Repetitive sequences have largely been ignored by genome-wide studies, and thus little is known about the DNA methylation profiles of different repetitive DNA elements types. As a step toward investigating epigenetic landscape of repetitive DNA, we have developed a repeat-specific DNA microarray called RepArray. The RepArray comprises 236 prototypic oligonucleotides that span the main repetitive elements families found in the human genome. Combined to a methylated DNA immunoprecipitation (MeDIP) approach, RepArray allows depicting simultaneously the global trends that affect multiple repeat classes through the analysis of a restricted number of targets. Here, we present the MeDIP-on-RepArray protocol as it was established in our laboratory to delineate DNA methylation changes after chemical or genetic disruption of DNA methyltransferase activity in cells. It might serve as a workflow guideline for screening DNA methylation changes on repetitive elements during development and aging, among tissues and in various types of stress or pathological situations.

Key words: DNA methylation, 5-Methylcytosine (5mC) antibody, DNA repetitive sequences, DNA repeats, Transposable elements, DNA satellite, Oligonucleotide microarray, MeDIP-on-chip, Epigenetics

1. Introduction

Aberrant DNA methylation is commonly observed in cancer progression and development, aging and also in autoimmune and neural disease (1). Carcinogenesis is frequently associated with promoter hypermethylation of specific genes concomitant to an overall demethylation. This hypomethylation largely affects the intergenic and intronic regions of the genome, particularly repeat

Yves Bigot (ed.), *Mobile Genetic Elements: Protocols and Genomic Applications*, Methods in Molecular Biology, vol. 859, DOI 10.1007/978-1-61779-603-6_16, © Springer Science+Business Media, LLC 2012

sequences and transposable elements. The significance of this hypomethylation of repetitive elements in cancer is still unclear, but it may play multiple roles in pathogenesis through enhanced chromosomal instability, increased mutation events and unwanted transcription by *cis*- or *trans*-effects (1–3). In this context, monitoring methylation level on repetitive sequences might be particularly important as it may pinpoint novel diagnostic or prognostic indicators.

Despite the large contribution of repetitive sequences to the human genome (≈50%), the majority of studies until recently have focused on regions where DNA methylation was assumed to have the greatest functional significance in the regulation of gene expression, such as CpG-rich promoters. In addition, although there are many different families of repetitive DNA elements in the genome, studies mainly focused on only few classes of repetitive elements (4–8). Therefore, the DNA methylation map in repetitive sequences remains poorly defined on large scale.

We have developed a microarray called RepArray for systematic and comprehensive analysis of all main families of DNA repetitive sequences found in the human genome (9). The RepArray comprises 236 prototypic oligonucleotides that span the four main categories of human repeats: tandem satellites found in centromeric, pericentromeric, and subtelomeric heterochromatin; interspersed transposable elements, such as LTR- and non-LTR retrotransposons as well DNA transposons (Table 1). Coupled to a methylated DNA immunoprecipitation approach (MeDIP), the RepArray platform provides information on the relative DNA methylation level of hundreds of repetitive sequences through the analysis of a restricted number of targets. Therefore, RepArray depicts relative prevalence of a particular DNA methylation pattern over a given repeat class and not the behavior of individual repetitive element in a given chromatin environment. MeDIP-on-RepArray is a relatively low cost, well-suited approach for rapid screening of DNA methylation changes occurring during development, aging as well as between diseased and healthy tissues or between genetically modified and unmodified control cells. As repeat-specific transcriptomic maps can also be established using RepArray (9), it enables a comprehensive epigenetic analysis on the repetitive compartment of the genome necessary to fully appreciate the roles of DNA repeats in disease initiation and progression.

In this chapter, we describe the MeDIP-on-RepArray assay as it is performed in our laboratory. The protocol we describe has been used successfully to study DNA methylation changes upon chemical or genetic disruption of DNA methyltransferase activity in cells (9). We provide detailed methods for DNA processing, MeDIP, labeling of enriched sequences and detection of enriched sequences using the RepArray as well as real-time PCR. Finally, we also provide a guide for primary data analysis of MeDIP-on-RepArray data.

Table 1
Human repetitive DNA sequences represented on RepArray microarray

Major classes	Subcategories	Oligos names
Satellite DNA		
	Centromeric	01A_003[a], 02A_005[a], 03A_012[a], 04C_005[a], 05I_005[a], 07D_004[a], 08A_008[a], 09A_002[a], 10E_002[a], 11A_003[a], 12A_003[a], 13D_003[a], 14A_001[a], 15J_003[a], 16A_004[a], 17M_005[a], 18C_008[a], 20C_002[a], 21H_008[a], 22B_002[a], XF06_004[a], chr_10_VERDIN[b], ALR, SATCONS[c], SN5
	Pericentromeric	48BP, BSR, CER, GSATX, HSATII, HSATI, SAT1, SAT3, PHUR98(99_159)[d], Phur98(1_60)[d], SatII_PUC1(101_161)[e]
	Sub- and telomeric	HEXA_TR_A1[f], REP522, TR_B1[g], TR_A6[g], TR_B2[g], TR_A16[g], TR_A19[g], TR_B5[g], TR_B6[g], TR_B9[g], TR_B13[g], TR_B16[g], TR_B19[g]
	Others	D4Z4, LSAU, MER122, MER22, MSR1, NBL2, SATR, R66, TETRA_NT_AATG[h], TETRA_NT_ACAG[i], HEPTA_TR_A4[j], NONA_TR_A17[k]
TEs—Class I		
Retroviruses and retrovirus-like	MaLR	MLT1A, THE1BR
	Others	HERV16, HERV18_2, HERV39, HERV3, HERVE, HERVH48I_1, HERVH48I_2, HERVI, HERVK, HERVL, HERVS71, HUERS_P1, HUERS_P2, MER21I_1, MER21I_2, MER41, MER4INT, MER57I, MER61I, MER65I, MER70I, MER89I
Long terminal repeats	MaLR LTRs	MLT1I, MSTA, THE1
	Other retrovirus LTRs	LOR1S, LTR10S, LTR12S, LTR13S, LTR16S, LTR17, LTR18_1, LTR18_2, LTR19, LTR1S, LTR21A_2, LTR22, LTR23, LTR24, LTR27, LTR29, LTR2, LTR30, LTR32, LTR33S, LTR34, LTR35, LTR36, LTR37, LTR3S, LTR40, LTR41, LTR42, LTR43, LTR44, LTR46, LTR47, LTR4S, LTR52, LTR53, LTR54, LTR55, LTR57, LTR62, LTR64, LTR66, LTR67, LTR68, LTR69, LTR6, LTR71, LTR72, LTR75, LTR7A, LTR9B, LTR9S, MER101, MER110, MER11, MER31, MER34, MER39, MER41E, MER48, MER49, MER4, MER50, MER51, MER61, MER65, MER66, MER67, MER68, MER70, MER73, MER76, MER83, MER84, MER87, MER89, MER90, MER92, MER93, MER95, MER9, MLT2E, PABL_A, TAR1
LINE	Non-LTR autonomous	CR1_HS, IN25, L1_3END, L1MC, L1MCC_5, L2
SINE	Non-LTR non-autonomous	ALU_ALL, MIR_MIR3, SVA

(continued)

Table 1 (continued)

Major classes	Subcategories	Oligos names
TEs—Class II		
DNA transposons		CHARLIE1, CHARLIE2, CHARLIE6, CHARLIE7, CHARLIE8, CHESHIRE, GOLEM, HSMAR, LOOPER, MADE1, MARNA, MER103, MER104, MER105, MER106, MER107, MER112, MER113, MER115, MER116, MER117, MER119, MER121, MER1AS, MER2, MER20, MER28, MER3, MER30, MER45, MER53, MER5C, MER5, MER63, MER69, MER6, MER75, MER80, MER81, MER82, MER85, MER91, MER94, MER96, MER97, MER99, ORSL, TIGGER1, TIGGER5, TIGGER6, TIGGER9, ZOMBI
Unclassified		MER120, MER109
Others		
Ribosomal DNA		RDNA_EST[l], RDNA_KOND[m]

Oligos names are listed according to their affiliation. Oligos names as used in Repbase Update except for sequences referring to some satellite DNA

[a]Chromosome-specific subsets of human alphoid DNA were designed with the first number referring to the chromosome

[b]Chromosome 10 alpha satellite DNA corresponding to a preferential integration site for HIV virus (15)

[c]Human alpha satellite DNA consensus as defined by Vissel and Choo (16)

[d]Clone pHuR 98, a variant satellite 3 sequence, specific to chromosome 9qh (17)

[e]Clone Sat_II Puc 1, a variant of satellite II from chromosome 1q (18)

[f]Telomere-specific hexanucleotide (TTAGGG)n (19)

[g]Subtelomeric minisatellites found at 1–840 kb of the chr2 telomeric tract

[h](AATG)n simple repeat

[i](ACAG)n simple repeat

[j](AACAAAAC)n simple repeat

[k](TTTTGTTTG)n simple repeat

[l]5′ end of the 28S component of ribosomal DNA (20, 21)

[m]ribosomal DNA (21)

2. Materials

Prepare solutions using ultrapure water and analytical grade reagents (unless otherwise indicated). All reagents can be stored at room temperature (unless otherwise indicated).

2.1. Purification of Human Genomic DNA

1. DNeasy Blood & Tissue Kit (Qiagen #69581).
2. 1× Dulbecco's PBS Ca & Mg free (PAA #H15-002).
3. RNAse A 100 mg/ml (Qiagen #19101), store at −20°C.

4. DNA resuspension Buffer: 10 mM Tris-HCl pH7.4.
5. Agarose DNA grade (Euromedex #D5-D).

2.2. Sonication of Genomic DNA

1. 0.2 M Na-phosphate buffer pH 7.0 (stock buffer): 39 ml 0.4 M monobasic sodium phosphate (NaH_2PO_4), 61.0 ml dibasic sodium phosphate (Na_2HPO_4). Add H_2O to 200 ml. Sterile filtrate.
2. The protocol was established using a Standard Bioruptor Waterbath Sonication device (Diagenod Bioruptor® UCD-200), (see Note 1).
3. Agarose DNA grade (Euromedex #D5-D).

2.3. Methylated DNA Immunoprecipitation

1. IP Buffer: (10× stock buffer) 100 mM Na-phosphate buffer pH 7.0, 1,4 M NaCl, 0,5% TritonX-100.
2. Rabbit anti-5mC polyclonal antibody (MegabaseResearch #CP 51000) (20 μg per IP point), (see Note 2).
3. Purified rabbit IgG (Sigma-Aldrich #I5006) or mouse IgG (Sigma-Aldrich #I5381).
4. 1× Dulbecco's PBS Ca & Mg free (PAA #H15-002).
5. Bovine serum albumin (BSA) 30% (Sigma #A9576). Store at 4°C.
6. Protein-A Sepharose CL4B beads (GE Healthcare #17-0780-01). Store at 4°C, (see Note 3).
7. 5 M NaCl (stock solution) = 292 g NaCl. Add H_2O to 1 L. Sterile filtrate.
8. 2× TE buffer = 20mM Tris pH 8.0, 2 mM EDTA pH 8.0. Sterile filtrate.
9. Proteinase K (Roche Applied Science #03115887001).
10. Proteinase K digestion buffer = 50 mM Tris–Cl pH 8.0, 10 mM EDTA, 0.5% SDS. Sterile filtrate.
11. UltraPure Phenol:Chloroform:Isoamyl Alcohol (Invitrogen #15593031).
12. 3 M Na-acetate pH 5.2 (stock solution) = Dissolve 408 g sodium acetate $3H_2O$ in 800 ml H_2O Add H_2O to 1 L. Adjust pH to 5.2 with 3 M acetic acid. Sterile filtrate.
13. Glycogen 20 mg/ml (Roche Applied Science #10901393001). Store aliquots at −20°C.
14. Ethanol, absolute for molecular biology.
15. DNA resuspension buffer: 10 mM Tris-HCl pH7.4.
16. NanoDrop spectrophotometer (Thermo Scientific).

2.4. Real-Time PCR Analysis of MeDIP

1. QuantiTect SYBR Green PCR kit (Qiagen #204143). Store at −20°C.
2. 25 mM $MgCl_2$ stock solution (Roche Applied Science #11 699 113 001). Store at −20°C.

3. Oligonucleotide primers to genomic regions of interest and control.
4. Real-time PCR machine: we used a Stratagene's Mx300OP™ qPCR System and the Agilent's MxPro qPCR software (Agilent Technologies Genomics), (see Note 4).
5. 96-well optical PCR plates and strip caps compatible with the Real-time PCR instrument (Agilent Technologies Genomics # 401333 and 401425).

2.5. Labeling of Immunoprecipitated DNA with Aminoallyl-dUTP

1. Random primers pd(N)6 (Roche Applied Science #11 034 731 001). Working solution 0.1 U/μl: dissolve 2 mg (50 A_{260} units) Random pd(N)6 Potassium Salt in 500 μl ultrapure water. Store in aliquots at −20°C.
2. Random prime reaction buffer (10× stock): 50 mM NaCl, 10 mM Tris–HCl pH 7.9, 10 mM $MgCl_2$, 1 mM Dithiothreitol (NEB #B7002S). Store at −20°C.
3. Klenow fragment (3′→5′ exo-) 50 U/μl (NEB # M0212M). Store at −20°C.
4. 100 mM dNTPs stock (NEB #N0446S). Store at −20°C.
5. 20× dNTPs working mix: 5 mM dATP, 5 mM dCTP, 5 mM dGTP, 2 mM dTTP.
6. 5-(3-aminoallyl)-dUTP: 10 mM aminoallyl ready-to-use solution (Ambion # 8438). Store frozen in aliquots below −70°C (see Note 5).
7. Qiaquick PCR Purification kit (Qiagen # 28104).
8. 1 M K-phosphate buffer pH = 8.0 (stock solution): mix 9.5 ml 1 M di-potassium hydrogen phosphate (K_2HPO_4) and 0.5 ml 1 M potassium di-hydrogen phosphate (KH_2PO_4). Sterile filtrate.
9. Column washing buffer: mix 20 ml 15 mM K-phosphate buffer pH 8.0 and 80 ml of absolute ethanol.
10. DNA elution buffer: 15 mM K-phosphate buffer pH 8.0.
11. DNA Speed Vacuum Concentrator (Thermo Scientific).

2.6. Coupling Monofunctional Reactive Cyanine Dyes to the Modified DNA

1. Fluorescent dye mono-reactive NHS-esters. We used Cy™ Dye Post Labelling Reactive Dye Pack (Amersham Bioscience #RPN 5661), (see Note 6).
2. Coupling buffer: dissolve 0.84 g $NaHCO_3$ in 90 ml ultrapure water. Adjust to pH = 9.0 with 1 M NaOH. Add water to 100 ml. Sterilize by filtration with a 0.45 μm filter and store aliquots at −20°C. Do not thaw and refreeze (see Note 7).
3. 4 M Hydroxylamine hydrochloride: dissolve 27.8 g hydroxylamine hydrochloride (Sigma Aldrich # H2391) in 100 ml ultrapure water.

4. NucAway™ Spin Column (Ambion # AM10070).
5. DNA Speed Vacuum Concentrator (Thermo Scientific).

2.7. Preparation of DNA Microarray

1. Library of long oligonucleotides DNA probes representative of a set of human repetitive DNA sequences, including satellites (centromeres, pericentromeres, telomeres, megasatellites, minisatellites), retrotransposons (LTR or non-LTR), DNA transposons and also rDNA intergenic region (Table 1), (see Note 8a). Additionally, unique sequences from *Arabidopsis thaliana*, *Drosophila melanogaster*, *Bacillus subtilis*, *Escherichia coli*, *Nicotiana tabacum*, *Photonis pyralis* were also added on the array as irrelevant (negative) controls. Long oligonucleotides are synthesized at 1 or 2 nmol scale and packaged in lyophilized form in 384-well plates by Eurogentec S.A., France, (see Note 8b).
2. Pronto!™ Universal Spotting Solution (Corning # 40027), (see Note 9).
3. UltraGAPS™ Coated Slides without Bar code (Corning #40016).
4. Arraying equipment and printing pins (see Note 9).

2.8. Array Hybridization

1. Hybridization chambers for 25 × 75 mm microarray slides (Corning #2551).
2. Bovine Serum Albumin 30% (Sigma #A9576). Store at 4°C.
3. Deionized Formamide (Sigma Aldrich #F9037).
4. 20% SDS stock solution (Euromedex # EU0660-A).
5. 20× SSC stock solution (Euromedex # EU0300-D).
6. Coverglass 24 × 60 mm (Corning # 2935-246).
7. Prehybridization solution (Fresh): 5× SSC, 0.1% SDS (w/v), 1% BSA(w/v).
8. 2× Hybridization stock solution: 50% formamide, 10× SSC, 0.2% SDS (see Note 10).
9. Waterbath at 42°C.
10. Post-hybridization wash solution 1 (fresh): 1× SSC, 0.2% SDS.
11. Post-hybridization wash solution 2(fresh): 0.1× SSC, 0.2%SDS.
12. Post-hybridization wash solution 3 (fresh): 0.1× SSC.
13. MilliQ Water.
14. Propanol-2 (VWR # 8.18766.1000).
15. Plastic containers for washing slides. BD Falcon™ conical tubes are ideal for individual washing of hybridized slides.
16. Scientific cleaning wipes (KIMTECH SCIENCE* Precision Wipes Tissue Wipers # 05511).

17. Tabletop centrifuge equipped with a swinging bucket rotor that can accommodate 50 ml tubes (Beckman Coulter # Allegra® X-15R).

2.9. Scanning and Analysis

1. Microarray scanner for imaging two-color microarrays: We opted for the robust and easy-to-use GenePix® 4000B microarray scanner that allows data acquisition at two wavelengths simultaneously. The GenePix® 4000B microarray scanner includes one license of GenePix® Pro Acquisition and Analysis software (Molecular Devices # GenePix 4000B).
2. RepArray specific array list (File extension ".gal").
3. Software for bioinformatic: Bioconductor (http://www.bioconductor.org/) and R packages (http://www.r-project.org/).

3. Methods

The complete protocol for MeDIP-on-RepArray starting from cell pellets up to data acquisition takes ≈7 days (Fig. 1).

3.1. Isolation of Genomic DNA

Typically, genomic DNA is purified from fresh or frozen cell pellets (≈ 5×10^6 cells) using DNeasy Blood and Tissue Kit according to manufacturer's recommendation. Briefly, adherent cells were grown to confluency, harvested after dispersion by trypsin treatment, washed in 1× PBS and pelleted by centrifugation ($300 \times g$).

1. Fresh or frozen cell pellets were resuspended in 200 µl 1× PBS. When using frozen pellets, before adding PBS allow cells to thaw until the pellets can be dislodged by gently flicking the tube.
2. Because the antibody also recognizes 5mC in the context of RNA molecules, it is important to completely remove RNA from the genomic preparation. Add RNase A to a final concentration of 300 µg/ml and incubate 10 min at room temperature.
3. Proceed to genomic DNA extraction according to manufacturer's instructions. Elution of purified DNA from DNeasy Mini spin column is performed in two successive steps using 150 µl of 10 mM Tris pH 7.4.
4. Determine DNA concentration with NanoDrop Spectrophotometer. Yields of genomic DNA will vary depending on cell type and quality of starting material. Approximately 6 µg of DNA should be obtained from 10^6 mammalian cells.
5. To assess DNA integrity, resolve 250 ng of DNA by 1% agarose gel electrophoresis.

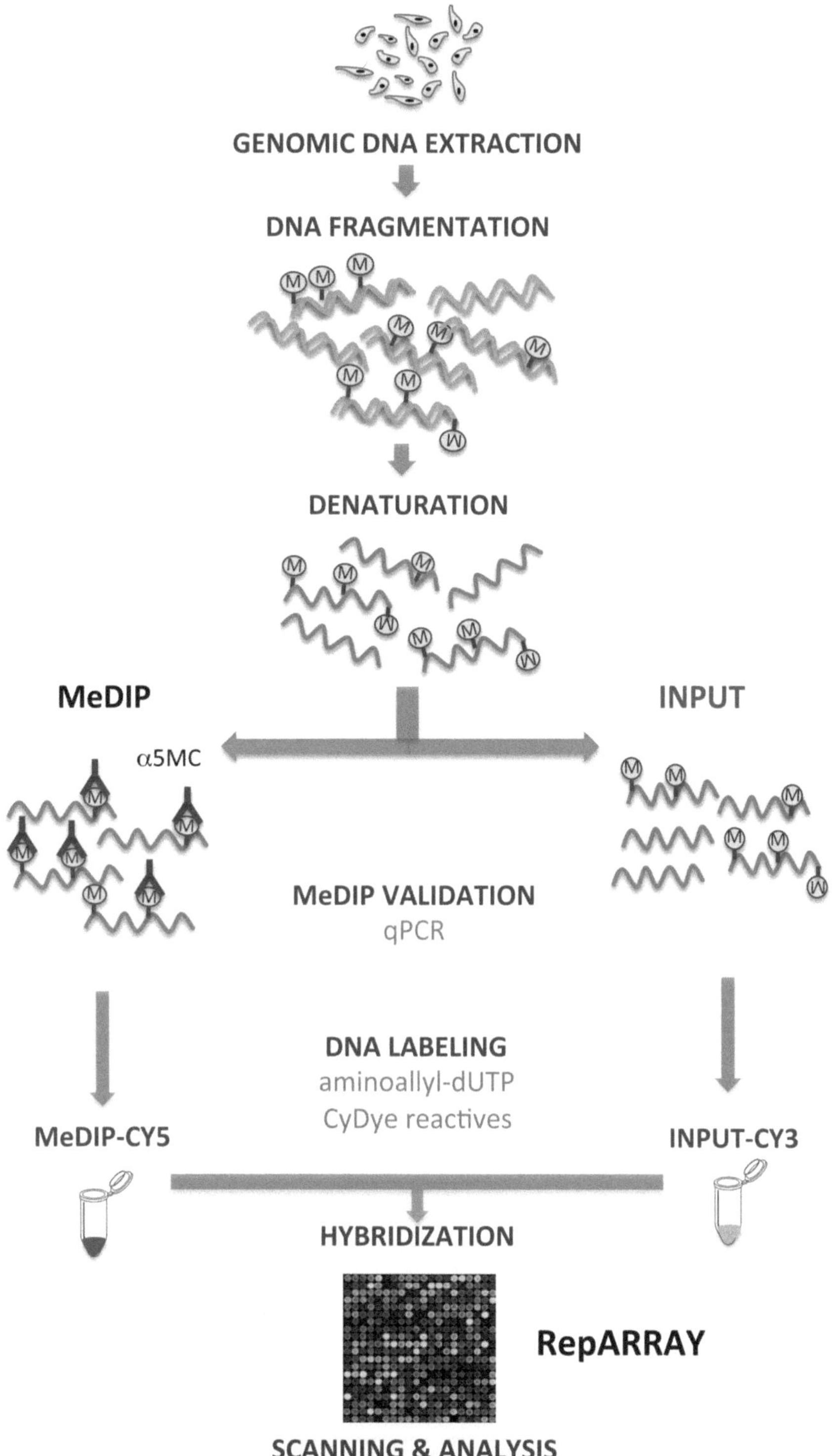

Fig. 1. Overview of the MeDIP-on-RepArray assay. Schematic diagram showing the steps involved in MeDIP from genomic DNA isolation, fragmentation by sonication, immunoprecipitation using anti-5 methylcytosine (α5mC) antibodies, real-time PCR validation of MeDIP reaction, labeling of MeDIP and Input DNA by random priming using aminoallyl-dUTP and coupling with CyDyes and hybridization to the repeat-specific oligo-array RepArray.

3.2. Fragmentation of Genomic DNA

DNA fragmentation is a critical step since it defines the resolution of MeDIP. The sheared DNA fragments should ideally range from 200 bp to 1,000 bp. Large DNA fragments should be eliminated for a better resolution and small DNA fragments below a size of 200 bp will be difficult to detect by PCR or microarray. The sonication protocol reported here is suitable for various cell lines (HCT116, HeLa, BJ, U2OS, NT2, human B lymphocytes), nevertheless optimization may be required for other cell lines, different amount of starting material or sonicator devices. Performing a sonication time course and products analysis on a 1% agarose gel should optimize sonication efficiency.

1. In a 1.5 ml microtube, dilute 8 μg of RNA-free genomic DNA in 300 μl of 10 mM NaP04 buffer pH 7.0, (see Note 11).
2. Sonicate for seven cycles 25 s ON/30 s OFF at the highest output level while cooling the samples to 4°C in the Bioruptor waterbath.
3. Resolve 5–10 μl by 1% agarose gel electrophoresis to check DNA fragmentation. A smear should be observed between 200 and 1,000 bp (Fig. 2). If necessary, sonicate for one or two additional pulses until the size of the DNA is 200–1,000 bp.

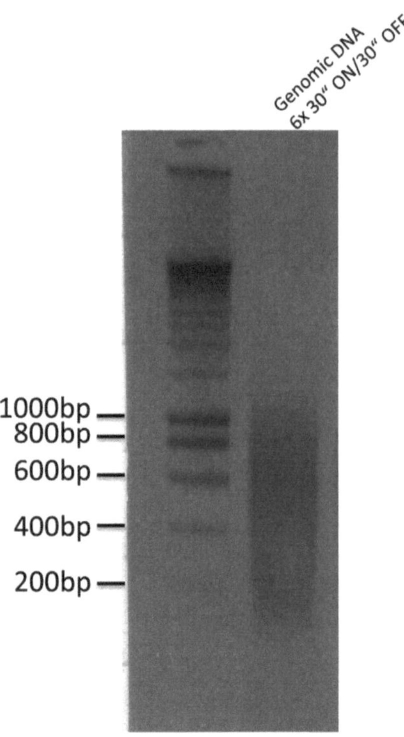

Fig. 2. DNA fragmentation by sonication. Example of gel electrophoresis analysis of sonicated genomic DNA. Typically, the sheared DNA fragments range from 200 bp to 1,000 bp.

4. Denature DNA by heating the solution 10 min at 94°C and immediately cool on ice.
5. Remove an Input sample: save 20 μl (≈ 500 ng) of the denatured DNA solution and keep at 4°C until step 18.
6. Divide the denatured DNA solution into 2× 140 μl aliquots (≈ 4 μg of DNA) in 1.5 ml microtubes, one that serves as the methylcytosine IP sample (5mC-IP) and the other as negative control (IgG-IP), (see Note 11).
7. Add 50 μl of 10× IP buffer and 300 μl of water to each DNA aliquot.
8. Add 10 μl of anti-5mC antibody (5mC-IP) or purified IgG (IgG-IP) and incubate O/N at 4°C with continuous gentle rotation. This step can be carried out in shorter time period (2–6 h) if more convenient.
9. Prepare a 50% slurry of preblocked Protein A-Sepharose beads. Into a 14 ml screw cap tube, wash the required amount of Sepharose beads in 10 volumes of 1× PBS so it swells up, then centrifuge at 700 × *g*, remove the supernatant and discard. Add 10 volumes of 1× PBS supplemented with 0.1% BSA and incubate O/N at 4°C with gentle overhead shaking. Remove the supernatant and make a 50% slurry (v/v) with 1× PBS-0.1% BSA. The Protein A-Sepharose slurry is now ready to use (see Note 12).
10. Into 1.5 ml microtubes, aliquot 80 μl of Protein A-Sepharose slurry, centrifuge at 650 × *g* using a 24-place microcentrifuge. Discard supernatant and continue with the pellet (beads), (see Note 13).
11. Capture the immune-complex by adding the 500 μl of immunoprecipitation mix from step 8 on the Protein A-Sepharose pellet (step 10) and incubate 2 h at 4°C with continuous gentle rotation.
12. Pellet the beads by centrifugation for 2 min at 650 × *g* using a 24-place refrigerated microcentrifuge. Transfer supernatant (unbound fraction) to a new tube without disturbing the pellet.
13. Wash the beads with 1 ml of cold 1× IP buffer for 5 min on rotating wheel at 4°C.
14. Repeat wash step 13 once. Recover the beads between washes by centrifugation for 2–3 min at 650 × *g* at 4°C using a 24-place refrigerated microcentrifuge and remove the supernatant nearly completely without disturbing the pellet.
15. Wash the beads with 1 ml of cold 1× IP buffer supplemented with 160 mM NaCl to give 1× IP buffer-300 mM NaCl for 5 min at 4°C with gentle rotation.

16. Wash the beads with 1 ml of 2× TE for 3 min at room temperature.
17. From that step, continue the procedure also with the Input sample (saved at step 5).
18. Elute DNA from the beads by adding 250 μl of proteinase K digestion buffer plus 7 μl of proteinase K (10 mg/ml stock) and incubate 3 h at 55°C while vigorously shaking (see Note 14). Add 230 μl of proteinase K digestion buffer to the Input sample.
19. Recover DNA by adding one volume (250 μl) phenol/chloroform/isoamyl alcohol, vortex vigorously, and centrifuge at 14,000 × *g* for 10 min in order to separate the phases. Carefully transfer the upper aqueous phase to a fresh microtube.
20. Precipitate DNA by adding 1/10 volume of 3 M Na-acetate pH 5.2, 1 μl of glycogen and 2.5 volume of ice-cold 100% ethanol. Mix and incubate for at least 2 h at −20°C. This step can be carried out O/N if more convenient.
21. Centrifuge at 14,000 × *g* for 20 min at 4°C to pellet the precipitated DNA. Wash pellet in 500 μl of ice-cold 70% ethanol. Centrifuge at 14,000 × *g* at 4°C to collect DNA.
22. Dissolve DNA in 40 μl of DNA resuspension buffer (10 mM Tris-HCl pH 7.4).
23. Measure DNA concentration with a NanoDrop spectrophotometer. The sample can be stored at −20°C.

3.3. Analysis by PCR

It is highly recommended to verify the MeDIP specificity by real-time PCR analysis on immunoprecipitated DNA using positive and negative control primers before starting labeling/hybridization procedure. Positive control primers typically amplify fragments spanning a region of the genome though to be methylated while negative control primers must be specific for a region expected to be unmethylated which is typically the promoter region from a housekeeping gene. Table 2 contains a set of control primers suitable for several human cell lines. qPCR using these primers is standardly carried out on MeDIP, IgG-IP, and Input samples. For repetitive sequences analysis, primers were designed to a consensus sequence for each repetitive element family and thus amplify a global pool of repetitive element rather than a single element or genomic locus.

1. Dilute part of the DNA samples in milliQ water to 8–10 ng/μl.
2. Prepare a master mix of A and B primers by mixing each primer at a final concentration of 5 mM in nuclease-free water.

Table 2
Control primers. For qPCR primers design we utilized Primer 3 (v.0.4.0) software (http://frodo.wi.mit.edu/primer3/) using the following specified constraints:

Product size range	100–300 bp		
Primer size	Min: 18 bp	Opt: 20 bp	Max: 25 bp
Primer Tm	Min: 57°C	Opt: 60°C	Max: 63°C

Locus		Sequences 5′–3′	Annealing	References
H4 promoter	Neg	AAATGGTGGGATCACAGACG CGAGCTTCTTGTTTCCGTGT	57°C	(9)
Gapdh promoter	Neg	CGTGCCCAGTTGAACCAG CGCCCGTAAAACCGCTAGTA	57°C	(9)
Beta Actin promoter	Neg	CCTGACTCCCCCAACACCAC TGAAGTCGGCAAGGGGGAGT	57°C	
TSH2B	Tissue-specific	CAGACATCTCCTCGCATCAA GGAGGATGAAAGATGCGGTA	57°C	(11)
MLH1	Cell-specific	CGAACCAATAGGAAGAGCGGA GGACAGCTTGAATGCCAGTCA	57°C	(10, 11)
RASSF1	Cell-specific	CCCAAGTGAAGGCTCGAGACT TGTTTCTGGAGGCCAGCTTTA	57°C	(10, 11)
RARB2	Cell-specific	GAGCAAACGAGTGCAGTCAA CTCTGTGCGCCTTTCTGTCT	57°C	(9, 22)
H19-ICR	Pos	GAGCCGCACCAGATCTTCAG TTGGTGGAACACACTGTGATCA	57°C	(10, 11)
Satellite 2	Pos	TCGCATAGAATCGAAGGAA GCATTCGAGTCCGTGGA	57°C	(20)
Alu-all	Pos	TGAAACCCCGTCTCTACTAAAAA GTCTCGCTCTGTCGCCCA	57°C	(9)
D4Z4 megasatellite	Pos	CTCAGCGAGGAAGAATACCG ACCGGGCCTTAGACCTAGAAG	57°C	(9)

Primers were optimized for real-time PCR by performing a standard curve with sheared genomic DNA. Amplification efficiency should be ≥95% (rate of amplification ≥1.9) and dissociation peak need to be checked

3. Combine the following in an optically clear 96-well PCR plates (see Note 15):

2× QuantiTec SYBRGreen PCR master mix	12.5 μl	1× final (contains 2.5 mM $MgCl_2$)
$MgCl_2$ 25 mM	1.5 μl	Provides a final concentration of 4 mM
Primers A+B Master Mix	1.5 μl	0.3 μM final each
UltraPure Water DNA sample (diluted)	7.5 μl 2 μl	

Each qPCR reaction for all samples is to be done in duplicate. Process also a "no DNA" control reaction without sample DNA to monitor possible contamination problems with foreign DNA (we recommend the use of filter-tips)

4. Set up and run the following PCR program (Mx300OP™, see Note 15):

Initial activation step	95°C	15 min
Denaturation	94°C	15 s
Annealing	57°C	30 s
Extension	72°C	30 s
	40 cycles	

5. To evaluate the enrichment in the MeDIP fraction we calculate the difference in cycles between the immunoprecipitated sample (MeDIP or IgG-IP) and the Input DNA for primer pair of interest and for negative control pair. The methylation enrichment of a particular sequence is then expressed as fold over the background methylation level at negative control locus (see Note 16).

3.4. Labeling of Immunoprecipitated DNA with Aminoallyl-dUTP

For our hybridization experiments, we compare the enrichment of the MeDIP sample relative to the Input DNA. For this, we label MeDIP sample and Input genomic DNA with different fluorescent dyes and co-hybridize them to the RepArray. We then generate a ratio of methylated versus Input signal for each sequence spotted on the array and use this as readout for the methylation level: a positive log2 ratio indicates hypermethylation, and a negative log2 ratio indicates hypomethylation. For labeling our DNA samples, we favor an indirect procedure. Indirect means of labeling incorporates a nonfluorescent nucleotide analogue, such as aminoallyl-dUTP (aa-dUTP), followed by cyanine dye coupling to the incorporate aminoallyl group. Because the small conjugate is incorporated in both the MeDIP and the Input samples equally, this method helps eliminating the incorporation biases that occur when using direct labeling methods.

1. In a microtube, bring 0.2–0.5 μg of MeDIP or Input DNA to 36.4 μl with milliQ water and mix with 4 μl of 10× random prime reaction buffer and 3 μl of 0.1 U/μl random primers pd(N)6 solution (see Note 17).
2. To denature DNA, incubate 10 min at 94°C and immediately place on ice afterward.
3. Add 2 μl of 20× dNTPs mix, 0.6 μl of 10 mM aa-dUTP, and 1 μl of 50 U/μl Klenow (exo-) polymerase. Incubate overnight at 37°C with gentle shaking.
4. Stop the reaction by adding 2 μl of 0.5 M EDTA pH 8.0.

5. Recover the labeled DNA using QiaQuick PCR Purification kit. We follow the Qiagen protocol except that we used column washing buffer (15 mM K-phosphate buffer pH = 8.0, EtOH 80%) for washing the silica column and then DNA is eluted using 2 × 40 µl of freshly prepared DNA elution buffer (15 mM K-phosphate buffer pH = 8.0) as the subsequent dye-coupling step is sensitive to amine buffer. It is very important to remove as much as possible of the unincorporated aminoallyl-dUTP as it will bind the NHS-ester cyanine dyes and thus reduce the labeling efficiency.
6. Check the quality of DNA after labeling step on a 1% agarose gel (5% of the final product). Determine DNA concentration and purity with a NanoDrop spectrophotometer using a DNA spectrum from 230 to 320 nm (see Note 18).
7. Spin-dry the prepared aminoallyl-DNA to ≈5 µl using a speed-vac centrifuge. Be careful not to overdry the DNA as it may become difficult to resuspend.

3.5. Coupling Monofunctional Reactive Cyanine Dyes to the Modified DNA

As Cy3 and Cy5 reactive dyes are easily photobleached, it is best to protect Cy dyes and all labeled products from light as much as possible during the procedure.

1. Bring the volume of aminoallyl-labeled DNA (from Subheading 3.4 step 7) to 40 µl by adding appropriate volume of 0.1 M sodium bicarbonate pH 9.0. Pipettes several times to perfectly mix the sample.
2. Transfer DNA samples to a freshly open vial of Cy3 or Cy5 reactive dye (see Note 19). For a given array, the MeDIP sample is coupled to one dye and the total Input to the other. Typically, we coupled MeDIP sample to the Cy5 dye and the Input reference to the Cy3 dye, but the reverse works as well.
3. Pipette up and down several times to resuspend completely the reactive dye/aminoallyl-labeled DNA mix. Bring all the contents at the bottom by briefly spinning.
4. Incubate at room temperature for 90 min in the dark to couple the NHS-ester cyanine dye to the DNA.
5. While coupling is going on, rehydrate NucAway™ Spin Column according to manufacturer's instructions. We have found that removal of the unincorporated dye is quite efficient
6. Add 15 µl of 4 M hydroxylamine and incubate at room temperature in the dark for 15 min. This will inactivate unreacted Cy dyes NHS-ester molecules
7. Bring the volume of labeling reaction to 85 µl by adding nuclease-free water.
8. Place the rehydrated NucAway™ Spin Column on top of a new collecting microtube, centrifuge 2 min at 750 × *g*. Discard the

flow-through. Slowly and carefully apply the labeled DNA directly to the center of the gel bed at the top of the NucAway™ Spin Column (avoid disturbing the gel surface).

9. Centrifuge at 750 × *g* for 2 min to collect the purified labeled DNA. The eluted dye-coupled product should be slightly colored (see Note 20).
10. Determine DNA yield and total dye incorporation by measuring the absorbance at A260, A550, and 650 nm using a NanoDrop spectrophotometer. The pmol incorporation of dye is calculated by using the formula: A550/0.15 (extinction coefficient for Cy3) or A650/0.25 (extinction coefficient for Cy5). Dye labeling density might be roughly estimated according to manufacturer's instructions manual (see Note 21).
11. Combine the respective Cy3- (MeDIP) and Cy5- (Input) labeled DNA and spin dry with heat the purple sample to ≈30 μl using a speed-vac centrifuge protected from light (be careful, this might take time). This is the probe solution ready for immediate hybridization (see Note 22).

3.6. Prehybridization/ Hybridization of the Array

Prehybridization should be done immediately preceding the application of labeled DNA to the target. Typically, we carried out the prehybridization of the microarrays while carrying concentration of probe solution (see Subheading 3.5 step 11). As general precautions, work in a clean and dust free environment and do not use powdered gloves when working with microarrays.

1. RepArray slides are placed for 1 h at 42°C in 50 ml of freshly prepared Prehybridization solution: 5× SSC, 0.1% SDS (w/v), 1% BSA (w/v). It is convenient to use 50 ml conical tubes to treat each slide individually.
2. Rinse briefly the slides five times in 50 ml milliQ water, then once in propanol-2. Place the slides in a new conical tube and dry by centrifugation for 4 min at 500 × *g* in a tabletop centrifuge at room temperature. Help the drainage of liquid from the slides by placing a piece of folded KimWipes paper at the bottom of the tube. The dried prehybridized slides are kept in dust-free container while completing the preparation of the hybridization solution.
3. Prepare the hybridization solution by mixing the concentrated probes (≈30 μl from subheading 3.5 step 11) with one volume of 2× hybridization stock solution (see Note 10). Tap gently to mix well and spin down.
4. Denature hybridization mixture at 95°C for 4 min. Briefly centrifuge to collect condensed water droplets and allow it to cool at room temperature for 2–3 min. Do not place the mixture on ice as this would cause SDS precipitation.

5. The denatured mixture is applied to the array by carefully adding the solution to one end of the slides and then a clean coverslip is carefully lower onto the array using forceps (see Note 23). Care must be taken to exclude formation of air bubbles and not to put pressure on the coverslip when handling the array.
6. Quickly place the microarray-cover assembly into the Corning hybridization chamber, coverslip side up, attach and seal the lid. Keep the chamber right side up and in a horizontal position all along the procedure.
7. Submerge the hybridization chamber in a humidified chamber consisting of a closed plastic box containing milliQ water prewarmed to 42°C (see Note 24).
8. Hybridizations are carried out in the dark for up to 16 h at 42°C in a water bath.
9. On day 2, prepare the post-hybridization wash solutions 1–3. Dispense all the necessary washing solutions in multiple 50 ml conical tube before starting the procedure. The post-hybridization wash solution 1 should be warmed at 42°C before use.
10. Remove array slides from humidified chamber and remove coverslip. The coverslip will slide off after a brief immersion in 50 ml of preheated post-hybridization wash solution 1. Do not agitate to avoid scratching the array with the loose coverslip. Gently move up the array using forceps and quickly transfer it into a new conical tube containing preheated post-hybridization wash solution 1
11. Wash arrays for 4 min at room temperature with gentle agitation on a platform shaker.
12. Transfer arrays in 50 ml of post-hybridization wash solution 2 for 4 min on a platform shaker.
13. Finally, to remove residual SDS wash arrays twice in 50 ml post-hybridization wash solution 3, each time for 4 min with gentle shaking.
14. Dry arrays by centrifugation for 4 min at 500×*g* in a 50 ml conical tube as described in step 2 (see Note 25). Store the arrays in a slide holder protected from light until ready to scan. It is recommended to proceed to scan as soon as possible, as the fluorescence will decay over time.

3.7. Scanning and Data Acquisition

The slides are scanned using an Axon 4000B scanner according to manufacturer's instructions.

1. Before scanning, warm up the 532 nm and 635 nm solid-state lasers for 30 min.
2. Load the slide into slide holder (for RepArray, the slide should be inserted with the printed side face down and label positioned

close to the user). Quickly scan the slides at low resolution to give a preview picture.

3. Set data acquisition area by selecting the area encompassing the spots. Set the laser intensity to 100%, adjust "pixel resolution" to 10 μm, "line to average" to 1, focus position 0 μm and optimize signal intensity by manually adjusting photomultiplier tube (PMT) gain values (see Note 26).
4. Full scan the selected area and save the image as a multiimage TIFF files.
5. Acquired images are analyzed using the GenePix Pro6.0 Microarray Analysis Software. For this, load the ".gal" file and adjust Alignment Setting (find circular features, do not resize the features during alignment, flag as "unfound" features that fail background threshold criteria). Find and align blocks manually so that the circular features of the .gal file are over the appropriate spots. Inspect all spots individually and flag as "bad" any feature affected by microarray imperfection, such as scratches or dust.
6. GenePix Pro 6.0 extracts numerical values and reports them in a table. Save the result file (.gpr file).
7. Preprocessing of the raw data is performed in R programming environment (http://cran.r-project.org/). Download the following tools: limma (http://www.bioconductor.org/packages/2.2/Software.html), marray (http://www.bioconductor.org/packages/2.2/Software.html), ade4 (http://pbil.univ-lyon1.fr/ade4html/00Index.html), gplots (http://cran.r-project.org/web/packages/gplots/index.html). Using the predictive model, a synthetic value for each spot on each individual array is computed and a data frame containing a set of nominal and numerical values ("Exp" "Lame" "Block" "Column" "Row" "Name""F635.Mean" "F635.SD" "B635.Mean" "B635.SD" "F532.Mean" "F532.SD" "B532.Mean" "B532.SD" "Flags" "Sens") is generated. Slides are kept in the analysis according to quality criteria, including signal-to-noise ratio inspection, control of the deviance from linear model. Features flagged "bad" or "unfound" and features that have not yielded data in minimum number of hybridizations (<75%) are excluded. Filtering for robust data may reduce the usable fraction but will ultimately result in greater confidence in data interpretation. Raw features intensities are used to determine $\log_2$ ratios MeDIP/Input (routinely $\log_2$ (F635/F532), see Subheading 3.5 step 2). Relative quantification of methylation level of individual sequence is inferred from the averaged log2 ratios (MeDIP/Input) of replicated features (see Note 27). Keep in mind that MeDIP enrichment at any target sequence depends both on the degree of methylation and on the density of CpG (10).

A modest enrichment can thus reflect an unmethylated state or the scarcity of CpG in the target region. Nevertheless, demethylation can affect repetitive sequences heavily or modestly methylated (9).

8. Validation of results: real-time PCR is a straightforward method for validation of methylation profiles at specific loci (see Subheading 3.3). Deeper analysis of methylation level can then be performed on individual identified repeat-class using bisulfite-PCR pyrosequencing for example (8).

4. Notes

1. Other sonication devices using direct sonication horns can be used, but conditions need to be adjusted for each apparatus. The Bioruptor™ sonicator allows the simultaneous processing of several samples (up to 6 × 1.5 ml tubes) in sealed tubes (no risk of contamination).
2. We mainly used the polyclonal 5-methylcytosine antibody from Megabase Research (#CP 51000), but mouse monoclonal 5-methylcytosine antibody from Eurogentec (#BI-MECY-0500) (20 μg per IP experiment) works well (as originally described by Weber et al. (11)). Since polyclonal antibodies are prone to batch-to-batch variability, each new immune-sera batch it is recommended to systematically check immunoprecipitation efficiency. Today, several companies (EUROMEDEX, DIAGENODE, etc...) propose similar poly- or monoclonal 5mC antibodies but DNA immunoprecipitation conditions must be optimized for each antibody. MeDIP performed with an unproven antibody must include real-time PCR validation step using appropriate controls. Antibodies must be aliquoted upon receipt and stored at −20°C to avoid freeze-thaw cycles.
3. Sepharose is a registered trademark of GE Healthcare for a cross-linked, beaded form of agarose. Several protocols have been developed using magnetic beads Dynabeads M-280 Sheep anti-mouse IgG (Dynal Biotech #112 01), for details see refs. (12, 13).
4. Methylation enrichment can be evaluated either by real-time PCR or by classical PCR (12).
5. Do not store aa-dUTP (or aminoallyl-labeled DNA) in a frost-free freezer. When stored properly, the reagent is stable for several months.
6. Do not store fluorescent components in a frost-free freezer. In addition, reactive dyes are sensitive to light and should be stored and used in dark as much as possible.

7. The CyDye reactive dye coupling to aminoallyl-containing DNA only occurs under alkali conditions between pH 8.5 and 9.5 which is set using 0.1 M sodium bicarbonate buffers. When stored properly, the reagent is stable for several months at –20°C or make the buffer fresh prior to setting up the labeling reactions.
8. (a) A full description of the probes selection procedure was previously described (9). Briefly, a conserved region specific for each repeat subfamily was inferred from multiple alignments of the most 50 similar DNA sequences identified using FASTA similarity search program in RepBase Update or GenBank databases. This region was used to design oligonucleotide sense probes using OligoArray 2.0 software (14). The probe search parameters were length range from 50 to 52 nucleotides, 25–65% CG percentage and melting temperature between 76 and 96°C. Oligonucleotides containing five consecutive A, C, G, or T were discarded. OligoArray 2.0 selects probes with the lowest cross-hybridization; the absence of secondary structure and the final oligonucleotide set is balanced in term of melting temperature. To limit overlapping between probes, an FASTA similarity search was conduct against RepBase and GenBank databases using the designed oligonucleotides. An antisense probe was then deduced from each designed sense probe. A full description of the oligonucleotides is hosted at: http://www.landesbioscience.com/supplement/HorardEPI4-5-Sup.pdf (see Supplemental Table 3).

   (b) Only oligonucleotides of the highest quality should be used for microarraying. Functionalization if the oligonucleotide with an amine or other reactive group is not necessary when using the UltraGAPS™ substrate.
9. RepArray slides were printed at The Transcriptome Plateform from Biology Department Genomic Service at ENS, Paris-FRANCE and information concerning microarray printing are hosted at: http://www.transcriptome.ens.fr/sgdb/. Briefly, dry oligonucleotides from were rehydrated in Pronto!™ Universal Spotting Solution (Corning # 40027) at a concentration of 0.5 ng/μl and printed on commercially prepared silinazed slides using a Microgrid TAS arrayer and pins from Biorobotics. Probes were arrayed as duplicates in a random order to minimize the risk of misinterpretation due to a geometric artifact during array hybridization. The printed arrays were kept at room temperature in a dessicator.
10. Prepare 2× hybridization stock solution, sterile filtrate, and make aliquots for storage at –20°C. Before use, 2× hybridization aliquot should be thawed completely and prewarmed at

68°C for few minutes to ensure complete dissolution of SDS crystals.

11. We recommended the use of 4 μg as starting DNA material per immunoprecipitation point. Smaller amounts down to 0.5 μg have been used successfully when combined with real-time PCR evaluation of enrichments. However, the yields of MeDIP DNA starting from small amount of material will be lower and thus not be adequate for immediate labeling and hybridization to microarray. Typical negative control reactions either use normal immunoglobulin (IgG) from the same species of the 5mC antibody or omit the antibody.
12. Protein A-coupled Sepharose beads are usually suitable when using polyclonal antibodies. If using a mouse monoclonal antibody combine (v/v) Protein A- and G-coupled matrix since some mouse isotypes do not bind to Protein A-Agarose. Lyophilized Sepharose should be swollen in 1× PBS (10 volumes) for at least 30 min at room temperature with gentle overhead shaking (1 g of powder typically swells to 3–4 ml of hydrated gel). Pre-swollen beads as slurry ready for use are also available from various manufacturers. Swollen Sepharose can be store at 4°C for few days in PBS 1×. To block nonspecific interactions with Protein A-Sepharose, it is important to use BSA 0.1% rather than salmon sperm DNA to avoid contaminations with unrelated repetitive DNA.
13. It is recommended to cut off the tip of the pipette when manipulating Protein A-Sepharose to avoid disruption of the beads. Beads will tend to stick to the sides of the tip so try to minimize the movement in the pipette.
14. We recommend using a shaking heat block at 750–900 rpm (Eppendorf Thermomixer R) to prevent sedimentation of the beads.
15. PCR program and cycling conditions might need to be adjusted according to qPCR reagents and machine used. It is best to premix all PCR components with the exception of the DNA template in a microtube (always prepare for one reaction more than you needed), aliquot the premix in the 96-well PCR plates and add the template.
16. The Ct values for each reaction within the duplicate are averaged if the duplicate does not fluctuate greater than or less than 0.5 cycles. It is recommended to repeat the PCR for samples where Ct ranges are >0.5. The rate of amplification is calculated for each PCR reaction. The theoretical optimum rate of amplification (R) is 2.0, but to make accurate and meaningful comparisons between primers pairs, the differences in amplification rates should be calculated. Amplification rate is determined by performing standard curves using serial dilutions

(0.01–30 ng) of sheared genomic DNA (=Input). The fold enrichment of a specific target sequence in the MeDIP sample relative to the Input sample is then calculated using the following equation: $R^{(\text{CtInput}-\text{CtMeDIP})}$. Finally, the methylation enrichment at a particular sequence relative to the background level is calculated according the following equation: $R^{(\text{CtInput}-\text{CtMeDIP})\text{Target}} / R^{(\text{CtInput}-\text{CtMeDIP})\text{Neg.cont}}$. The IgG-IP samples allow demonstrating that enrichment is specific for methylated DNA.

17. With the described conditions, the MeDIP procedure generally yields sufficient amount of material to allow direct labeling by Klenow random priming, thus bypassing the need for a PCR amplification step that might introduce bias in sequences representation. We confirm using qPCR that specific enrichments of several control loci detected in MeDIP samples versus Input samples are the same before (DNA from Subheading 3.2 step 22) and after the random priming reaction (DNA from Subheading 3.4 step 5). Typically, the larger the amount of DNA template used, the greater the yield of probe. Start labeling procedure from the same amount of MeDIP and Input DNA samples.
18. Successful labeling results in a smear from 100 to 1,000 bp with an average of size of 300–400 bp (i.e. slightly shorter than the smear detected after sonication). This is due to random incorporation of pd(N)6 oligonucleotide into DNA. For pure DNA, optimal absorbance ratios are A260/280≈1.8 and A260/230≈1.8. Typically, Klenow (exo-) reactions produce between 4 and 6 μg of aminoallyl-labeled DNA.
19. Place reactive dye pouch at room temperature for 15 min before opening the vials.
20. The color of the purified probe is a good indicator for the procedure success. If the eluted probe is colorless, the labeling was probably not efficient. If eluted product has a strong pink (Cy3) or blue color (Cy5), the dye removal procedure most likely failed.
21. According to manufacturer's instructions, the labeling density of one dye per X unlabeled nucleotide can be roughly estimated by using the formula:

$$\text{Dye ratio } (X) = ((\text{total DNA yield (ng)} \times 1{,}000) / 324.5) / \text{pmol incorporation of dye}$$

where: Total DNA yield (ng) = A260 × 37 μg/μl × total volume, dye yield (pmol) = A550 (or 650)/dye coefficient × total volume, 324.5 is the average molecular weight of nucleotide in DNA. 1,000 is the DNA length average assumed. Best hybridization results are obtained when one dye per every 30–60 unlabeled nucleotides is targeted.

22. Because cyanine dyes are extremely sensitive to degradation (by air humidity, ambient light, ozone), we recommend to proceed immediately to hybridization and not to store the labeled DNA for further hybridization.
23. Rinse cover glass in milliQ water then store them in 95% ethanol. Before use, allow cover glass to air dry.
24. We prefer that hybridization chambers do not directly contact the bottom of the water bath since it might cause nonhomogeneous heat transfer.
25. Transferring the slides to fresh conical tubes between wash steps helps to minimize SDS (which is autofluorescent) carry-over. As liquid evaporating slowly from the slides can lead to wash artifacts, proceed quickly with the final centrifugation step. Transfer the arrays from wash solution to conical tube placed in the centrifuge and immediately start the centrifuge.
26. Because some repeats account for 50 copies in the human genome while other can account for 5×10^7 copies, large amplitude of signal intensity is observed on a scan. Therefore, each array is scanned at three different photomultiplier tubes (PMTs) settings (low, median, and high). In order to correlate the measurements over a wide range of incidental light intensities, we developed a linear predictive model that generates a synthetic value of intensities. A full description of the predictive model is available at: http://www.landesbioscience.com/supplement/HorardEPI4-5-Sup.pdf (see Supplemental Method).
27. At least three independent MeDIP reactions are performed and hybridized to total genomic DNA (Input). The hybridized slides are then grouped together accordingly for the analysis. Friedman's analysis of variance can be performed on the whole dataset to check the coherence between the different microarrays hybridized. Both sense and antisense features populations might display similar $\log_2$ ratios in methylation experiments (Supplementary figure 1 at: http://www.landesbioscience.com/supplement/HorardEPI4-5-Sup.pdf), analysis can be performed on mean $\log_2$ ratios from one strand (sense or antisense) or on averaged $\log_2$ ratios from both strands (sense + antisense). Once DNA methylation status of each element on the array has been determined, methylation pattern can be investigated across samples. Principal component analysis and unpaired test analysis might facilitate methylome comparisons (9). To identify aberrant DNA methylation, it is imperative to have a "reference" map for comparison and thus experimental strategies should include control and test samples. For example, comparison a comparison of DNA methylation from tumor and nonmalignant samples is a common strategy in cancer research.

Acknowledgments

We are grateful to Dr N. Vassetzky, Dr G. Fourel, Pr. P. Barbry, Dr. J. Puechberty, Dr G. Roizes, Pr. C. Vourc'h, Dr F. Devaux, and Pr C. Gautier for their contribution in developing the oligoarray RepArray and the computational tools. This work was supported by European Union 6th framework program grant RISCRAD; ARECA and EpiPro framework programs from Canceropole Lyon Auvergne Rhône Alpes and the Association pour la Recherche sur le Cancer and the Ligue Nationale contre le Cancer ("Equipe Labellisée").

References

1. Wilson AS, Power BE, Molloy PL (2007) DNA hypomethylation and human diseases. Biochim Biophys Acta 1775 (1):138–162. doi:S0304-419X(06)00056-4
2. Hedges DJ, Deininger PL (2007) Inviting instability: Transposable elements, double-strand breaks, and the maintenance of genome integrity. Mutat Res 616 (1–2):46–59. doi:S0027-5107(06)00333-2
3. Eymery A, et al. (2009) A transcriptomic analysis of human centromeric and pericentric sequences in normal and tumor cells. Nucleic Acids Res 37 (19):6340–6354. doi:gkp639
4. Rodriguez J, et al. (2008) Genome-wide tracking of unmethylated DNA Alu repeats in normal and cancer cells. Nucleic Acids Res 36 (3):770–784. doi:gkm1105
5. Yang AS, et al. (2004) A simple method for estimating global DNA methylation using bisulfite PCR of repetitive DNA elements. Nucleic Acids Res 32 (3):e38. doi:10.1093/nar/gnh032/3/e38
6. Weisenberger DJ, et al. (2005) Analysis of repetitive element DNA methylation by MethyLight. Nucleic Acids Res 33 (21):6823–6836. doi:33/21/6823
7. Roman-Gomez J, et al. (2008) Repetitive DNA hypomethylation in the advanced phase of chronic myeloid leukemia. Leuk Res 32 (3):487–490. doi:S0145-2126(07)00307-4
8. Choi SH, et al. (2009) Changes in DNA methylation of tandem DNA repeats are different from interspersed repeats in cancer. Int J Cancer 125 (3):723–729. doi:10.1002/ijc.24384
9. Horard B, et al. (2009) Global analysis of DNA methylation and transcription of human repetitive sequences. Epigenetics 4 (5):339–350. doi:9284
10. Weber M, Schubeler D (2007) Genomic patterns of DNA methylation: targets and function of an epigenetic mark. Curr Opin Cell Biol 19 (3):273–280. doi:S0955-0674(07)00063-4
11. Weber M, et al. (2005) Chromosome-wide and promoter-specific analyses identify sites of differential DNA methylation in normal and transformed human cells. Nat Genet 37 (8): 853–862. doi:ng1598
12. Mohn F, et al. (2009) Methylated DNA immunoprecipitation (MeDIP). Methods Mol Biol 507:55–64. doi:10.1007/978-1-59745-522-0_5
13. Sorensen AL, Collas P (2009) Immunoprecipitation of methylated DNA. Methods Mol Biol 567:249–262. doi:10.1007/978-1-60327-414-2_16
14. Rouillard JM, Zuker M, Gulari E (2003) OligoArray 2.0: design of oligonucleotide probes for DNA microarrays using a thermodynamic approach. Nucleic Acids Res 31 (12):3057–3062
15. Jordan A, Bisgrove D, Verdin E (2003) HIV reproducibly establishes a latent infection after acute infection of T cells in vitro. EMBO J 22 (8):1868–1877. doi:10.1093/emboj/cdg188
16. Vissel B, Choo KH (1987) Human alpha satellite DNA – consensus sequence and conserved regions. Nucleic Acids Res 15 (16):6751–6752
17. Moyzis RK, et al. (1987) Human chromosome-specific repetitive DNA sequences: novel markers for genetic analysis. Chromosoma 95 (6):375–386
18. Cooke HJ, Hindley J (1979) Cloning of human satellite III DNA: different components are on different chromosomes. Nucleic Acids Res 6 (10):3177–3197

19. Moyzis RK, et al. (1988) A highly conserved repetitive DNA sequence, (TTAGGG)n, present at the telomeres of human chromosomes. Proc Natl Acad Sci U S A 85 (18):6622–6626
20. Espada J, et al. (2004) Human DNA methyltransferase 1 is required for maintenance of the histone H3 modification pattern. J Biol Chem 279 (35):37175–37184. doi:10.1074/jbc.M404842200
21. Kondo Y, et al. (2004) Chromatin immunoprecipitation microarrays for identification of genes silenced by histone H3 lysine 9 methylation. Proc Natl Acad Sci USA 101 (19):7398–7403. doi:10.1073/pnas.0306641101 (pii)
22. Jacinto, F.V., et al. (2007) Discovery of epigenetically silenced genes by methylated DNA immunoprecipitation in colon cancer cells. Cancer Res 67(24): p. 11481–6

Chapter 17

Novel Approach for the Development of New Antibodies Directed Against Transposase-Derived Proteins Encoded by Human Neogenes

Ahmed Arnaoty, Bruno Pitard, Benoit Bateau, Yves Bigot, and Thierry Lecomte

Abstract

Molecular domestication of several DNA transposons has occurred during the evolution of the primate lineage, and has led to the emergence of at least 42 new genes known as neogenes. Because these genes are derived from transposons, they encode proteins that are related to certain recombinases, known as transposases. Consequently, they may make an important contribution to the genetic instability of some human cells. In order to investigate the role of these neogenes, we need to be able to study their expression as proteins, for example in tumours, which often provide good models of genetic instability. In order to perform such studies, polyclonal antibodies directed against the proteins expressed by neogenes are obtained using a recently developed new method of Nanospheres/DNA immunisation in laboratory mammals. In this chapter, we describe a fully integrated process of producing antibodies that consists of a series of steps starting with the preparation and synthetic formulation of plasmids encoding neogenes, and culminating in the final production and confirmation of the quality of these polyclonal antibodies.

Key words: DNA transposons, Genetic instability, Neogenes, Nanopheres/DNA immunisation, Antibody

1. Introduction

Transposable elements (TEs) are discrete segments of DNA that are able to catalyse their own ability to move from one genomic location to another. They are powerful forces for genetic change, and have made a significant contribution to the evolution of many genomes. TEs are divided into two classes depending on the transposition intermediate they use: class I RNA transposons (retrotransposons) function through the reverse transcription of an

Yves Bigot (ed.), *Mobile Genetic Elements: Protocols and Genomic Applications*, Methods in Molecular Biology, vol. 859, DOI 10.1007/978-1-61779-603-6_17, © Springer Science+Business Media, LLC 2012

RNA intermediate, whereas class II DNA transposons move by a cut-and-paste mechanism in which the DNA segment corresponding to the transposon is excised from one location and reinserted at another location (1). At least 45% of the human genome is made of TE-derived sequences, with retrotransposons forming the dominant TE type, and DNA transposons contributing about 3% of the genome (2). A DNA transposon encodes at least one protein, known as a transposase, which shows many homologies with the family of DD35E integrases (3, 4). DNA transposons can potentially influence the evolutionary trajectory of their host in three distinct ways: (1) by altering gene function as a result of their insertion; (2) by producing chromosomal rearrangements; (3) by providing a source of coding and noncoding material that permits genetic novelties to appear (5).

TEs are generally present in genomes as discrete DNA repeats that are epigenetically silenced by the host genome to prevent their transcription and subsequent transposition (6, 7). This silencing means that maintaining TE mobility is not entirely subject to selection. Consequently, over time TEs accumulate mutations in their DNA sequence that inactivate their abilities to encode functional proteins and to move about within chromosomes. Nevertheless, it has recently been shown that a few TEs have escaped this fate as a result of having been "domesticated" by the host genome, resulting in the creation of novel genes, known as neogenes (7, 8). Some of domesticated genes originating from retrotransposons or DNA transposons are known to be implicated in the cellular and developmental functions involved in placental development, viral resistance, chromatin structure, DNA recombination and repair, gene regulation, apoptosis, and brain development (8). This shows that complex evolutionary interactions occur between the genomic silencing of TEs to prevent their proliferation and the exaptation of genes encoding transposon proteins to provide novel cellular functions in higher eukaryotic genomes (5, 9, 10). RAG1 and RAG2, proteins that are involved in the recombination processes of immunoglobulin and T-cell receptor genes in lymphocytes, are among the neogenic proteins that have received most investigation (11). The domestication of these two proteins is ancient since it occurred into the agnathan ancestor genome of all vertebrates about 500 million years ago.

The creation of neogenes has occurred in all eukaryotic lineages from the most remote aeons of evolution to our own time. One case of a "recent" domestication is the *Setmar* gene that encodes the SETMAR (the so-called METNASE) proteins. The *Setmar* gene is a modified transposase gene that originated from an ancient *mariner* transposon, *Hsmar1*, which was domesticated when the primate lineage originated (12). The SETMAR protein is in fact a fusion of two protein moieties: an SET domain that displays histone lysine methyltransferase activity, and an *Hsmar1* transposase that displays specific DNA binding, DNA looping, and DNA cleavage activities (13). This gene contains three exons that extend over 13.8 kbp, and is located in human chromosome 3p26 (14). SETMAR has been shown to be involved in the mechanisms of genome repair and to play important roles in genetic disorders and gene instability (14). Analysis of deletions in the 3p26 region that contains *Setmar* has revealed that this gene may be involved in neuron stability and mental retardation (15). *Setmar* is highly transcribed in many different human tissues and cancers, and could play a role in DNA repair by enhancing the non-homologous end joining (NHEJ)DNA repair pathway, promoting exogenous random DNA integration, enhancing topoisomerase II α function (16), and suppressing chromosomal translocations (17). From a clinical standpoint, several cytogenetic abnormalities involved in human malignancies are located in the 3p26 chromosomal region: non-Hodgkin's lymphoma, acute leukaemia, hereditary prostate cancer, myeloma, and myelodysplastic syndromes. Recently, it was reported that in vitro METNASE may be involved in the mechanisms of resistance to topoisomerase II inhibitors in a cell lineage model of breast cancer treated by anthracycline as well as in cell lineage models of acute leukaemia treated by VP-16 (18, 19).

Overall, the features of human neogenes and the accumulated findings about the recombinogenic activities of the RAG and SETMAR proteins suggest that neogenic proteins that have kept their DNA binding and catalytic activities could be involved in human genetic instability. Of the 42 neogenes so far identified in the human genome (5), at least 23 could potentially have these properties (Table 1). To make it possible to study these putative neogenic recombinases, we need tools that make it possible to track their cellular expression as proteins. One basic requirement is the availability of recombinant neogenes produced in vitro for animal immunisation or antibodies or sera that target these neogenic proteins, thus making it possible to monitor the presence of their various isoforms in biopsies or cell extracts.

2. Principle Underlying the Production of Polyclonal Antibodies by Nanospheres/ DNA Immunisation

Producing polyclonal antibodies directed against neogenic recombinases is a difficult task in this area of research. Different strategies can be used to generate these new antibodies but genetic immunisation is currently the only method capable of overcoming the difficulties of in vitro neogene production. The existing methods, including immunising animals with recombinant viruses encoding antigens, neogenic proteins, or peptides corresponding to the predicted epitopes, are not suitable for the rapid generation of highly functional antibodies targeting native neogenic recombinases.

Table 1
List of the human neogenes against which Nanospheres/DNA immunisation were performed to obtain antibodies

Family of transposon	Name of neogene	Domain	Chromosomal location
IS630-Tc1-mariner/pogo	JRK: Jerky	DBD + CatD	8q24.3
IS630-Tc1-mariner/pogo	TIGD2: Tigger TE-derived 2	DBD + CatD	4q22.1
IS630-Tc1-mariner/pogo	TIGD3: Tigger TE-derived 3	DBD + CatD	11q13.1
IS630-Tc1-mariner/pogo	TIGD4: Tigger TE-derived 4	DBD + CatD	4q31.3
IS630-Tc1-mariner/pogo	TIGD5: Tigger TE-derived 5	DBD + CatD	8q24.3
IS630-Tc1-mariner/pogo	TIGD6: Tigger TE-derived 6	DBD + CatD	5q32
IS630-Tc1-mariner/pogo	TIGD7: Tigger TE-derived 7	DBD + CatD	16p13.3
IS630-Tc1-mariner/Tc2	POGK	DBD + CatD	1q24.1
IS630-Tc1-mariner/Tc2	POGZ	DBD + CatD	1q21.3
IS630-Tc1-mariner/ Hsmar1	SETMAR* (Metnase)	DBD + CatD	3p26.1
piggyBac	PGBD1: PiggyBac-derived 1	DBD + CatD	6p22.1
piggyBac	PGBD2: PiggyBac-derived 2	DBD + CatD	1q44
piggyBac	PGBD3: PiggyBac-derived 3	DBD + CatD	10q11
piggyBac	PGBD4: PiggyBac-derived 4	DBD + CatD	15q14
piggyBac	PGBD5: PiggyBac-derived 5	DBD + CatD	1q42.13
Hobo-Ac-Tam	ZBED1 (hDREF)	DBD + CatD	Xp22.33 Yp11
Hobo-Ac-Tam	ZBED5 (Charlie 1 ; Buster 1)	DBD + CatD	11p15.3
Hobo-Ac-Tam	ZBED6	DBD + CatD	1q32.1
Hobo-Ac-Tam	Charlie 8 (GTF2IRD2)	DBD + CatD	7q11.23
PIF-Harbinger	HARBI1	DBD + CatD	11p11.2
PIF-Harbinger	NAIF1	DBD	9q34.11
P	THAP4	DBD + CatD	2q37.3
P	THAP9	DBD + CatD	4q21.22

Moreover, of the eight antibodies available in suppliers' catalogues (Jerky, TIGD3, POGK, PGBD1, PGDB3, SETMAR, ZBED1, KRBA), we found that only those directed against the *Hsmar1* moiety of SETMAR were active and specific under our experimental conditions. This highlights the need for new technologies to make it possible to use quick and easy methods to produce high-titre

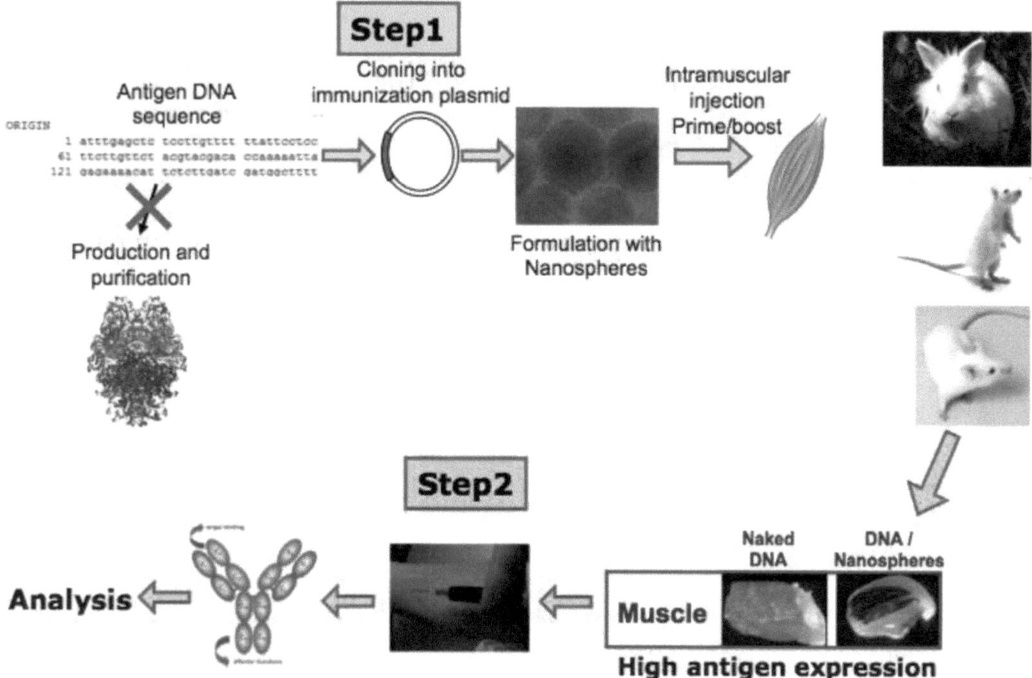

Fig. 1. The polyclonal antibody production process using the ICANtibodies technology.

functional antibodies against the 24 selected neogenes. Recently, our group has designed a new adjuvant that dramatically enhances the immunogenicity of plasmid DNA encoding antigens, including tumour- or pathogen-derived antigens. Here, we explain this process in detail (Fig. 1). Our DNA synthetic formulation based on the use of an amphiphilic polymer is a breakthrough, and corresponds to an important step forward towards the production of polyclonal antibodies via a DNA immunisation process, because it is suitable for producing virtually any antigen that is encoded by a plasmid. This novel DNA synthetic formulation makes it possible to generate highly functional antibodies against any natively expressed nuclear, cytoplasmic, secreted or membrane proteins. Our approach to generating polyclonal antibodies is a fully integrated process, starting from an in silico antigen sequence to synthesise a gene encoding the antigen, DNA formulation and genetic immunisation in various species including mice, rats, rabbits and goats, to antibody quality controls. This recent innovation in the field of in vivo plasmid DNA formulation technology has made it possible to produce more than 100 functional different antibodies. Indeed, ever since the serendipitous discovery some 20 years ago that naked plasmid DNA could be used as a vector for immunisation, numerous attempts have been made to develop a gene-based approach to antibody production. Most recent DNA approaches have either involved ways of increasing cell entry by electroporation or other means (tattooing, patches...) or particle-based delivery (e.g. Vical's vaxfectin or Pfizer/PowderMed gold particles). Electroporation or other naked DNA approaches generally require large amounts of DNA, and have still to demonstrate their real efficacy in every animal as the devices used restrict entry to a very limited space. Particle-mediated methods generally require much less DNA (e.g. gold particles covered with microgram quantities of DNA). However, the development of gold nanoparticles has the drawback of requiring a complex and costly device. Other approaches, such as vaxfectin, which consists of cationic lipids like those used in cell cultures, have also been tested for their capacities to produce transfection in vivo. DNA-based nucleic acid immunisation faces critical limitations: the DNA not only has to cross the outer cell membrane, but also has to enter the nucleus in order to be transcribed before translation of the targeted antigen can take place within the cytoplasm. In addition, entry generally occurs via endosomes, and therefore the DNA has to escape from endosome entrapment and then undergo degradation in order to be active. Recently, our group has developed a breakthrough delivery technology, named ICANtibodies technology, that makes it possible to deliver low doses of nucleic acid molecules directly into the cell cytoplasm, thus by-passing the limiting endosomal step used by cationic vectors, such as positively charged lipids, polymers, or proteins (20–22). This results in dramatic antigen expression and stimulation of the innate immune system. This new synthetic vector with this revolutionary mechanism of action has been successfully applied to the treatment of the mouse hepatocellular carcinoma using the tumoral antigen AlphaFetoProtein in an extremely relevant autochthonous HepatoCarcinomaCellular model (23). Similarly, an immune response against a self-antigen was also obtained with mouse erythropoietin, leading to an autoimmune form of anaemia.

Here, we describe this potent "one-way" technology, which involves several steps leading from the in silico sequence of the neogene to the production of the final functional polyclonal antibody directed against neogenic recombinases. This technology has been used in attempts to synthesise antibodies directed against proteins encoded by 24 neogenes derived from DNA transposons.

3. Materials

3.1. General Material

1. HeLa cell line.
2. DMEM medium.
3. PVDF membrane.

4. Whatmann paper.
5. Primary polyclonal antibodies (serum).
6. Anti-actin antibody.
7. Secondary antibodies (Donkey Anti Mouse Abcam).
8. Mini-Gel apparatus (Mini-PROTEAN® II-Bio-Rad).
9. Semi-dry electrotransfer apparatus (Bio-Rad).

3.2. Enzymes

1. Protease inhibitors.
2. Restriction enzymes selected to control plasmid integrity.

3.3. Molecular Biology Products and Chemicals

1. *N,N,N′,N′*-Tetramethylethylenediamine (TEMED).
2. Protein assay reagent (Bradford kit, Bio-Rad).
3. Transfection reagent JetPEI (Poly+transfection).
4. A mammalian expression plasmid encoding green florescent protein (GFP).

3.4. Solutions and Kits

All solutions are prepared with ultrapure water.

1. 20% Acrylamide/bis-acrymalide 15/0.425.
2. Buffer SDS pH 8.8 (4×=1.5 M Tris–HCl, 0.4% SDS).
3. Buffer SDS pH 6.8 (4×=0.5 M Tris–HCl, 0.4% SDS).
4. Ammonium persulfate (APS) 20%.
5. 10× TBS=Tris–HCl 0.2 M pH 7.4, NaCl 1.5 M.
6. Loading buffer SDS-PAGE (6×=Tris–HCl 2 M pH 6.8, Glycerol, SDS 5%, β-mercaptoethanol 10 mM, Bromophenol blue).
7. Methanol.
8. SDS-PAGE buffer (Tris 0.26 M-glycine 1.86 M, SDS 10%).
9. Transfer buffer (Tris 0.05 M-glycine 0.38 M, pure methanol, SDS 10%).
10. 1× TBS–Tween–milk (5 g non-fat dry milk in 100 ml 1× TBS–Tween 0.5%).
11. Enhanced Chemoluminescence Kit (Amersham).
12. Endotoxin-free, plasmid DNA purification kit (Maxi-Macherey Nagel).
13. Buffer lyses (SDS 20%, NaCl 50 mM, 10 mM β-mercapto ethanol, protease inhibitor cocktail).

3.5. Animal Husbandry

1. ICANtibodies technology (In Cell Art, Nantes, France).
2. Breeding of BALB/c and/or Swiss mice (see Note 1).

4. Methods

4.1. Plasmid DNA Design and Production

The information needed to start the production of polyclonal antibodies directed against neogenic protein consists of the DNA and protein sequences, which are available from private or public databases. Several features of the gene sequence are optimised:

1. The codon usage is adapted to optimum expression in rodent species. Regions with very high or very low GC content are mitigated when possible.
2. The following *cis*-acting motifs are avoided where applicable: (a) internal TATA-boxes, chi-sites, and ribosomal entry sites (inner Kozak'box), (b) AT-rich or GC-rich sequence stretches, (c) RNA instability motifs, (d) repeated sequences and RNA secondary structures, and (e) cryptic splice donor and acceptor sites found in the genes of higher eukaryotes.

The optimised sequence was synthesised and subcloned in a plasmid specifically designed for use in the development of DNA vaccines that contains a CMV promoter and a kanamycin resistance gene. The synthesis of the synthetic genes of the 24 neogenes and their subsequent cloning in the immunisation plasmid is performed by ATG Biosynthetic (Germany). Plasmids are amplified in *Escherichia coli*, and then purified with EndoFree plasmid purification columns (Qiagen). The quality control of the plasmid is monitored by electrophoresis on an agarose gel to confirm that the supercoiled form predominated. Analyses of restriction enzyme profiles are also performed to check the identity and integrity of the plasmid.

4.2. DNA Formulation and Animal Injection

Mouse experiments must be performed in accordance with the State guidelines (in our case, those of the French *Institut National de la Santé et de la Recherche Médicale*).

1. Each purified plasmid at the desired concentration is formulated using an aqueous solution of the synthetic vector by equivolumetric mixing. Nanospheres are assembled using the ICANtibodies technology, as recommended by the Supplier (InCellArt, Nantes, France).
2. Adult Swiss and BALB/c mice are used for the immunisations.
3. Intramuscular injections are administered into shaved muscles at a single site in five anaesthetised mice of each strain.
4. 21, 42, and 63 days after the primo injection, the animals are given a booster with a formulation containing the same amount of DNA formulated with the synthetic vector.
5. Serum samples are collected on days 0, 21, 42, 63, and in some cases at day 84, and analysed for their antibody titre.

4.3. Antibody Analysis

Analysis of the serum quality (recognition specificity and activity) is performed by Western blot using HeLa cell extracts over-expressing (positive control) or not expressing (negative control) a neogenic protein. HeLa cell extracts are obtained as follows:

1. HeLa cells are cultured in DMEM supplemented with 10% foetal bovine serum (FBS) at 37°C and with 5% CO_2.
2. Approximately 4×10^5 cells are seeded onto each 6-well plate 1 day prior to transfection.
3. Cells are transfected with jetPEI, according to the manufacturer's instructions (Poly + transfection). Briefly, plasmid DNA (2 μg: 1.8 μg of the plasmid expressing the neogenetic protein, and 0.20 μg of a plasmid expressing GFP (see Note 2)), and PEI (4 μl) are each diluted in 50 μl of 150 mM NaCl, and then mixed gently.
4. After being incubated for 15 min, the mixture is diluted with OPTIMEM medium to a final volume of 1 ml.
5. Cells are then incubated with 0.2 ml of the complexes for 2–4 h. The transfection solution is then discarded, and replaced by fresh DMEM supplemented with 10% FBS before being incubated for 24 h at 37°C.
6. The culture medium is removed and the HeLa cells are washed twice with 1× PBS.
7. The HeLa cells are then lysed by adding 500 μl of the lysis buffer (Tris–HCl 2 M pH 6.8, SDS 20%, glycerol 80%, DDT 1 M, bromophenol blue) to each well.
8. Cell extracts (viscous solutions) are recovered by gentle pipeting, and then transfered into a 1.5- or 2.2-ml microtube.
9. To reduce the viscosity of the solution, the genomic DNA is broken by shearing with a sonicator (40 W, twice for 10–15 s).
10. The isolated protein is quantified by means of a commercially available modified Bradford assay (Bio-Rad Laboratories; Note 3).
11. Protein extracts are then kept at –20°C until use.
12. Before being used for gel electrophoresis, sample 10-μg protein extracts are denatured by heating to 95°C for 5 min, cooling with ice for 5 min, and then centrifuging at 5,500 × g for 5 min to pellet the debris.

4.4. Western Blot Analysis

1. Before being used for gel electrophoresis, sample 10-μg protein extracts are denatured by heating to 95°C for 5 min, cooling with ice for 5 min, and then centrifuging at 5,500 × g for 5 min to pellet the debris.
2. The proteins are then separated by SDS-PAGE on a 10% polyacrylamide mini-gel.

3. Proteins are transferred onto a polyvinylidene difluoride (PVDF) membrane (Bio-Rad, Richmond, USA).
4. Post-transfer, the membranes are blocked for 1 h at room temperature with 5% non-fat dry milk in 1× TBS, adjusted with 0.5% Tween 20.
5. After removing the blocking solution, the primary antibodies (a serum directed against a neogenic protein) is incubated with the membrane for 2 h at room temperature in 1× TBS, 0.5% Tween 20, 5% non-fat dry milk.
6. The membrane is washed three times for 10 min with 1× TBS, 0.5% Tween 20, at room temperature.
7. The membrane is incubated with the secondary antibody (horseradish peroxidase–conjugated antibody directed against mouse IgG) for 1 h at room temperature.
8. The membrane is then washed three times for 10 min with 1× TBS, 0.5% Tween 20, at room temperature.
9. The proteins recognised by the primary antibodies are finally visualised with the enhanced chemoluminescence kit (Amersham Biosciences) according to the manufacturer's instructions. Pictures of the membranes are obtained by autoradiography or using an accurate chemoluminescence imager.
10. Serum samples collected at 0, 21, 42, 63 days post the primo injection are analysed in order to track the antibody production, and in some cases to find out whether a time of less than 84 days could be better for recovering active serum (see Note 4).

4.5. Efficiency of the Polyclonal Antibody Production

Table 2 reports the 14 neogenic proteins for which we have obtained polyclonal antibodies with high specificity (see Note 5). For nine cases in which antibody production was negative in Western blot analysis, ELISA assays on lysat of transfected HeLa cells with the corresponding plasmid encoding the neogene, allowed us to confirm that conformational antibodies were in fact produced in four of them, which indicated that antibodies directed against both linear and conformational antigens can be produced using this new ICANtibodies method. We successfully obtained specific antibodies for the eight neogenic proteins for which commercial antibodies are available and claimed to be efficient, with the exception of the KRBA proteins The positive signals that obtained with our antibodies when tested with Western blot analysis and ELISA assays on lysat of transfected and non transfected HeLa cells with the corresponding plasmid encoding the neogene compared with the commercial antibodies from different supplier that never show any signals neither in transfected nor in non transfected Hela cells. That's mean, our antibodies have better quality and efficiency than commercial antibodies. Overall, our results indicate that we produced polyclonal antibodies usable for

Table 2
Neogene proteins against which antibodies were obtained by Nanospheres/DNA immunisation[a]

Neogene	Calculated MW (Da)	MW of the detected proteins (Da)
JRKL	57,402	59,911
TIGD2	59,623	59,623
TIGD3	52,027	52,026
TIGD5	64,228	49,333
TIGD7	63,236	63,236
POGK	69,444	69,443
Hsmar1	76,669 SETMAR	40,286 HSMAR1
PGBD1	92,515	92,515
PGBD3	67,567	67,595
PGBD4	67,033	67,003
ZBED6	109972.89	110,000
Charlie 8	107,233	107,233
NAIF	35,164	35,164
THAP9	103,411	103,411

MW molecular weight
[a]All serum were verified by Western Blot and ELISA

Western blot analyses and ELISA for 61% and 78% of the neogenic proteins assayed, respectively. The absence of produced polyclonal antibodies for the other neogenes can be attributed to the non-immunogenicity of these neogenic proteins in the mouse immune system, as we confirmed the high protein expression after transfection of their corresponding plasmids.

5. Notes

1. Inbred mouse lineages and mouse strains "close to a wild population" have a reactivity to immunisation with formulated Nanospheres/DNA vaccination that depends on the antigen, and which can vary dramatically from the absence of any response to very efficient immunisation. This is why it is recom-

mended that DNA immunisation with ICANtibodies technology, assays should be carried out using two mouse strains with very different genetic backgrounds.

2. The GFP expression marker can be used to evaluate transfection quality control (the proportion of cells transfected) by epifluorescence microscopy, FACS, or Western blot analysis using an antibody directed against GFP as the primary antibody.
3. The relative amount of protein in each sample blotted onto the PVDF membrane can be evaluated by probing the membrane with an antibody directed against actin (a "house-keeping" protein), which then makes it possible to compare the relative amounts of a protein of interest in different samples.
4. For a few neogenic proteins, we found that withdrawal dates 42- or 63-days post primo immunisation work better than 84 days for the recovery of sera with optimal antibody levels from mice.
5. In general, all five mice of the same strain are found to react against a given antigen. If only some of the mice are reactive, antibody production is not stable over time, and so the optimum production date will vary from individual to individual.

Acknowledgments

We thank E. Goudeau (INSERM U915) for providing excellent technical expertise in production of plasmids DNA. This work was funded by a Research Program grant from the Cancéropôle Grand-Ouest, and grants from Amgen and the French National Society of Gastroenterology.

References

1. Muñoz-López M, et al. (2008) Transposition of Mboumar-9: identification of a new naturally active mariner-family transposon. J Mol Biol 382:567–72
2. Lander ES, et al. (2001) International Human Genome Sequencing Consortium. Initial sequencing and analysis of the human genome. Nature 409:860–921
3. Rice P, Craigie R, Davies DR (1996) Retroviral integrases and their cousins. Curr Opin Struct Biol 6:76–83
4. Doak TG, et al. (1994) A proposed superfamily of transposase genes: transposon-like elements in ciliated protozoa and a common "D35E" motif. Proc Natl Acad Sci USA. 91:942–946
5. Feschotte C, Pritham EJ (2007) DNA transposons and the evolution of eukaryotic genomes. Annu Rev Genet 41:331–368
6. O'Donnell KA, Boeke JD (2007) Mighty Piwis defend the germline against genome intruders. Cell 129:37–44
7. Miller WJ, et al. (1999) Molecular domestication – more than a sporadic episode in evolution. Genetica 107:197–207
8. Volff JN (2006) Turning junk into gold: domestication of transposable elements and the creation of new genes in eukaryotes. Bioessays 28:913–922
9. Kalitsis P, Saffery R (2009) Inherent promoter bidirectionality facilitates maintenance of sequence integrity and transcription of parasitic

DNA in mammalian genomes. BMC Genomics. 10:498

10. Sinzelle L, Izsvk Z, Ivics Z (2009) Molecular domestication of transposable elements: from detrimental parasites to useful host genes. Cell Mol Life Sci 66:1073–1093
11. Agrawal A, Eastman QM, Schatz DG (1998) Transposition mediated by RAG1 and RAG2 and its implications for the evolution of the immune system. Nature 394:744–751
12. Cordaux R, et al. (2006) Birth of a chimeric primate gene by capture of the transposase gene from a mobile element. Proc Natl Acad Sci USA 103:8101–8106
13. Beck BD, et al. (2008) Human Pso4 is a metnase (SETMAR)-binding partner that regulates metnase function in DNA repair. J Biol Chem 283:9023–9030
14. Lee SH, et al. (2005) The SET domain protein Metnase mediates foreign DNA intégration and links integration to nonhomologous end-joining repair. Proc Natl Acad Sci USA. 102: 18075–18080
15. Higgins JJ, et al. (2004) Candidate genes for recessive non-syndromic mental retardation on chromosome 3p (MRT2A). Clin Genet 65: 496–500
16. Shaheen M, et al. (2010) Metnase/SETMAR: a domesticated primate transposase that enhances DNA repair and chromosome decatenation. Genetica 138:559–566
17. Wray J, et al. (2010) The transposase domain protein Metnase/SETMAR suppresses chromosomal translocations. Cancer Genetics and Cytogenetics 200:184–190
18. Wray J, Williamson EA, et al. (2009) Metnase mediates resistance to topoisomerase II inhibitors in breast cancer cells. PLoS One 4:e5323
19. Wray J, et al. (2009) Metnase mediates chromosome decatenation in acute leukemia cells. Blood 114:1852–1858
20. Pitard B, et al. (2004) Negatively charged self-assembling DNA/poloxamine nanospheres for in vivo gene transfer. Nucleic Acids Res 32:e159
21. Chèvre R, et al. (2011) Amphiphilic block copolymers enhance the cellular uptake of DNA molecules through a facilitated plasma membrane transport. Nucleic Acids Res 39:1610–1622
22. Le Bihan O, et al. (2011) Probing the in vitro mechanism of action of cationic lipid/DNA lipoplexes at a nanometric scale. Nucleic Acids Res 39:1595–1609
23. Cany J, et al. (2011) AFP-specific immunotherapy impairs growth of autochthonous hepatocellular carcinoma in mice. J Hepatol 54:115–21

Index

Yves Bigot (ed.), *Mobile Genetic Elements: Protocols and Genomic Applications*, Methods in Molecular Biology, vol. 859, DOI 10.1007/978-1-61779-603-6,

If you have any concerns about our products,
you can contact us on
ProductSafety@springernature.com

In case Publisher is established outside the EU,
the EU authorized representative is:
Springer Nature Customer Service Center GmbH
Europaplatz 3, 69115 Heidelberg, Germany

Printed by Libri Plureos GmbH
in Hamburg, Germany